List of Symbols

A = area
C = discharge coefficient
d = distance
D = diameter
e = voltage
E = energy
f = friction factor
F = force
g = gravitational acceleration
h = height (distance) or head
H = relative humidity
I = electrical current
k = adiabatic process exponent
K = surface roughness factor
L = length (distance)
m = mass
N = rotational speed
p = pressure
P = power
q = volume flow rate
Q = mass or weight flow rate

r = radius
R = gas constant
Re = Reynolds Number
s = stroke length
S = specific gravity
t = time
T = torque
T = temperature
v = velocity
V = volume
w = work
W = weight
x = horizontal distance
y = vertical distance
γ = gamma = specific weight
η = eta = efficiency
ν = nu = kinematic viscosity
π = pi = 3.14159
ρ = rho = mass density
ω = omega = frequency of rotation

About the cover:

(Left) Hydraulic systems are unsurpassed in generating the incredibly large forces required of equipment used for heavy manufacturing, construction, and agriculture. (*John Deere Construction Equipment Co., Moline, Ill.*)

(Right) Pneumatic systems, like the one used to operate this robotic gripper, are capable of fast, precise, and repetitive motions needed for the mass production of small or delicate objects.

Fluid Power Technology
Second Edition

Robert P. Kokernak

Fitchburg State College

Prentice Hall
Upper Saddle River, New Jersey   Columbus, Ohio

Library of Congress Cataloging-in-Publication Data

Kokernak, Robert P.
   Fluid power technology / Robert P. Kokernak. — 2nd ed.
      p. cm.
   Includes index.
   ISBN 0-13-912487-X (hc)
   1. Fluid power technology.   I. Title.
TJ843.K637 1999
620.1'06—dc21   98-21880
                                                      CIP

Cover photo: ©John Deere Construction Equipment Company
Editor: Stephen Helba
Production Editor: Patricia S. Kelly
Design Coordinator: Karrie M. Converse
Text Designer: Linda M. Robertson
Cover Designer: Brian Deep
Production Manager: Deidra M. Schwartz
Marketing Manager: Frank Mortimer, Jr.

This book was set in Times Roman and Gill Sans by Carlisle Communications, Ltd. and was printed and bound by R.R. Donnelley & Sons Company. The cover was printed by Phoenix Color Corp.

©1999 by Prentice-Hall, Inc.
Simon & Schuster/A Viacom Company
Upper Saddle River, New Jersey 07458

Earlier edition © 1994 by Macmillan College Publishing Company, Inc.

Printed in the United States of America

10 9 8 7 6 5 4 3 2 1

ISBN: 0-13-912487-X

Prentice-Hall International (UK) Limited, *London*
Prentice-Hall of Australia Pty. Limited, *Sydney*
Prentice-Hall of Canada, Inc., *Toronto*
Prentice-Hall Hispanoamericana, S. A., *Mexico*
Prentice-Hall of India Private Limited, *New Delhi*
Prentice-Hall of Japan, Inc., *Tokyo*
Simon & Schuster Asia Pte. Ltd., *Singapore*
Editora Prentice-Hall do Brasil, Ltda., *Rio de Janeiro*

This one's for Jean,
a loving helpmate who has shared so many projects
with me during the past three decades.

Preface

A Balanced Approach

This textbook is aimed specifically at students enrolled in two-year and four-year college-level programs both in engineering technology and industrial technology. Its primary objective is to strike an appropriate *balance* between theory and application.

Traditionally, engineering students have used courses in fluid mechanics, hydraulics, and compressible fluid flow to examine the theoretical behavior of liquids and gases without regard to hardware or applications. At the other extreme, conventional fluid power courses for both technicians and technologists often emphasize components and the "plumbing" aspects of hydraulic or pneumatic systems with little discussion of the basic physical principles governing the operation of these systems. However, because many technology students enroll in only a single fluid power course, broad coverage of this subject can provide future benefits to graduates employed in a wide variety of technical industries.

In addition, the objective of *Fluid Power Technology*—to offer a balance of theory and practice—supports the accepted role of the technologist in industry. Figure 1 illustrates that within the spectrum of technical occupations, the optimum distribution of knowledge between theory and application is 60–40 for the technologist and 50–50 for the technician.

Presenting a balanced approach to any technical discipline, however, has drawbacks and limitations. As we shall see, one of the most appealing aspects of fluid power is its tremendous diversity. From dentists' drills to giant earth movers, the number of applications for compressed gases and liquids is staggering. Unfortunately, this same diversity encourages a high degree of industrial specialization; technologists and technicians tend to work in narrow occupational "compartments" and are often unaware of the activities taking place in other specialized areas in different industries. For example, personnel involved with compressed air systems used in continuous processes such as cleaning, drying, painting, or combustion might have no knowledge of, and little immediate interest in, basic hydraulic circuits applied to robots and other automated machines.

Since most undergraduates do not know in what specific areas or industries they will find employment during their lifetimes, the value of a broad-based knowledge of fluid power should be apparent. Yet many students consider irrelevant any subject matter that does not deal specifically with their own limited area of interest. It is unlikely that any single textbook can provide in-depth coverage for each of these diverse areas of fluid power; indeed, complete volumes have been written about most of the topics discussed in *Fluid Power Technology*.

The goal, then, of producing an effective text of reasonable length, which presents a balanced and comprehensive view of fluid power at a level suitable for the average technology

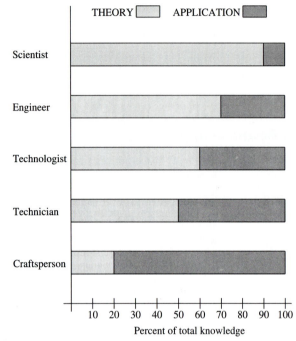

Figure I Desired skills for various technical occupations. (After R. J. Pond, *Introduction to Engineering Technology,* 4th ed. Upper Saddle River, NJ: Prentice Hall, 1999.)

student, imposes certain constraints on the scope and depth of material covered. A broad-based text, however, allows instructors to select and emphasize material important to their students, while giving cursory treatment to topics of secondary value. Realistically, no textbook can be all things to all people, and it is doubtful that any two instructors would use all of the material presented in *Fluid Power Technology* to the same degree.

Another area of concern to educators is that of laboratory activities. Although most instructors agree that hands-on problem solving provides excellent reinforcement of theoretical concepts, reviewers of this text have pointed out that there is a tremendous diversity in the lab equipment found at schools across the country. Although some institutions have extensive facilities and are able to offer a rigorous laboratory experience, others operate under budget constraints that severely limit the availability of equipment and restrict the allowable range of laboratory work.

Such conditions also limit the type and complexity of hands-on exercises that may reasonably be presented in a textbook on fluid power. For this reason, suggested activities that appear at appropriate points in *Fluid Power Technology* are intentionally generic, and require little or no specialized equipment. Most schools having well-equipped facilities will likely follow a specific sequence of experiments or exercises. Although the activities suggested here, then, may be used to supplement such a laboratory program, they should prove especially useful in schools having limited resources and equipment. Although some exercises may appear to be quite simple and unsophisticated, both students and instructors are often surprised at the educational value offered by the activities involved.

Features of This Book

To help students develop an understanding both of fluid behavior and its relevance to hydraulic/pneumatic components and systems, this text features the following:

- *Basic Theory: Keeping It Simple.* Compared with other areas of science and technology—computers and electronics, for example—whose theoretical bases have advanced significantly during the past several decades, the fundamental laws governing the behavior of fluids have exhibited quite remarkable durability and versatility. Many of these basic principles were formulated in the seventeenth and eighteenth centuries by men such as Pascal, Torricelli, Pitot, and the Bernoullis. Even those concepts presented in Archimedes's *Treatise on Floating Bodies,* written before 212 B.C., have not been superseded by more sophisticated theories. This text emphasizes the inherent simplicity of fluid power systems and their underlying principles of operation. Because most components are mechanical in nature, flow paths and system behavior are easily visualized. Students are encouraged to develop an understanding of how and why components and systems behave as they do, with quantitative evaluation a secondary concern.

- *Computational Techniques: Levels of Learning.* Mathematically, *Fluid Power Technology* should contain something for everyone. Although this text requires only a knowledge of basic algebra, students may seek their own level of learning as determined by their individual math skills. To aid the marginal student, frequent references are made to tabulated values presented in the text and exercises. Although "number crunching" is not a primary objective of this book, the process does develop a student's confidence and, like a good lab experiment, aids in his or her understanding of the concepts presented. On the other hand, students who can function at a higher level of mathematics should find problems that challenge their abilities, and students with access to computers are encouraged to generate their own spreadsheets and BASIC programs following the examples of section 1.5. Problems that lend themselves to machine computation are designated by a computer monitor and keyboard symbol. Throughout the book, an attempt has been made to achieve a balance in the use of U.S. customary and SI Units.

- *Applications: Learning by Example.* This book discusses systems and devices with which the student, as a consumer, is already familiar, as well as applications most likely to be encountered by technologists in industry. Specific examples—from automobile brake systems and hummingbird feeders through automated machines and hot-air balloons—are integrated with the text to transform each theory from merely an abstract concept into a concrete image in the reader's mind.

- *Illustrative Problems and Examples.* Where applicable, worked numerical examples are presented in the text, and problems for student assignment follow each of these major sections. Comprehensive review problems appear at the end of appropriate chapters, and answers to odd-numbered problems appear at the end of the book. Examples and problems have been carefully selected to expand the theories and applications presented, while bringing together concepts discussed elsewhere in the text. For example, virtually every college student knows that the buoyant force on an object immersed or floating in a fluid may be computed as the weight of fluid displaced by that object. Ask students to explain what causes this force, however, and only a few can relate this phenomenon to the pressure difference between the top and bottom surfaces of the object. Although Archimedes's principle finds little application in fluid power systems, the principle makes an excellent example for the discussion of fluid statics and pressures existing at various depths in a column of liquid. In addition, the principle relates directly to the operation of a hydrometer, a device that is mentioned early in the book as a means of measuring specific gravity of a liquid. The concept of buoyant force, then, is used by example to introduce the phenomenon and unify ideas presented in several other areas of the text.

- *End-of-Chapter Questions.* Instructors may find these questions useful for stimulating class discussion. Students can test both their qualitative understanding of material presented in each chapter and their own abilities to apply these concepts to new situations or conditions.

- *An Emphasis on Graphics: Seeing Is Believing.* This book makes liberal use of photographs and line drawings to "open up" the text, maintain reader interest, and present specific information not contained in the text itself. Although a substantial amount of the artwork used has been provided by manufacturers and users of fluid power components, I have attempted to supplement this material with clear, detailed photographs of unusual and interesting applications of fluid power. For this reason, race cars, liquid cutting jets, wheel crushers, sphygmomanometers, aerial buckets, and the Jaws of Life® all appear in this book.

- *Safety Sidebars.* Scattered throughout this text are sidebars that deal with the safe use of various fluids, components, and systems. Experience has shown that safety is an area in which students and other fluid power users tend to be very lax. It is hoped that highlighting this material through the use of boxed sidebars may increase each reader's awareness of the constant dangers present in *any* system that uses a compressed liquid or gas.

- *Suggested Activities.* Simple activities appropriate for group participation or class demonstrations are listed at the end of chapters 1, 2, 3, 5, 6, and 11. While the activities described require minimal equipment, they are intended to demonstrate basic principles presented in each chapter and identify some of the discrepancies that may exist between theory and practice.

As a final note, this text purposely uses a somewhat narrative style. Over the years I have written many articles for newspapers and consumer magazines on such nontechnical subjects as horse racing, private detectives, train travel, and wildlife photography. Although many textbooks "talk down" to their readers, trade publications require a writing style that catches the attention of readers, draws them into the subject, and makes them feel that they are experiencing a shared adventure with the author. To a certain extent I have tried to achieve the same results with this book, in the hope that anyone who reads the text—whether or not they ever perform a numerical calculation—will come away with an understanding of how fluids and fluid power systems behave. I hope that you will find it clear and concise.

Acknowledgments

Many people contributed to the preparation of this book by supplying factual data and illustrations, as well as explaining, demonstrating, or posing for photographs of various pieces of equipment in action. I appreciate all of their efforts and especially thank the following: Lt. Steve Cormier and the members of the Gardner Fire Department; Paul Guerin of Nypro, Inc.; Ashburnham physician Dr. Curtis Clayman; the personnel at Consentino Salvage Co.; Barbara and Gene Columbus of Blackstone Valley Balloon Co.; Gene and Jamie Fleck of Gene's Service Center; Ron Roberts of Parker's Maple Barn; the workers at Tuttle's Auto Parts; David Mei of the Ashburnham Municipal Light Plant; Denis Bland of the John Deere Company; Nadine Pinter and Deb Hendricks from Hydro-Line, Inc.; automotive photographer Joan Eident; and Glenn Danuser of the Danuser Machine Company.

I am particularly indebted to my parents, John and Frances Kokernak, for a lifetime of support and encouragement; and to the other members of my immediate family—Neal and Christine Smith and Jean, Susan, and Jim Kokernak—for gracing the pages of this book. Additional thanks go to Jim Kokernak, Ph.D., for his assistance with various computer programs and spreadsheets, as well as for composing so many of the intricate equations that appeared in the initial manuscript.

I also thank Doris Decicco, Bob Greene, and Bill Charpentier for supplying the computer and word processing expertise necessary to this project; Richard Gould for his assistance with the equipment and components featured in many of the figures; Jamie Roger and his staff at the Fitchburg State College Press for their help and advice on various areas of the text preparation; and to Professor Bob Tapply of the English Department at Fitchburg State College for his suggestions that, through the years, have proven so helpful to me in this business of writing.

I also thank the following reviewers, who offered helpful suggestions: John D. Coluccini, University of Massachusetts, Lowell; Douglas B. Cook, Owens Community College; Wendell Johnson, University of Akron; and Kenneth A. Seidel, Columbus State Community College.

Finally, I extend my appreciation to the staff at Prentice Hall Publishing Company for their patience and cooperation during the preparation and production of this book. Special thanks to editor and friend Steve Helba for the patience, advice, and encouragement that have meant so much to me during our several projects together.

Robert Kokernak

Contents

What This Book Is All About

This book has two very simple objectives: (1) to provide students with a working knowledge of the *components* and *circuits* found in many basic fluid power systems, and (2) to introduce students to the handful of *physical principles* that govern the behavior of these systems.

Traditional texts in this subject area tend to present only one side of the picture. Some are little more than parts catalogs or "how-to" books that fail to explain even the simplest of fluid properties and behavior; others feature rigorous mathematical analyses without regard to actual hardware. *Fluid Power Technology* strives to remedy this imbalance between theory and practice, to develop in each student an understanding of why fluids behave as they do, and how this behavior may be controlled by the proper selection of components.

The need for a balanced approach to the study of fluid power is obvious: Although the form and function of specific components may evolve with time, basic fluid behavior will not. In fact, many of the fundamental laws upon which the operation of fluid power systems is based were formulated during the sixteenth and seventeenth centuries. Even the concepts presented by Archimedes prior to his death in 212 B.C. have not been superseded by more sophisticated theories through the intervening centuries. Although our understanding and application of fluid power have advanced with time, the basic principles continue to exhibit a durability and versatility that is quite remarkable when compared with other areas of science and technology.

1.1 Fluid Power Defined

The expression *fluid power* may be used to describe any process, device, or system that converts, transmits, distributes, or controls power through the use of a pressurized liquid or gas. *Hydraulic* systems use a *liquid* as the working fluid, while *pneumatic* systems operate using a *gas*.

Because of their versatility, fluid power systems are common in industry, where they may be employed in a wide variety of processing or manufacturing operations. They also abound in daily life. The water supply systems that provide us with liquid for drinking, showering, and fighting fires are simple fluid power systems, as are the automatic transmission, hydraulic brakes, air conditioning, ventilating, engine lubrication, and cooling systems on the motor vehicles that we drive. Even the service vehicles used to plow snow from our highways or to pick up and compact our trash generally contain auxiliary fluid power systems that allow them to perform their designated functions.

Figure 1–1   Powerful hydraulic systems are used to produce the enormous forces required on many types of construction equipment. Cylinders such as the ones shown on these wheeled loaders may be used to lift, position, extend, and retract the various blades, buckets, and booms commonly found on such equipment.

Fluid power systems have several distinct advantages over electrical or conventional mechanical systems.

- Unlike most electromechanical systems, fluid power systems are amazingly straightforward. Just a handful of physical principles govern their operation, which can be visualized easily because all their components are mechanical. This feature also simplifies the troubleshooting and repair of malfunctioning systems.

- Because they act as hydraulic levers, single- and multiple-piston hydraulic systems are absolutely unsurpassed in the production of sustained, controllable forces at incredibly high levels. For applications in construction (Figure 1–1), agriculture, and manufacturing (Figure 1–2) that require tremendous forces to lift, bend, shear, clamp, or press, these forces are invariably provided by a pressurized liquid within a closed hydraulic system. Comparable forces may also be produced by a free flowing stream of liquid; the cutting jet shown in Figure 1–3 consists of a high-velocity liquid whose dynamic force is used to slice through materials such as

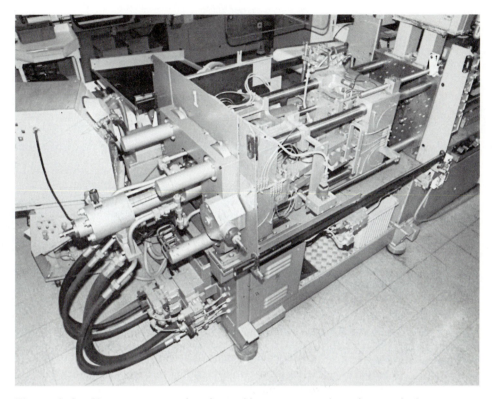

Figure 1–2   Plastic parts are often formed by injecting molten plastic at high pressure into a mold consisting of two or more sections. Molding machines such as the one shown here contain hydraulic pistons that exert clamping forces to prevent the mold from opening during the injection process. The forces provided by these pistons are multiplied mechanically by a toggle mechanism; although clamping force varies with mold size (90 tons for the machine pictured), total forces in excess of two million pounds are not unusual.

granite, steel, and titanium alloys with both accuracy and precision. Examples such as these truly illustrate the absolute "power" available with fluid power systems.

• Their simplicity, durability, and compact size also make individual components and whole systems well suited to mobile operation. Each jaw of the portable rescue tool shown in Figure 1–4, for example, produces a 35-ton cutting force, yet the entire unit, including fluid and fittings, weighs less than 40 lb. Capable of cutting through automobile bodies, this hand-operated device is part of a hydraulic system including rams, spreaders, and power supply that is used by fire departments and rescue squads to extricate accident victims who have been trapped inside their vehicles.

Figure 1–3 Water containing an abrasive grit is pressurized to 45,000 psi (311 MPa) to produce the cutting jet shown here slicing through a 1-in.-thick stone slab. (*Waterjet Industries, Inc., Brooklyn, N.Y.*)

- Fluid power systems are able to deliver smooth, uniform power concurrently to multiple sites and offer the convenience of remotely controlling this power from a central location. A good example of this ability is the hydraulic brake system on any modern automobile (Figure 1–5). Activated at one point (the brake pedal), the system amplifies this input, splits it into separate outputs for front/rear or diagonal brake combinations, and delivers these outputs simultaneously to each of the four wheels. Imagine the complexity of an equivalent system of mechanical linkages and levers that would produce the same result and the difficulty of keeping such a system adjusted for uniform braking!

- Because of the ability of a fluid to move and change shape, hydraulic and pneumatic systems allow for a variety of power conversions with a minimum of mechanical hardware. The power of a motor's rotating shaft may readily be converted to linear form using a pump/compressor connected to a single- or double-acting cylinder, and similar conversions are possible for various combinations of rotary-rotary, linear-rotary, and linear-linear motion.

- Fluid power systems, particularly pneumatic systems, are well suited to the automation of simple, repetitive tasks. Robots such as the one shown in Figure 1–6 are commonly used in the plastics industry to load/unload, position, and assemble

Figure 1–4   This portable rescue tool, commonly known as the Jaws of Life®, uses hydraulic power to shear through the windshield and door posts of an automobile with ease, yet may be operated and maneuvered by one person. (*Jaws of Life® is a registered trademark of Hurst Emergency Products, a division of Hale Products, Inc., Conshohocken, Pa.*)

    high volumes of parts. Often such parts are small or delicate, and frequently they must be handled under sterile or clean-room conditions.

- Most fluid power components are able to withstand considerable abuse and function well under extreme conditions such as those involving shock and vibration. They are especially well suited to operation in explosive environments, and the development of fire-resistant and nonflammable hydraulic liquids has extended their use in high-temperature applications as well.

    Like anything else, however, fluid power systems are not perfect. They are subject to *leaks,* which can disrupt manufacturing operations and degrade clean-room conditions. Behavior of a system may also be affected by *contamination* from external sources. And like high voltages, high pressures can constitute a *safety* hazard unless adequate precautions are taken. Despite these potential disadvantages, fluid power systems continue to provide the motive power for a wide range of industries. During the coming decade, integration of electronic and computer process controls with

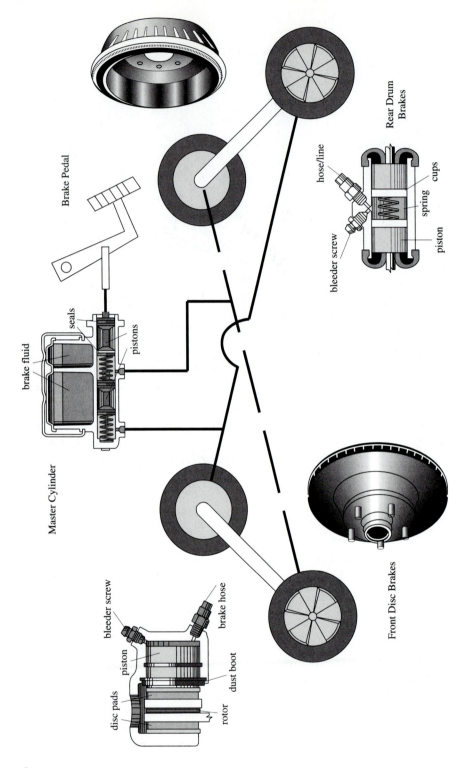

Figure 1–5 The components in a typical automobile braking system. When activated by brake pedal movement, dual pistons in the master cylinder each pressurize a separate circuit consisting of one front disc brake and its diagonally positioned rear drum brake. These two independent circuits deliver smooth, steady braking forces to all four wheels simultaneously and can provide the vehicle with some degree of braking stability even if one circuit fails.

Figure 1–6 Pneumatic robots may be designed to provide a range of movements that will successfully automate simple repetitive tasks. One example involves production of the plastic cases used by the recording industry to package compact discs. During each work cycle of an injection molding machine, the two robots shown remove bodies and lids for four complete cases (bottom center) and transfer them to another automated system for assembly and packaging. In a 24-h period, each robot handles 30,000 components.

conventional fluid power systems will increase dramatically, thus enhancing the versatility of these systems.

1.2   The Industry and Its Acronyms

Both nationally and internationally, the fluid power industry is substantial. In the United States alone, sales of fluid power equipment grew from a meager $1 million at the end of World War II to $8.9 billion in 1991. Figure 1–7 shows that the United States is a global leader in the manufacture and sales of fluid power products, outstripping its nearest competitors by a wide margin. Within the array of American industries, fluid power is larger than both mining machinery and machine tools, and is one of the few manufacturing areas able to maintain a trade surplus. In 1996, U.S. fluid power exports exceeded imports by $22 million.

As Figure 1–8 illustrates, the U.S. fluid power industry has grown steadily in the past decade and a half. Sales of hydraulic equipment exceed those of pneumatic equipment by

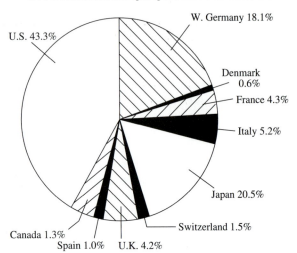

1995 North America/Europe/Japan Fluid Power Production

U.S. 43.3%

W. Germany 18.1%

Denmark 0.6%

France 4.3%

Italy 5.2%

Japan 20.5%

Switzerland 1.5%

Canada 1.3%

Spain 1.0%

U.K. 4.2%

Figure 1–7 1995 international fluid power market. (*National Fluid Power Association, Milwaukee, Wis.*)

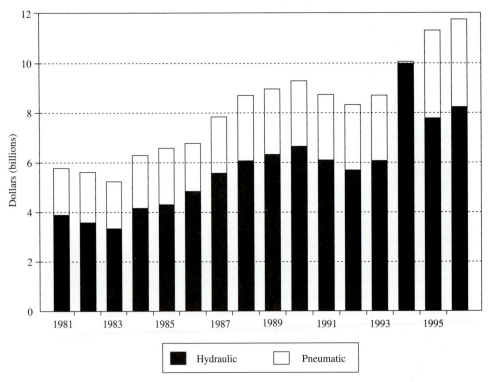

U.S. Fluid Power Industry Shipments

■ Hydraulic   □ Pneumatic

Figure 1–8 U.S. fluid power shipments. (*Sources: Data from U.S. Department of Commerce, MA35N; NFPA, CSS Program. Illustration from National Fluid Power Association, Milwaukee, Wis.*)

Table 1–1 Leading Markets for Hydraulic and Pneumatic Equipment in 1996 (*National Fluid Power Association, Milwaukee, Wis.*)	**Market**	**Percent**
	Hydraulics:	
	Aerospace	20.0
	Construction equipment	24.3
	Farm equipment	12.1
	Highway vehicles	11.3
	Pneumatics:	
	Aerospace	20.0
	Factory automation	15.5
	Food products machinery	3.0
	Packaging machinery	5.0

a ratio of approximately 3 to 1. The four top markets for hydraulics and pneumatics are shown in Table 1–1.

A number of organizations provide input to the fluid power industry, much of it aimed at establishing design and performance standards for hydraulic and pneumatic products. (See Figure 1–9.) Since references to these organizations, usually in acronym form, are often found in the textbooks, catalogs, manuals, and specification sheets of the industry, a brief description of the major associations and their functions is in order.

The National Fluid Power Association (NFPA) is a not-for-profit trade organization whose 180+ member companies account for 75% of all U.S. fluid power products. (This NFPA is often confused with the National Fire Protection Association, which uses the same acronym and itself establishes standards for fire-fighting apparatus such as water pumps.) The NFPA seeks to establish technical guidelines, or standards, for the fluid power industry and collects statistical data pertaining to sales and use of equipment. It works closely on the development of standards with both the American National Standards Institute (ANSI) and the International Organization for Standardization (ISO). A federation of professional and scientific societies, trade associations, and individual company members, ANSI reviews and approves national standards in a wide variety of technical areas. ISO, composed of national standards bodies from 83 nations, acts in a similar fashion.

One of NFPA's most important functions is the dissemination of information through its fluid power publications, including individual volumes of the NFPA standards. These standards, many of which are approved by ANSI and ISO as well, establish guidelines ranging from the use of graphic symbols and preferred SI units, through the design, testing, and presentation of results for individual components.

A similar organization is the Hydraulic Institute (HI), which is an association of pump manufacturers whose objectives include the development and publication of standards for pumps. Their book of standards for centrifugal, rotary, and reciprocating pumps provides valuable information pertaining to the classification, installation, and operation of all types of pumps. In addition, an engineering data book published by the Institute contains detailed

American National Standards Institute(ANSI)
11 West 42nd Street
New York, NY 10036
(212) 642-4900
www.ansi.org

American Petroleum Institute (API)
1220 L Street, N.W.
Washington, DC 20005
(202) 682-8000
www.api.org

American Society for Testing and Materials (ASTM)
1916 Race Street
Philadelphia, PA 19103
(610) 832-9500
www.astm.org

Compressed Air and Gas Institute (CAGI)
1300 Sumner Avenue
Cleveland, OH 44115
(216) 241-7333
www.taol.com/cagi

Department of Transportation (DOT)
400 Seventh Street, S.W.
Washington, DC 20590
(202) 366-4000
www.dot.gov

Fluid Power Distributors Association
P.O. Box 1420
Cherry Hill, NJ 08034
(609) 424-8998
www.fpda.org

Fluid Power Educational Foundation (FPEF)
c/o National Fluid Power Association
(see address below)
(414) 778-3364
www.fpef.org

Fluid Power Society (FPS)
2433 North Mayfair Road
Milwaukee, WI 53226
(414) 257-0910
www.ifps.org

Hydraulic Institute (HI)
9 Sylvan Way
Parsippany, NJ 07054
(973) 267-9700

International Organization for Standardization (ISO)
Case Postale 56
CH-1211
Geneva, Switzerland
(Country Code 41) 227490111

Figure 1–9   Major organizations with links to the fluid power industry. Where available, Web page addresses on the World Wide Web are given.

information on complete hydraulic systems, covering such topics as friction factors for pipe and pressure losses through valves and fittings.

On the pneumatic side, the Compressed Air and Gas Institute (CAGI) is a nonprofit organization comprised of companies that manufacture air and gas compressors, pneumatic machinery, and air- and gas-drying equipment. CAGI prepares educational literature and films used to promote the industry, develops and publishes standards and engineering data for compressors and related equipment, and participates in cooperative educational and research activities.

Another acronym often seen in fluid power literature is that of the former Joint Industry Council (JIC). This organization was originally formed almost three decades ago by manufacturers concerned with standardizing the use of hydraulic, pneumatic, electrical, and electronic equipment for uses in manufacturing. Membership included General Motors Corporation, Ford Motor Company, and the Aluminum Company of America. The JIC documents produced between 1967 and 1975 all pertain to the design, installation, performance, maintenance, and safety of fluid power, electrical, and electronic systems.

National Fire Protection Association
(NFPA)
One Batterymarch Park
P.O. Box 9101
Quincy, MA 02269-9101
(617) 770-3000
www.nfpa.org

National Fluid Power Association (NFPA)
3333 North Mayfair Road
Milwaukee, WI 53222-3219
(414) 778-3344
www.nfpa.com

National Machine Tool Builders Association
(NMTBA)
7901 Westpark Drive
McLean, VA 22102
(703) 893-2900
www.mfgtech.org

Naval Publications and Forms Center
(NPFC)
5801 Tabor Avenue
Philadelphia, PA 19120
(215) 697-2000

Occupational Safety and Health Administration
(OSHA)
Department of Labor
200 Constitution Avenue N.W.
Washington, DC 20210
(202) 523-1452
www.osha.gov

Penton Publishing
1100 Superior Avenue
Cleveland, OH 44114
(216) 696-7000
www.fpweb.com

Society of Automotive Engineers (SAE)
400 Commonwealth Drive
Warrendale, PA 15096-0001
(412) 776-4841
www.sae.org

Society of Tribologists and Lubrication Engineers
(STLE)
(formerly the American Society of Lubrication
Engineers)
840 Busse Highway
Park Ridge, IL 60068
(847) 825-5536
www.stle.org

Shortly thereafter, JIC became inactive, and for a short time its role was carried on by the National Machine Tool Builders Association (NMTBA). However, in 1981 NMTBA asked the NFPA to update JIC's standards for hydraulic/pneumatic machinery, and the National Fire Protection Association to do the same for electrical/electronic machinery. The resulting fluid power documents are entitled *NFPA/JIC Pneumatic Systems* and *NFPA/JIC Hydraulic Systems,* and establish standards for systems of industrial machinery. In addition, the JIC designation is still often used to denote individual components whose design or performance was originally specified by this group (see Figure 7–72).

Although the American Petroleum Institute (API) is best known for its statistical data on the production and supply of petroleum-based products, this group also develops standards related to the hydraulic and pneumatic equipment found in all areas of the refining process. API also makes available to both companies and individuals a series of educational videos on topics ranging from basic fluid mechanics to the operation of specific components such as pumps, compressors, and oil-and-gas separators.

In Chapter 2 we examine the important physical properties of a number of hydraulic liquids. Standardized methods for the experimental measurement of these properties have been established by the American Society for Testing and Materials (ASTM), and may be found in the ASTM standards published annually by this organization. The 1998 edition consists of 72 separate volumes describing standard test methods for a wide variety of materials, including the petroleum liquids most often used in hydraulic systems. The standards, which are also available on CD-ROM, describe in detail not only tests for such properties as viscosity, flash point, thermal conductivity, water content, corrosiveness, and oxidation stability, but also standard practices for cleaning, flushing, and purification of hydraulic systems.

Minimum required values for various properties of hydraulic liquids are often specified either by professional or governmental organizations. The Society of Automotive Engineers (SAE), for example, has established ratings for many liquids used in automotive applications, including brake fluids and engine lubricating oils. (Oil designations such as "SAE 10W-30" use this system.) The National Highway Traffic Safety Administration's Department of Transportation (DOT), which also imposes stringent requirements on brake fluid properties, is one of several governmental agencies that specify characteristics of liquids or other fluid power products found in automotive applications. Lubrication characteristics and recommended practices concerning the use of fluids and seals in hydraulic equipment are available from the Society of Tribologists and Lubrication Engineers (STLE). Products used by America's armed forces must also satisfy rigorous military specifications, details of which are available from the Naval Publications and Forms Center (NPFC). Manufacturers' catalogs and specification sheets often refer to these individual rating systems to describe the minimum level at which a product is guaranteed to perform. Of the many companies that actually manufacture or distribute hydraulic/pneumatic components, approximately 1000 are members of the Fluid Power Distributors Association.

Two other organizations also deserve mention because they deal not with products or standards, but with people. The Fluid Power Educational Foundation (FPEF) is administered by NFPA and strives to develop fluid power courses in school curricula. In addition, FPEF awards scholarships and provides informational resources to students, educators, and industry.

The Fluid Power Society (FPS), often referred to in print as "FPS—the International Organization for Fluid Power and Motion Control Professionals," strives to provide educational information to engineers, technicians, specialists, mechanics, and other professionals involved in the area of fluid power technology. Besides publishing a series of reference and self-study manuals, as well as the bimonthly *Fluid Power Journal,* FPS administers certification tests for fluid power technicians at various levels of proficiency.

Finally, two periodicals available through Penton Publishing are the monthly *Hydraulics & Pneumatics Magazine,* written for technical personnel involved with the design, manufacture, and maintenance of fluid power and control systems, and *Fluid Power Service Center,* a quarterly publication aimed at the staff of hydraulic maintenance and repair facilities.

Addresses and telephone numbers for all of the major organizations discussed above are listed in Figure 1–9.

1.3   Levels of Learning

To a certain extent, you may select the level at which you are introduced to the field of fluid power.

- If you are interested in a general overview of the principles and hardware, you will find a heavily illustrated text written in readable prose style and supplemented by questions and suggested activities where appropriate.

- If you are seeking to reinforce the technical concepts through numerical examples, you may do so at a level and pace determined by your own mathematical skills. Students having a general knowledge of algebra will find numerous worked examples to help in the solution of those problems presented at the end of individual sections, while advanced students should find the comprehensive problems at the end of each chapter both challenging and informative.

- If you have access to a computer, you can take advantage of suggested activities and problems presented throughout the text.

Although "number crunching" is not the primary objective of this book, it should develop your self-confidence and aid in your understanding of the material. The next section, then, establishes some ground rules for this process.

1.4   Numerical Considerations

Throughout this book, numerical examples are used to illustrate the physical principles that govern the behavior of fluid power systems and components. Although most U.S. industries continue to describe their products quantitatively in *U.S. customary units* (formerly referred to as British or English units), other nations generally employ *SI units* (for Système International d'Unités, or International System of Units), a revised form of the metric system. Since the typical workplace often contains a mix of foreign and domestic products, a familiarity with both unit systems is essential to those working in the area of fluid power. For this reason, examples and problems found in the text are approximately balanced between U.S. customary and SI units.

Many of the numerical computations presented involve the process of *conversion,* either between systems of units or to an equivalent set of units within the same system. As shown in the following examples, conversions may be made one of two ways: by *direct substitution* or through the use of *conversion factors.*

EXAMPLE 1

Convert 75 mm to an equivalent number of inches.

Solution   The distance $d = 75$ mm may be written as $d = 75 \times (1$ mm$)$. From tables inside the back cover of this book, we find that:

$$1 \text{ in.} = 25.4 \text{ mm}$$

Dividing both sides of this equality by 25.4 *without units* yields:

$$\frac{1 \text{ in.}}{25.4} = \frac{25.4 \text{ mm}}{25.4}$$

or:

$$0.03937 \text{ in.} = 1 \text{ mm}$$

Since these quantities are equal, direct substitution yields:

$$d = 75 \times (1 \text{ mm}) = 75 \times (0.03937 \text{ in.})$$

or:

$$d = 2.953 \text{ in.}$$

Although the method of substitution is suitable for direct conversions, those involving multiple steps are best approached through the use of conversion factors. Let us repeat the problem above beginning with the equality:

$$1 \text{ in.} = 25.4 \text{ mm}$$

Dividing both sides by 25.4 mm *including units* yields:

$$\frac{1 \text{ in.}}{25.4 \text{ mm}} = \frac{25.4 \text{ mm}}{25.4 \text{ mm}}$$

Since the units, or dimensions, in a fraction may be canceled according to the rules for numbers, our division results in:

$$\frac{1 \text{ in.}}{25.4 \text{ mm}} = 1$$

The quantity on the left, known as the conversion factor, has a numerical equivalent of unity with no associated units. Therefore, multiplying any quantity by the conversion factor is equivalent to multiplying by unity, a process that has no algebraic effect on the original quantity. Using this factor, our conversion becomes:

$$d = 75 \text{ mm} \times 1 = 75 \text{ mm} \times \frac{1 \text{ in.}}{25.4 \text{ mm}} = \textbf{2.953 in.}$$

Notice that, for this problem at least, the actual computational operations are identical with those performed using substitution and that some dimensional analysis is necessary to determine whether the factor should contain units of inches over millimeters or millimeters over inches. The advantage of this method for multiple conversions is demonstrated by the next example.

EXAMPLE 2

The liquid in a jet of water is traveling at 30 mi/h. Find its equivalent velocity, *v*, in ft/s.

Solution   The appropriate equalities from tables inside the back cover are:

$$1 \text{ mi} = 5280 \text{ ft}$$

$$1 \text{ h} = 60 \text{ min}$$

$$1 \text{ min} = 60 \text{ s}$$

Forming the appropriate conversion factors and multiplying gives:

$$v = 30 \frac{\cancel{\text{mi}}}{\cancel{\text{h}}} \times \frac{5280 \text{ ft}}{1 \cancel{\text{mi}}} \times \frac{1 \cancel{\text{h}}}{60 \cancel{\text{min}}} \times \frac{1 \cancel{\text{min}}}{60 \text{ s}} = \textbf{44.0 ft/s}$$

Our final consideration here deals with computational accuracy as indicated by the number of *significant digits* found in our input data. What are significant digits in a given number? Any nonzero digit is considered significant; zeros that appear at the left and right of a number and are used to locate the decimal point are not significant. Thus the numbers 0.00345, 3.45, 3450, and 3,450,000 all contain three significant digits. Zeros bounded by significant digits are also considered significant; 0.0005067, 5.067, 50.67, and 506,700 each contain four significant digits. Although it is not a common practice, an overbar may be used to indicate that one or more trailing zeros is significant; 123,0̄00 contains four significant digits. Notice that the number of significant digits should not be confused with the number of *decimal places* contained in any particular numerical value.

Significant digits are important because numbers themselves may be *exact* or *approximate*. In real applications, numbers are exact only if they are obtained by *counting,* or as part of a *definition*. For example, a freight train may consist of an engine and 38 boxcars. The number 38 is exact because it is determined by counting, and the counted units (boxcars) are not divisible. Similarly, we know that a linear distance of 6 ft is exactly equal to 72 in., because 1 ft, by definition, contains 12 in.

Most other numbers, however, are *approximate,* because they are obtained by *measurement*. In fact, all measured values are assumed to contain some amount of error, resulting either from *fluctuations* in the measured quantity itself, systematic or random errors due to the measuring *instruments* used, or *operator* errors introduced by reading and recording the indicated values.

Whatever the sources of error, computations performed using approximate numbers can produce only approximate results. For this reason and to maintain consistency in such calculations, measured values are generally *rounded* to a specific number of significant digits. To round any number, examine the digit immediately to the right of the last significant digit desired.

- If this digit is 5 or greater, add 1 to the last significant digit.
- If this digit is less than 5, leave the last significant digit unchanged.
- In both cases, either discard all digits following the last significant digit or change them to zeros if they are necessary to maintain decimal point location.

EXAMPLE 3

Round these measured values to the indicated number of significant digits:

(a) 107.1968 ft to 4 significant digits

(b) 3.294 psi to 3 significant digits

(c) 86.5 ft/s to 2 significant digits

Solution   Inspect the digit (underlined) immediately to the right of the last desired significant digit and follow the given rules:

(a) 107.1968 ft   becomes   107.2 ft

(b)   3.294 psi   becomes   3.29 psi

(c)   86.5 ft/s   becomes   87 ft/s

(In instances where the number after the last significant digit is a 5, some industries utilize a variation known as the *round-even rule;* here, the last significant digit is rounded up only if this process yields an even (0, 2, 4, 6, 8) digit. Under this rule, for example, the given value of 86.5 ft/s would become 86 ft/s, whereas a value of 81.5 ft/s would become 82 ft/s. The reason for using such a procedure is quite simple: When the number following the last significant digit is a 5, it is just as reasonable to round the significant figure *down* as it is to round it *up.* If such numbers are always rounded up, it may have the effect of *skewing* a calculation or distribution of numbers to the high side. Because a significant figure is just as likely to be odd as even, rounding to an even digit should take place in about half the measured values, thus minimizing the likelihood of skewing.)

For approximate quantities, *accuracy* is defined simply as the number of significant digits contained in a particular value. Thus, 54,380,000 (with four significant digits) is considered to be more accurate than 0.921 (with only three significant digits).

Precision, on the other hand, is used to indicate the "fineness" to which a quantity is known. For instance, consider a piece of tubing whose diameter is measured and found to have an approximate value of 8.76 mm. Although this figure is accurate to only three significant digits, its level of precision is said to be 1 one-hundredth of a millimeter, because the final significant digit is located two places to the right of the decimal point. By contrast, if the length of this tubing is given as 1075.4 mm, the numerical value is accurate to five significant digits, but its level of precision is only 1 tenth of a millimeter. A simple method for comparing the precisions of numbers is to write the numbers in a column with their decimal points aligned; the number whose final significant digit is furthest to the right is the most precise.

The distinction between accuracy and precision is an important one, as indicated by these algebraic procedures recommended for computations involving measured values:

- If approximate numbers are *multiplied or divided,* the final result should be rounded to the *accuracy* of the least accurate number used in the computation.

- If approximate numbers are *added or subtracted,* the final result should be rounded to the *precision* of the least precise number used in the computation.

The following example demonstrates how and why these rules are applied.

EXAMPLE 4

Perform the indicated calculations, rounding the results to an appropriate number of significant digits.

(a) Standard pipe designated *nominally* (in name only) as half-inch pipe actually has an inside diameter of 0.622 in. Compute the cross-sectional area for an opening of this size.

(b) A reservoir containing hydraulic fluid weighs 997 lb. If a filter weighing 6.09 lb is installed, calculate the total weight of the unit.

(c) For a cubical storage tank of volume *V,* side length, *s,* may be computed as:

$$s = \sqrt[3]{V}$$

Find the dimensions of a storage tank whose volume is 235 in.^{3}.

Solution   (a) The area, *A,* of a circle of diameter *D* is computed as:

$$A = \frac{\pi}{4} \times D^2$$

Pi (π) is the Greek letter that represents a constant whose accuracy is known to several *million* significant digits and whose value has such importance in technology that it is given its own key on all scientific calculators. The value of π shown on a typical 10-digit calculator is 3.141592654. When this value is divided by the constant 4 from the preceding formula, the resulting numerical value is (3.141592654/4) = 0.785398163, so the given formula may be rewritten as

$$A = 0.785398163 \times D^2$$

Because the numerical value of diameter contains only three significant digits, the constant itself could be rounded to 0.785 or to the more conservative 0.7854 (a value that you will see frequently throughout this book). Computing *A* for these three values of the constant gives:

$$A = 0.785398163 \times (0.622 \text{ in.})^2 = 0.303857983 \text{ in.}^2$$

$$A = 0.7854 \times (0.622 \text{ in.})^2 = 0.303858694 \text{ in.}^2$$

$$A = 0.785 \times (0.622 \text{ in.})^2 = 0.303703940 \text{ in.}^2$$

Because hand calculators will produce as many "significant" figures as there are display spaces on the screen, all three answers imply an accuracy of nine digits. In fact, these answers agree only to the accuracy of the given diameter (three significant digits), regardless of the number of digits retained in our constant value. Because multiplication is the only operation used here, our result should be rounded to the same accuracy as the least accurate factor:

$$A = 0.304 \text{ in.}^2$$

(b) Simple addition yields a combined weight of:

$$
\begin{array}{r}
997. \quad \text{lb} \\
+ \quad 6.09 \text{ lb} \\
\hline
1003.09 \text{ lb}
\end{array}
$$

Because our original weights were *added,* this answer should be rounded to the *precision* of our least precise value (the 1-lb. digit in the 997-lb. value), or 1003 lb. Note that if the given weights had themselves been obtained using standard procedures for rounding, their actual values could be anywhere in these ranges:

	Minimum Weight	Maximum Weight
Reservoir	996.5 lb	997.4 lb
Filter	6.085 lb	6.094 lb
Combined	1002.585 lb	1003.494 lb

Each of the combined values agrees to the same level of precision (1003 lb) as that obtained from simple addition of the given weights.

(c) Computing the cube root of 235 in.^{3} as well the roots of its extreme un-rounded values gives:

$$s = \sqrt[3]{234.5 \ \text{in.}^3} = 6.166626085 \text{ in.}$$

$$s = \sqrt[3]{235 \ \text{in.}^3} \ = 6.171005793 \text{ in.}$$

$$s = \sqrt[3]{235.4 \ \text{in.}^3} = 6.174505088 \text{ in.}$$

These answers agree to the third significant digit (6.17 in.), suggesting that roots should be rounded to the *accuracy,* not the precision, of their radicand (the number under the radical sign).

Even if these rules are faithfully applied (as in all sample calculations throughout this text), values of the final significant digit may vary slightly. In practice, computations are often performed without rounding either initial or intermediate values. (For example, once the quantity ($\pi/4$) has been obtained and displayed on a hand calculator, it is more cumbersome to clear the display and enter a rounded value such as 0.785 or 0.7854 than it is to proceed using the full displayed value of 0.785398163.) Final results are then rounded to either three or four significant digits, allowing for possible errors of 1% or less. (Note that a three-digit number that rounds to 101 instead of 100 contains a 1% error, while a four-digit number that rounds to 1001 instead of 1000 contains a 0.1% error.) Except in very unusual circumstances, such accuracies are more than adequate for fluid power applications. As the examples of Chapter 4 will illustrate, systems are often designed with capacities increased by factors ranging from 10% to 30% to accommodate intermittent overloads, and anyone who has ever tried to measure pressures within a dynamic or flowing system realizes that fluctuations of the gauge readings seldom remain within 1% of their actual values.

Finally, please note that an attempt has been made to follow the common or accepted usage of all symbols that appear in *Fluid Power Technology*. Although each symbol is defined in the text near that formula or equation in which it is contained, a summary list of symbols may also be found inside the front cover of the book.

Problem Set I

Note: Answers to odd-numbered problems appear at the end of the book.

1. Convert 400 in.^{3} to cm^3.

2. Convert 100,000 cm/min to km/h.

3. Knowing that 1 m = 100 cm = 1000 mm, convert 87.2 cm^2 to mm^2.

4. If 1 ft^3 of water weighs 62.4 lb, how much does 1 gal weigh?

5. Electric power is generally measured in watts (W), whereas mechanical power is often measured in horsepower (hp). If 1 hp = 745.5 W:
   (a) Find the horsepower of a 60-W lightbulb.
   (b) How many watts are equivalent to a 210-hp automobile engine?

6. United States customary units for pressure are given in pounds per square inch, abbreviated as psi. Because force is measured in SI units of newtons (N), SI pressures are often given in N/m^2. (These units are usually given as pascals (Pa) in honor of the French philosopher and mathematician who was a pioneer in the field of hydraulics.) Using the following equivalents, convert 65.0 psi to Pa.

$$1 \text{ lb} = 4.448 \text{ N}; \qquad 1 \text{ in.}^2 = 6.452 \text{ cm}^2 \qquad 1 \text{ cm} = 0.01 \text{ m}$$

7. Three pieces of steel pipe are measured and found to have lengths of 42.6 ft, 3.15 ft, and 157 ft. Based upon the number of significant figures given, what is the possible range of values for:
   (a) The 42.6-ft piece?
   (b) The 3.15-ft piece?
   (c) The 157-ft piece?
   (d) Add the given values and round off the answer to an appropriate number of significant figures.
   (e) Add the three minimum dimensions for the pipes and round to the same number of significant figures as part (d). How do the answers compare?
   (f) Repeat (e) using the maximum dimensions.

8. The side of a cube measures 28.7 cm. Compute the volume of this cube. Recompute the volume using the probable maximum and minimum values implied by the significant figures.

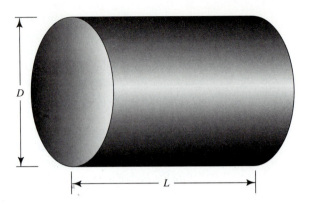

Figure I–10   A right circular cylinder of diameter *D* and length *L.* (See problem 9.)

9. A right circular cylinder is shown in Figure 1–10. The total surface area, *A,* for such a cylinder is given by the formula:

$$A = \pi(D \times L + D^2/2)$$

Compute the area of a cylinder whose diameter is 30.5 cm and whose length is 102 cm using two different methods:

(a) Round the *intermediate products* ($D \times L$) and ($D^2/2$); the *sum* obtained by adding these two products; and the *final result* obtained after multiplication by π.

(b) Carry as many significant digits as your calculator will allow, rounding *only* the final answer.

How do these two computed values compare?

10. Add the following two measured lengths, rounding your answer to an appropriate number of significant digits:

$$2.30 \text{ ft} + 6.50 \text{ in.}$$

11. The *dynamic pressure head* for a moving liquid may be calculated as:

$$\frac{v^2}{2g}$$

where *v* is the liquid velocity and *g* is a measured quantity known as the *acceleration of gravity.* If *v* for a jet of water is approximately 76.4 ft/s and the average value of *g* is 32.1740 ft/s², compute the dynamic pressure head for this liquid, rounding your result appropriately. What units are associated with the numerical value?

12. The *moment,* or *torque,* produced by a force is a numerical value assigned to the *twisting effect* that the force creates about a particular point. If the force, *F,* and a line of length *d,* drawn from the point to the force, are perpendicular (Figure 1–11), the moment about this point (center of bolt) is defined as:

$$\text{Moment} = F \times d$$

(Notice that the *direction* of such a twisting effect must either be *clockwise* or *counterclockwise.*) When a force of $F = 257.8$ lb is applied to a *moment arm* whose length is $d = 9.63$ ft, compute the torque produced by this force about *A,* rounding your answer appropriately.

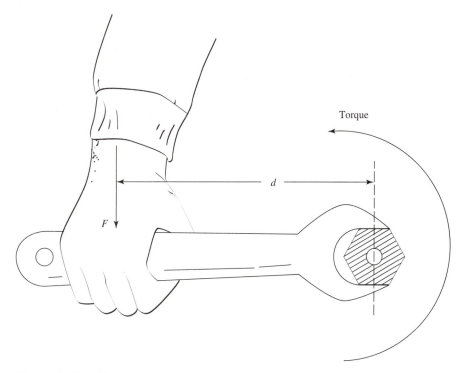

Figure 1–11   The twisting effect of a force about a point is called the moment, or torque, of the force. (See problem 12.)

1.5   Using Computers

As in other areas of technology, some familiarity with computers is desirable for anyone whose work involves fluid power. Not only are these machines well suited to repetitive calculations, but many manufacturers today offer free software that allows for circuit design, component selection, and pricing of systems utilizing their specific products.

Although it is not the intent of this book to teach computer applications, two useful options—*BASIC programming* and *spreadsheets*—appear on virtually all modern personal computers and are discussed briefly in this section. Use of these computational tools is demonstrated in examples 5 and 6; other computer problems and suggested exercises appear throughout the text, each denoted by a computer monitor and keyboard symbol, as shown on the two following examples.

EXAMPLE 5

 Repetitive numerical calculations may be performed using a simple computer programming language called *Beginner's All-Purpose Symbolic Instruction Code.* Known universally by its acronym, *BASIC,* this language (or one of its several variations) provides the computer

Table 1–2   Some Common Algebraic Operations Represented In BASIC

Operation	BASIC Equivalent
$5x + 4$	$5 * x + 4$
$\dfrac{9x}{8}$	$9 * x / 8$
$\dfrac{x - 7}{6}$	$(x - 7) / 6$
$3x^2$	$3 * (x \char94 2)$ or $3 * x * x$
2^x	$2 \char94 x$
$\sqrt{x}$	$x \char94 0.5$ or SQR (x)

with a set of directions that specify the exact algebraic operations to be carried out. Although some versions of BASIC may differ slightly in their representations for specific algebraic operations, Table 1–2 lists a few of the more common computational equivalents, which are essentially the same for all versions of this programming language.

Write a BASIC computer program that generates a table of circular areas calculated for given diameters. Program inputs should consist of desired units, initial diameter, final diameter, and incremental diameter.

Solution   The following program yields a table of diameters and their corresponding areas. In Step 40, the user is asked to select dimensional units of inches, feet, millimeters, centimeters, or meters. Likewise, numerical values for initial diameter, final diameter, and incremental diameter are requested in Steps 70, 100, and 130, respectively. In Step 160, the computer is already organizing its output in tabular form, but the actual calculations begin with Step 190. Here, the program computes an area for initial diameter, DI, specified in Step 80 using the formula ($A = 0.7854D^2$) introduced in Example 4(a). After Steps 200 and 210 print the corresponding values of diameter and area, Steps 220 through 240 then compare the current value of diameter (DI) to the final value desired (DF). If DI has reached (Step 220) or exceeded (Step 230) the specified value for DF, the program jumps to Step 270 and stops operating. If DI has not yet reached DF (Step 240), the program creates a *new value* of DI (called *indexing*) by adding incremental diameter D to the current DI value (Step 250). Step 260 sends the machine back to Step 190, where it repeats the process by calculating an area for the new diameter (DI). This programming sequence—calculation, comparison to a limiting value, indexing, and calculation of a sequential value—is called *iteration*. Such a program is known as an *iterative loop* because it performs a repeated computational cycle within the specified range of values.

```
 10 REM   CIRCULAR AREA
 20 REM
 30 PRINT
 40 PRINT "ENTER UNITS (SELECT in, ft, mm, cm, or m)"
 50 INPUT "UNITS = ";UNITS$
 60 PRINT
 70 PRINT "ENTER INITIAL DIAMETER (";UNITS$;")"
 80 INPUT "Di = ";DI
 90 PRINT
100 PRINT "ENTER FINAL DIAMETER (";UNITS$;")"
110 INPUT "Df = ";DF
120 PRINT
130 PRINT "ENTER INCREMENTAL DIAMETER (";UNITS$;")"
140 INPUT "d = ";D
150 PRINT
160 PRINT "DIAMETER (";UNITS$;")"    AREA (";UNITS$;"^2)"
170 PRINT "*****************************";
180 PRINT
190 A = .7854*DI*DI
200 PRINT USING " ##.##"; DI;:PRINT "               ";
210 PRINT USING "####.##";A
220 IF DF = DI THEN 270
230 IF DF < DI THEN 270
240 IF DF > DI THEN 250
250 DI = DI + D
260 GOTO 190
270 END
```

When this program is RUN, the following questions appear on the screen. If the user types in the italicized responses shown, a table of diameters and areas is generated.

```
ENTER UNITS (SELECT in, ft, mm, cm, or m)
UNITS = ? mm

ENTER INITIAL DIAMETER (mm)
Di = ? 10

ENTER FINAL DIAMETER (mm)
Df = ? 20

ENTER INCREMENTAL DIAMETER (mm)
d = ? 1
```

```
DIAMETER (mm)          AREA (mm^2)
* * * * * * * * * * * * * * * * * * * * * * * * * * * * *
        10.00              78.54
        11.00              95.03
        12.00             113.10
        13.00             132.73
        14.00             153.94
        15.00             176.71
        16.00             201.06
        17.00             226.98
        18.00             254.47
        19.00             283.53
        20.00             314.16
```

Note that with only slight modification, the generic iteration program presented here can be used to construct tables of values for any two related quantities. (See problem 2 in problem set 2 of this chapter.)

EXAMPLE 6

Another powerful tool used to generate tables of data is a type of computer program known as a *spreadsheet.* Such prepackaged software is available for all types of personal computers, and, except for a few minor differences in program commands, its use and operation are relatively consistent between the various brand names available.

As it appears on the computer screen, a spreadsheet has two important elements: an array of boxes (called *cells*) arranged in rows and columns such as those found in the business ledgers used by accountants and bookkeepers and some type of *editing bar* (usually just a separate line or box located at the top of the computer screen), by which entries on the spreadsheet may be specified or defined. As shown in the abbreviated spreadsheet of Figure 1–12, columns are commonly *lettered* from left to right, whereas rows are *numbered* from top to bottom. Thus the location, or *address,* of any cell may be indicated by its column letter and row number; the addresses A5, C2, and D4 are shown within the cells to which they correspond.

Manipulating spreadsheet data—such as specifying numerical values and mathematical operations for individual cells or groups of cells—is usually achieved by a combination of *keyboard commands* and *cursor movements* involving a *mouse.* The cursor is simply a movable indicator, often in the shape of an arrowhead or cross, that appears on the computer screen. The mouse, a small hand-held box that contains a ball or wheel, is wired to the computer; when rolled around a flat surface (usually a cushioned *mousepad*), the mouse generates electrical signals that control the cursor's position on the screen.

Using spreadsheets, generate a table of data for the volumes (in gallons) of right circular cylinders (see Figure 1–10) for diameters, *D,* from 2 in. to 6 in. and lengths,

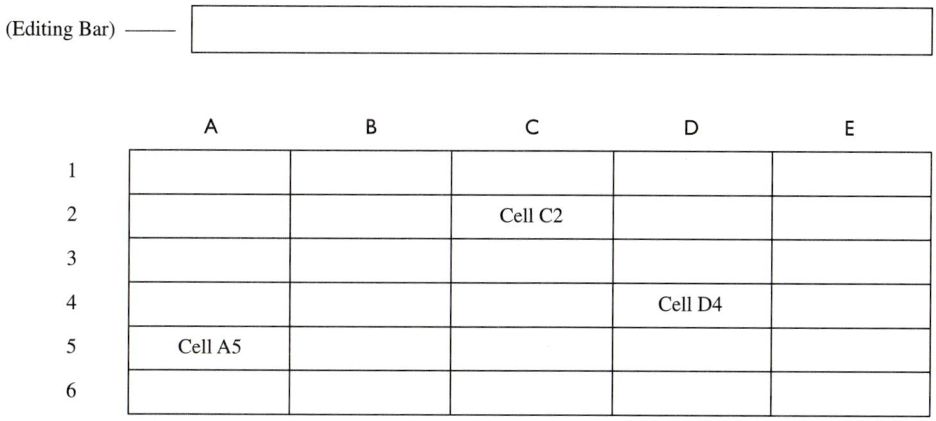

(Editing Bar) ————

	A	B	C	D	E
1					
2			Cell C2		
3					
4				Cell D4	
5	Cell A5				
6					

Figure 1–12   The screen displayed for a typical spreadsheet contains two important elements: an *editing bar* and *cells,* or boxes, arranged in *lettered columns* and *numbered rows.* As shown by the three entries on this partial spreadsheet, the location, or *address,* of any cell is designated by its column letter and row number. (See example 6.)

L, from 12 in. to 24 in. For *D* and *L* values in inches, the volumes, *V,* may be computed as:

$$V = A \times L = \frac{\pi}{4} \times D^2 \times L \times \frac{1 \text{ gal}}{231 \text{ in.}^3} = 0.0034 D^2 L$$

Solution   A good first step is to label the columns for our *D* and *L* values. To do so, we might follow the *typical* sequence of steps given next, again noting that *specific* procedures may differ slightly among spreadsheet software from various manufacturers.

1. Use the mouse to move the cursor to cell A1.
2. With the left-hand mouse button, click "on" to highlight this cell. (See shaded cell in Figure 1–13(a).)
3. Use the mouse to move the cursor up to the editing bar.
4. Use the keyboard to type " = D" into the editing bar space.
5. Click the cursor off this cell by left-clicking onto another cell or by using the arrow keys on the keyboard. The heading D should now appear in cell A1.
6. Repeat Steps 1 to 5 to label cell F1 for L: highlight cell F1; move to the editing bar and type in " = L" (Figure 1–13(b)); click the cursor off. At this point, the spreadsheet should appear as shown in Figure 1–13(c).

(Note that for short entries, such as column headings or numerical values, these procedures can often be streamlined by simply highlighting a desired cell and typing the entry directly into that cell rather than moving to the editing bar. With some spreadsheet packages, this happens automatically.)

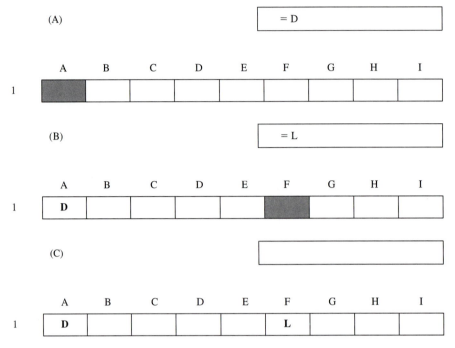

Figure 1–13   To label a column, highlight the top cell in that column; then move to the editing bar and type in the heading using the computer keyboard. (See example 6.)

Our next step is to establish a series of values for both *D* and *L,* but we need not type each value individually. Instead we can follow these steps:

7. Highlight cell A4 and enter an initial *D*-value of 2 in this cell.

8. Highlight cell A5, move to the editing bar, and enter the expression " = A4 + 0.5." This step will generate our second *D*-value by adding 0.5 to the previous value. (If diameter increments of 1 in. or 2 in. are preferred, our typed entries in the editing bar should read " = A4 + 1" and " = A4 + 2" respectively.)

9. The remaining cells in our *D*-column can generally be filled by using either *fill down* or *copy* and *paste* options (available from the *pull-down* menus located somewhere on the computer screen) to copy our directions from Step 8 into the remaining cells of column A. Typically, this is executed by clicking onto the entry of cell A5, holding the mouse key on, and dragging the cursor down the column. Through this procedure, the machine fills in the remaining A-cells, as shown literally in Figure 1–14(a), to produce the resulting numerical values listed in Figure 1–14(b).

10. Similarly, we type an *L*-value of 12 in cell C2, specify that cell D2 should be " = C2 + 2," and copy the contents of cell C2 to fill the remaining cells in row 2, and produce the spreadsheet of Figure 1–15.

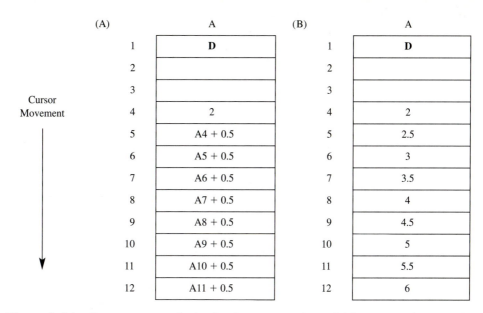

Figure 1–14 Once a numerical value has been entered in cell A4, the entry for cell A5 can be defined in terms of this initial value. "Fill" procedures for specific software generally involve dragging the cursor down the column to copy this relationship into the remaining A-cells. The computer sees the entries in (A) but displays the numerical values in (B) on its screen. (See example 6.)

	A	B	C	D	E	F	G	H	I
1	**D**					**L**			
2			12	14	16	18	20	22	24
3									
4	2								
5	2.5								
6	3								
7	3.5								
8	4								
9	4.5								
10	5								
11	5.5								
12	6								

Figure 1–15 The spreadsheet of example 6 with incremental values of *D* and *L* established in column A and row 2, respectively.

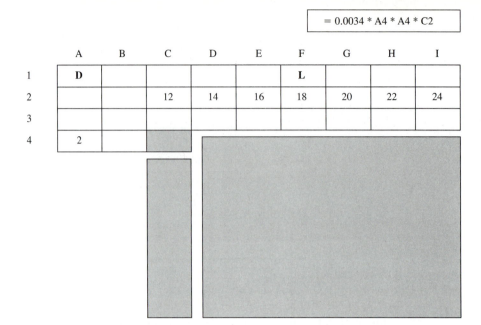

$$= 0.0034 * A4 * A4 * C2$$

	A	B	C	D	E	F	G	H	I
1	**D**					**L**			
2			12	14	16	18	20	22	24
3									
4	2								

Figure 1–16 The entry for cell C4 is defined in the editing bar by a relationship expressed using addresses of the initial values for *D* and *L*. The contents of cell C4 must then be copied into each of the shaded regions separately. Normally this is a two-step process, which involves copying down to fill the column and then copying across to fill the rows of cells under columns D through I. (See example 6.)

	A	B	C	D	E	F	G	H	I
1	**D**					**L**			
2			12	14	16	18	20	22	24
3									
4	2		0.163	0.190	0.218	0.245	0.272	0.299	0.326
5	2.5		0.255	0.298	0,340	0 383	0.425	0.468	0.510
6	3		0.367	0.428	0.490	0.551	0.612	0.673	0.734
7	3.5		0.500	0.583	0.666	0.750	0.833	0.916	1.000
8	4		0.653	0.762	0.870	0.979	1.088	1.197	1.306
9	4.5		0.826	0.964	1.102	1.239	1.377	1.515	1.652
10	5		1.020	1.190	1.360	1.530	1.700	1.870	2.040
11	5.5		1.234	1.440	1.646	1.851	2.057	2.263	2.468
12	6		1.469	1.714	1.958	2.203	2.448	2.693	2.938

Figure 1–17 The completed spreadsheet of example 6 lists volumes (in gallons) for right circular cylinders (see Figure 1–10) of various diameters, *D* (in inches), and lengths, *L* (in inches).

Once the range of values for D and L has been established, we compute the desired volumes as follows:

11. Highlight cell C4 and enter the equation by which this entry is to be computed: "= 0.0034 * A4 * A4 * C2." Note that the operations are specified in terms of the cell addresses for our initial values for D and L. Using appropriate *fill* or *copy* and *paste* procedures, we now fill the shaded areas shown in Figure 1–16 to produce the completed table of Figure 1–17.

Spreadsheets have several advantages over programmed calculations. Unlike the program of example 5, which provides values of only two related quantities (D and A), spreadsheets are able to generate tables involving several related quantities (V, D, and L in example 6). To achieve this same result with a computer program would require one iterative loop within another (called a *nested* loop). Such a program would set an initial value for L, iterate through the specified values of D to compute one column in our table, and then index the L-value and repeat the process for the set of D-values. This procedure would continue until the final L-value has been reached. Writing such a program and organizing the printout into usable tabular form involves considerably more work than that required in setting up a spreadsheet.

Also, from example 6 you should note that all cell entries on a spreadsheet may be defined in terms of the initial values of the variable quantities and the equation that relates them. Thus, changing the entry in one cell will cause an automatic change in all cells whose contents depend on the data contained in that first cell. (In example 6, for instance, this characteristic allows any change in the initial or incremental D- and L-values to automatically generate a revised table of volumes.) For this reason, it is relatively simple to examine the real or potential effects that might result from changes in the physical conditions for any given problem.

Problem Set 2

1. Using the BASIC program from example 5, generate a table of areas for circle diameters, D, in each range.
   (a) From 0 in. to 2.5 in., in increments of $\frac{1}{16}$ in.
   (b) From 2.5 in. to 10 in. in increments of $\frac{1}{4}$ in.

2. In North America, where temperatures are often measured using the traditional Fahrenheit scale, many people still have some difficulty working with values given in Celsius degrees. Modify the BASIC program from example 5 to generate a table that allows conversion of:
   (a) Celsius temperatures, C, to Fahrenheit temperatures, F, for $0° \leq C \leq 100°$ at intervals of $1°C$, where:

$$F = \frac{9}{5}\,C + 32$$

   (b) Fahrenheit temperatures, F, to Celsius temperatures, C, for $32° \leq F \leq 212°$ at intervals $2°F$, where:

$$C = \frac{5}{9}\,(F - 32)$$

3. Many hydraulic pumps require that a specified torque, T, be applied to their input shaft in order for the pump to operate at a particular speed, N. If T is in units of inch-pounds and N is in units of revolutions per minute (rpm), the amount of horsepower, P, required to drive the pump is given as:

$$P = \frac{N \times T}{63,000}$$

Using the following format, create a spreadsheet that lists the input horsepower required by a pump for various combinations of operating speed and torque.

N				T				
	50	**100**	**150**	**200**	**250**	**300**	**350**	**400**
500								
1000								
1500								
2000								
2500								
3000								

4. Portable air compressors are generally powered by four-cycle gasoline engines. In an ideal engine, *thermal efficiency, E* (the engine's effectiveness in converting chemical energy of the fuel into mechanical energy of its reciprocating pistons), is related to a fixed operating characteristic known as the *compression ratio,R,* by the equation:

$$E = 1 - \frac{1}{R^{0.4}}$$

Using spreadsheets, create a two-column table that lists ideal engine efficiencies for compression ratios in the range from 6 to 12 at intervals of 0.2. (Modern automobile engines typically operate at compression ratios in the range 7.5 to 10.)

Questions

1. List several advantages of fluid power systems over other electromechanical power systems.

2. Swimming pools, refrigerators, and vacuum cleaners are common household devices that utilize either a liquid or a gas. Answer the following for each system.
   (a) Identify the working fluid. Is the same fluid used repeatedly, or is it used once and then discarded?
   (b) What is the main purpose of each system? Can the principal use of its working fluid be neatly classified as the distribution, transmission, conversion, or control of power?

   What other types of systems found in the home involve the movement of a liquid or a gas?

3. Which three countries represent more than 80% of the total international market for fluid power products and equipment?

4. Identify the two largest market segments for hydraulic equipment; for pneumatic equipment.

5. In 1996, U.S. fluid power industry shipments totaled just under $12 billion. Approximately what percentage of that amount was for hydraulic equipment? For pneumatic equipment?

6. Because most relationships between mathematical and scientific quantities are strictly defined, their conversion factors remain essentially constant. By contrast, the equivalences between various international currencies, such as U.S. and Canadian dollars, Swiss francs, German marks, and Japanese yen, fluctuate on a daily basis. Such conversion factors, or *exchange rates,* are generally listed to at least four significant figures in most financial publications. Can you think of any other quantities whose relative values and appropriate conversion factors also vary?

7. When converting one physical quantity to another, care must be taken not to confuse *amounts* or *changes* in these quantities with their *actual values.* For example, the temperature *difference* between the freezing point of water (0°C or 32°F) and the boiling point of water (100°C or 212°F) may be expressed as *either* 100°C or 180°F. From this, we may conclude (correctly) that 1°C = 1.8°F. However, when we attempt to use this equivalence to convert a temperature of 40°C to °F, we obtain:

$$40°C \times \frac{1.8°F}{1°C} = 72°F$$

Is this the correct temperature? (*Hint:* See problem 2 in problem set 2.) Why doesn't this method of direct conversion produce the desired result? (As we see in Chapters 3 and 5, a similar discrepancy can exist between *pressures* and *pressure differences.*)

Suggested Activities

1. Form a three-member group. Give one member of the group a tape measure, give the second member a yardstick, and give the third member a 1-ft ruler. Then have each member, working alone, measure the same three linear distances: a distance less than 1 ft (such as the height of a textbook cover); a distance greater than 1 ft but less than 3 ft (such as the width of a small desk or table); and a distance greater than 3 ft. Compare these independent measurements for each distance, and discuss the following questions:
   (a) How much variation is there between the three values obtained for each distance measured?
   (b) Which measurement do you think is most accurate for each distance? Why?
   (c) What is your group's best estimate of the probable error for each of the three values selected in (b)? How might you use the number of significant figures in your final values to reflect this probable error?

2. Place an automobile's spare tire on the ground some distance away from the vehicle. Have members of your group go singly, in turn, to the tire, and use a tire-pressure gauge to determine the inflation pressure. Each member can make as many measurements as desired but should record all readings and not reveal any values obtained to other members of the group. All members should make their measurements using the same pressure gauge.
   (a) Have each group member submit a *single* estimate of the actual inflation pressure. How do these values compare?
   (b) How did each group member determine his or her best estimate for part (a)? Did any group members obtain one or more readings that were abnormally high or low?
   (c) Did the readings show a trend over time? In other words, were the third group member's readings lower than the second member's readings, and the second member's readings lower than those of the first member? If so, what could this indicate?
   (d) As a group, estimate the actual inflation pressure of the tire. Also indicate the probable error in this value, and adjust the number of significant digits in your estimate to reflect this uncertainty.

2

Liquid Properties and Their Measurement

The performance of a hydraulic system may be significantly affected by the properties of its working liquid. Although most fluid power systems use petroleum-based liquids, synthetic liquids and invert emulsion hydraulic fluids are also common, particularly where specific properties are needed. Invert emulsion fluids, for example, are a mixture of mineral oil and water, and are used for conditions requiring a fire retardant liquid or compatibility with water-based metalworking liquids.

Because of its availability, cost, and low environmental impact, water is the primary liquid used for those applications in which the liquid passes through the system once and is then discarded. This universal liquid has relatively uniform physical properties that are often used as a standard for those of other liquids.

In this chapter, we examine some of the more important physical characteristics of liquids, as well as the methods used in their measurement. We also discuss the important and frequently overlooked subject of fluid contamination. We include numerous references to pertinent ANSI and ASTM standards for students who desire additional information on a given topic. New technology graduates should also find these standards helpful in determining recommended practices for use in industry.

2.1   Density, Specific Weight, and Specific Gravity

Density of a material refers to the amount of actual matter contained in a specified volume of that material. Balsa wood, for example, is quite porous and therefore has a low density, while granite, which is quite compressed, or dense, has a high density. Although the densities of such dissimilar materials may easily be compared *qualitatively,* a more accurate, *quantitative* method of measurement is necessary for materials that are similar to each other. This is particularly true for most common liquids—from lubricating oil to milk to liquid soap to rubbing alcohol—whose densities all lie within a narrow range.

How can we determine the amount of matter contained in a particular volume of material so that we may assign a numerical value of density to that material? If it were possible to examine the material on an atomic scale, we could count the number of protons, neutrons, and electrons that form a particular mass of the material and present a sort of atomic density based upon our counting process. Such time-consuming measurements would not be practical, however, and the resulting values would be of little use to most engineers and technologists.

On the other hand, this mass of atomic particles that make up a material has a collective *weight,* which is easily determined simply by weighing a particular volume of the ma-

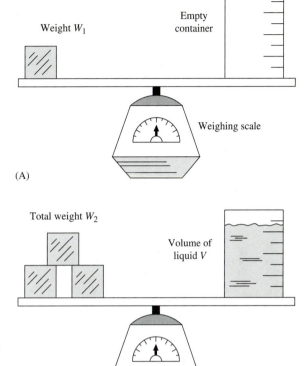

Weight W_1

Empty container

Weighing scale

(A)

Figure 2–1   Specific weight is determined by weighing an empty graduated container (A) and then reweighing the container after a measurable volume of liquid has been added (B). The net weight of the liquid is the difference in scale readings.

Total weight W_2

Volume of liquid V

(B)

terial. Dividing the weight of a material by its volume yields its *weight density,* or, as it is more commonly known, its *specific weight.* Defined mathematically, specific weight, γ, is simply the weight, W, of a material divided by its volume, V, or:

$$\gamma = \frac{W}{V} \tag{2–1}$$

The specific weight of a liquid is most easily determined as shown in Figure 2–1; a graduated container is weighed empty and then weighed again after some volume of liquid has been added to the container. The net weight of liquid, $W_2 - W_1$, is then divided by the liquid's volume to obtain the numerical value of specific weight.

This fluid property is most commonly used in the U.S. customary system of measurement. Since fresh water is such a universal liquid and has relatively constant properties, its specific weight of 62.4 lb/ft^3 is often used as a reference value and appears frequently in fluid power calculations. The unit of weight in the metric system is the newton (N); thus, specific weight in that system of measurement is in N/m^3.

EXAMPLE I

Compute the specific weight of water in (a) lb/in.^{3}, (b) N/m^3, and (c) N/cm^3.

Solution   (a) Since 1 ft^3 = 1728 in.^{3},

$$62.4\,\frac{\text{lb}}{\text{ft}^3} \times \frac{1\ \text{ft}^3}{1728\ \text{in.}^3} = \textbf{0.0361 lb/in.}^3$$

(b) Since 1 N = 0.2248 lb and 1 ft = 0.3048 m,

$$62.4\,\frac{\text{lb}}{\text{ft}^3} \times \frac{1\ \text{N}}{0.2248\ \text{lb}} \times \frac{(1\ \text{ft})^3}{(0.3048\ \text{m})^3} = \textbf{9800 N/m}^3$$

(c) Since 1 m = 100 cm,

$$9800\,\frac{\text{N}}{\text{m}^3} \times \frac{(1\ \text{m})^3}{(100\ \text{cm})^3} = \textbf{0.00980 N/cm}^3$$

Specific weight is known as a *gravimetric* property because its value is dependent upon the force of gravity. Although the effect of gravity varies by approximately 0.5% worldwide, resulting changes in specific weight are within the limits of accuracy required for most applications in engineering and technology.

However, scientists seeking a nongravimetric property to describe quantities of matter developed the concept of *mass,* which is based upon the fact that changes in the gravitational field cause proportional changes in the apparent weight of a given volume of material. Doubling the gravitational field, for example, doubles the measured weight of an object, even though the amount of matter in that object remains constant. Mass, then, is defined as the ratio of weight to gravitational acceleration, *g,* or:

$$m = \frac{W}{g} \tag{2–2}$$

In the U.S. customary system, *g* has an average value of 32.2 ft/s^2 or 32.2 ft/s/s. If weight is in pounds, the unit of mass is the *slug,* a unit seldom seen in contemporary technical work. Assuming a gravitational acceleration of 32.2 ft/s^2, an object weighing 32.2 lb would have a mass of 1.00 slug.

Far more common is the SI unit of mass, either the *gram* or the *kilogram.* Since the average value of *g* in this system is 9.81 m/s^2, an object weighing 9.81 N is defined as having a mass of 1.00 kg. (The italic symbol *g* used to represent the gravitational acceleration is different from the roman g used to indicate units of grams; similarly, the italic *m* used to represent mass differs from the roman m representing distances in meters. Take care to differentiate *m,* m, *g,* and g when using these quantities in numerical calculations.)

In order to measure mass experimentally, an object must be carefully weighed. The gravitational field at that exact location is then determined using a sensitive instrument called a *gravity meter* or *gravimeter,* and the results substituted into equation (2–2). Except for highly precise scientific work, this procedure is seldom followed. Mass is computed using the weight of an object and an average value for *g;* the results find wide use, however, in the analysis of technical problems.

The *mass density,* ρ, of a material may now be defined as:

$$\rho = \frac{m}{V} \tag{2-3}$$

where *m* is the mass of the object and *V* is its volume. Whenever the term *density* is used alone, it is always assumed to mean mass density; conversely, if weight density is intended, it is always designated as *specific weight.* The density of water at 60°F is 1.938 slug/ft^3, and at 15°C (59°F) it is 0.9991 g/cm^3. This latter value is often rounded to 1.00 g/cm^3.

Generally, the concepts of specific weight, density, and mass are not difficult for most people to understand. The units, however, particularly those for mass and weight, are often a bit more difficult to grasp. This confusion is increased by the fact that units for mass and weight, U.S. customary and SI, are used indiscriminately in today's society: bathroom scales that measure in both pounds and kilograms and grocery containers whose labels indicate both ounces and grams are just two cases that illustrate this point, as does the following example.

EXAMPLE 2

A bag of flour at the local market is labeled as follows: Net Wt. 5 lb (2.26 kg). Determine the numerical values for consistent sets of units by filling in the following blanks:

(a) 5 lb (_____ slugs)

(b) _____ slugs (2.26 kg)

(c) 5 lb (_____ N)

(d) _____ N (2.26 kg)

Solution

(a) From equation (2–2), a 5-lb weight has a mass of:

$$m = W/g = (5 \text{ lb})/(32.2 \text{ ft/s}^2) = \textbf{0.155 slugs}$$

(b) Because 0.155 slugs and 2.26 kg both have an apparent weight of 5 lb, they are equivalent.

(c) Rearranging equation (2–2) yields:

$$W = m \times g = (2.26 \text{ kg}) \times 9.81 \text{ m/s}^2) = \textbf{22.2 N}$$

(d) Because 2.26 kg has an apparent weight of 22.2 N, these units are consistent within the SI.

Perhaps the most useful property relating to the density of a material is its *specific gravity.* This dimensionless quantity is defined as the ratio of density or specific weight of a material to the corresponding property of water, or:

$$S = \frac{\gamma_{material}}{\gamma_{water}} = \frac{\rho_{material}}{\rho_{water}} \qquad\qquad (2\text{--}4)$$

where S is the specific gravity of the material. This definition applies to both *solids* and *liquids; S* is customarily specified at a particular temperature and pressure since these conditions, especially through thermal expansion and contraction, can affect the volume of a material and thus its density or specific weight. As we shall see in later chapters, the specific gravity of a *gas* is determined by comparing its density or specific weight to that of dry air at standard conditions, typically atmospheric pressure and 68°F.

Notice from equation (2–4) that if identical units are used in both numerator and denominator of the fraction, all units cancel and S becomes a pure number. This number, in fact, is a simple factor that tells us how much lighter or heavier (less dense or denser) than water a particular material is. From the definition, water has a specific gravity of 1.00; materials with S values less than 1.00 are lighter than water, while those with values greater than 1.00 are heavier than water. Table 2–1 gives typical values for solids, and Table 2–2 lists average values for some common liquids.

EXAMPLE 3

(a) Compute the density of gasoline in g/cm^3.

(b) Calculate the specific weight of mercury in lb/ft^3.

Solution

(a) Rearranging equation (2–4) yields:

$$\rho_{gasoline} = S_{gasoline} \times \rho_{water} = 0.68 \times 1.0 \ g/cm^3$$

$$= \mathbf{0.68 \ g/cm^3}$$

(b) Similarly,

$$\gamma_{mercury} = S_{mercury} \times \gamma_{water} = 13.6 \times 62.4 \ lb/ft^3$$

$$= \mathbf{849 \ lb/ft^3}$$

A quick glance at Table 2–2 reveals that the specific gravities of most liquids fall within a rather narrow range of values. With few exceptions, carbon-based liquids lie between 0.6 and 1.0, while water-based liquids hover in the vicinity of 1.0 to 1.2. Mercury, an unusual material that remains liquid at room temperature, establishes the upper limit with an S value of 13.6, but specific gravities are uncommon in that wide span bounded by 1.2 and 13.6.

Specific gravity may be determined by measuring the specific weight of a liquid and comparing this property to the published values for water. Direct measurements are possible through the use of an instrument called a *float tube hydrometer,* several of which are shown in Figure 2–2. As explained in Chapter 5, the depth at which a hydrometer floats in a particular liquid is determined by the weight and volume of the device, as well as the specific weight of the liquid; the hydrometer floats higher in a dense liquid than it does in a lighter

Table 2–1 Specific Gravities for Common Solids	**Material**	**Average Specific Gravity**
	Aluminum	2.7
	Brick	1.8
	Concrete	2.8
	Glass, window	2.5
	Gold	19.3
	Ice	0.92
	Nylon	1.2
	Polystyrene	1.1
	Rubber, synthetic	1.5
	Steel	7.8
	Wood, oak	0.68
	Wood, pine	0.44

Table 2–2 Specific Gravities of Common Liquids*	**Liquid**	**Average Specific Gravity**
	Acetone	0.79
	Alcohol, denatured	0.80
	Ammonia	0.83
	Carbon tetrachloride	1.6
	Ethylene glycol	1.1
	Fuel oil, heavy	0.91
	Fuel oil, medium	0.85
	Gasoline	0.68
	Glycerine	1.26
	Hydraulic fluid, petroleum[†]	0.88
	Hydraulic fluid, synthetic[†]	1.2
	Hydraulic fluid, water-glycol[†]	1.1
	Kerosene	0.82
	Linseed oil	0.93
	Lubricating oil, SAE 10W[‡]	0.87
	Lubricating oil, SAE 30W[‡]	0.89
	Mercury	13.6
	Seawater	1.03
	Turpentine	0.87

*Values at 14.7 psia and 77°F except as noted.

[†]At 68°F

[‡]At 60°F

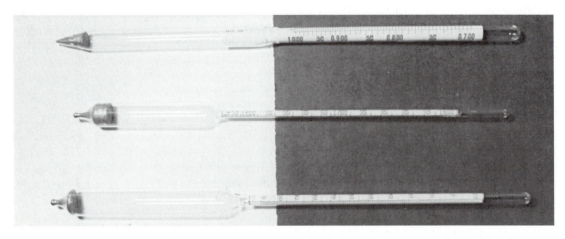

Figure 2–2   Float tube hydrometers measure specific gravities of liquids directly. Characteristics such as stability and depth of float are controlled by relative volumes and diameters of *stem* (narrow upper section containing scale) and *bulb* (lower section containing weight), as well as the overall weight of the device. The hydrometer at top is 12 in. long and measures specific gravities from 0.7 to 1.0.

liquid. By calibrating a vertical scale on the instrument, it is possible to accurately relate flotation depth to specific gravity of the liquid.

Besides providing information that allows us to compute the density and specific weight of a liquid, specific gravity may also be used to measure the properties of a liquid mixture. When soluble material, either solid or liquid, is added to a pure liquid, the specific gravity and boiling point of the resulting solution are generally higher than that of the pure liquid, while the freezing point is lowered. Changes in these properties are directly related to the amount of solute material contained in the mixture. Hydrometers are frequently used to measure the concentration of a solution; the information may then be used to predict boiling and freezing points for the mixture. Common examples include the salinity of water, acid content of solution in an automobile battery, and the concentration of antifreeze in a radiator. New Englanders also use these devices to determine the sugar content of maple syrup before it is "drawn off" from its final stages in the evaporator (see Figure 2–3). (Additional information concerning the construction and use of hydrometers can be found in ASTM Standards D 1298 and E 100.)

Problem Set I

1. Find the specific weight of water in lb/gal.

2. When a solid object is placed in a liquid, the object will float if its specific gravity is less than the specific gravity of the liquid. For one particular material, a 6-in. cube weighs 7.0 lb. Will this material float in water? In an oil for which $S = 0.85$?

3. A particular water-glycol liquid has a specific weight of 69.9 lb/ft^3. Find its density in g/cm^3.

Figure 2–3   Approximately 40 gal of sap must be boiled down to produce 1 gal of a liquid whose sugar content allows it to be labeled as maple syrup. Hydrometers are used as shown here to determine the concentration of sugar in the liquid.

4. A 55-gal drum contains hydraulic oil whose specific gravity is 0.84. Find the weight of liquid in the drum.

5. A cylindrical tank 3 m in diameter and 5 m long is filled with liquid industrial waste. If net weight of the liquid is 783 kN, find its specific gravity.

2.2   Viscosity

Viscosity is a property that measures the internal resistance of a liquid to flow. Thick liquids have a high viscosity and move slowly (Have you ever been as "slow as molasses"?), while thin liquids such as water or alcohol have low viscosities that allow them to flow freely. This internal friction is related to the shape of the molecules in any given liquid, as well as the interaction between these molecules. Since increasing the temperature of a liquid reduces this molecular interaction, viscosity of a liquid drops as the temperature of the liquid rises.

The viscosity of a liquid and its changes with temperature can have significant effects on the system in which the liquid is used. Lubricating oil, for example, must maintain a low enough viscosity at low temperatures so that the oil does not congeal and is not prevented from flowing into the tiny clearances where lubrication is required. On the other hand, at

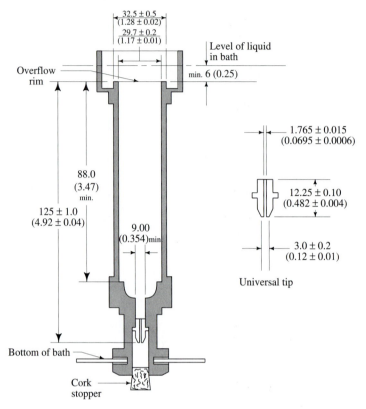

Figure 2–4   Configuration of Saybolt viscometer and universal orifice. Dimensions are in millimeters (dimensions in parentheses are in inches). (*American Society for Testing and Materials, Philadelphia, Pa.*)

high temperatures the oil must retain a high enough viscosity to provide that thin film of liquid necessary between sliding surfaces.

Similar conditions exist in hydraulic systems where the liquid not only transmits power but lubricates components as well. Low-viscosity liquids are prone to leakage and promote higher wear rates on moving parts, whereas high-viscosity liquids suffer higher frictional losses, thus generating heat and consuming power. To a certain extent, liquid temperature can be controlled in a hydraulic system. Heaters can be used to maintain a minimum temperature for the working fluid; fans, cooling fins, a larger reservoir, or external heat exchangers can all be used to cool a liquid. Such devices, however, are normally used only where extreme operating conditions exist but not to compensate for poor quality hydraulic fluid that may be ill suited to the particular application at hand.

Although several different types of viscosities and measurement techniques have been defined and developed, the most common test utilizes a *Saybolt viscometer.* This device, whose configuration is specified by ASTM Standard D 88, consists of a liquid holding tube and calibrated orifice as shown in Figure 2–4. The viscometer itself is placed inside test

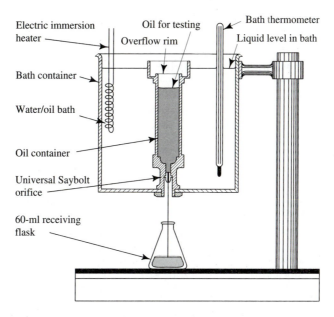

Figure 2–5   Typical arrangement of test apparatus for measuring Saybolt viscosity. *(American Society for Testing and Materials, Philadelphia, Pa.)*

apparatus such as that shown in Figure 2–5. A liquid sample is then placed in the viscometer and brought to a specified temperature (100°F and 210°F are most commonly used) by the water or oil bath, and the stopper is removed (see Standard D 88 for exact test procedure). Saybolt Universal viscosity is defined as the time, in seconds, required for 60 milliliters (ml) of test liquid to flow through the calibrated orifice. The values of viscosity may be given either in Saybolt Universal Seconds (SUS) or Seconds Saybolt Universal (SSU). Table 2–3 compares Saybolt viscosity with SAE viscosity ratings for automobile lubricating oils, and Table 2–4 presents typical SUS values for various hydraulic fluids and specialty oils.

Another commonly used measure of fluid viscosity is *kinematic viscosity*. The test method (D 445) and subsequent calculations produce units of mm^2/s, which, by definition, are also called *centistokes* (cSt). Conversions between kinematic and Saybolt viscosities may be made using standardized formulas or tabulated values (D 2161).

Notice that Table 2–4 also includes values for the viscosity index (V.I.). This property measures the change in viscosity for a given change in temperature. Liquids having high viscosity indices experience smaller changes in viscosity over a given range of operating temperatures than do liquids having low viscosity indices. This property is not necessarily related to the initial viscosity of the liquid. Figure 2–6 illustrates the behaviors of four liquids, all of which have a viscosity of 90 SUS at 210°F. Liquid A, which exhibited a viscosity of 2115 SUS at 100°F, undergoes a greater viscosity change at 210°F than liquid D, which had an initial viscosity of only 534 SUS. Liquid D, then, has the higher viscosity index of the two fluids. (For standard practices in computing viscosity indices, see ASTM Standard D 2270.)

Table 2–3 Saybolt Viscosities for SAE-Rated Lubricating Oils
(*Copyright ©1993, Society of Automotive Engineers, Warrendale, Pa. Reprinted by Permission.*)

| | Saybolt Viscosity (SUS) | | | |
| | At 0°F | | AT 210°F | |
SAE Rating	Minimum	Maximum	Minimum	Maximum
5W		4000		
10W	6000[*]	12000		
20W	12000[†]	48000		
20			45	58
30			58	70
40			75	85
50			85	110

[*]Waived if viscosity at 210°F is not below 40 SUS.

[†]Waived if viscosity at 210°F is not below 45 SUS.

Table 2–4 Viscosity and Viscosity Index for Various Liquids

Liquid	Type	Use	Specific Gravity	Viscosity, SUS (100°F)	Viscosity, SUS (210°F)	Viscosity Index
Sun 2105[*]	Paraffin base oil	General-purpose hydraulic fluid	0.865	206	52.4	167
Sunsafe F[*]	Invert[†] emulsion	Fire-resistant hydraulic fluid	0.922	410		
Royco 717[‡]	Mineral oil	General purpose hydraulic fluid		124[§]	57.5[§]	
Royco 756[‡]	Petroleum base	Aircraft, ordnance, missiles	0.854	64.6[§]		
Royco 782[‡]	Synthetic hydrocarbon	Fire-resistant, aircraft, ordnance, missiles	0.851	74.7[§]	37.8[§]	
Sunoco Ultra Super C Gold[*]	Petroleum oil	SAE 15W-40 motor oil			81.6	135
Sunoco Ultra[*]	Petroleum oil	SAE 30 motor oil			66.0	111
Sunoco Type F[*]		Automobile transmission fluid			52.3	174
Sunep 68[*]		Gear lubricant		353	55.7	100

[*]Data provided by Sun Refining and Marketing Company, Lubes Division, Philadelphia, Pa.

[†]Water in mineral base oil.

[‡]Data provided by Royal Lubricants Co., Inc., East Hanover, N.J.

[§]Converted from kinematic viscosities provided by Royal Lubricants Co., Inc., according to ASTM Standard D 2161.

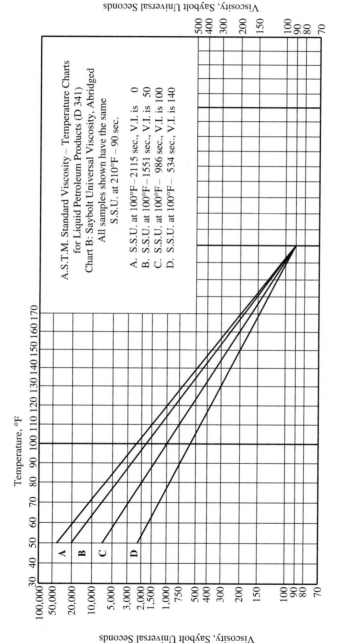

Figure 2-6 Viscosity behavior for several liquids of different viscosity indices. (*USX Corporation, Pittsburgh, Pa., and American Society for Testing and Materials, Philadelphia, Pa.*)

2.3   Important Thermal Points

The behaviors of most hydraulic liquids are characterized by several experimentally determined thermal points. *Pour point* is the lowest temperature at which motion of the liquid
can be detected visually. The test method (D 97) for petroleum oils is often applied to other
types of liquids as well and consists of heating a liquid and then cooling it at a prescribed
rate (see test apparatus in Figure 2–7). A sample of the liquid is examined at intervals of
3°C until a temperature is attained at which no movement of the liquid is observed. The
pour point represents an extreme limit that is generally well below the minimum temperature at which a hydraulic fluid may safely be used in a fluid power system.

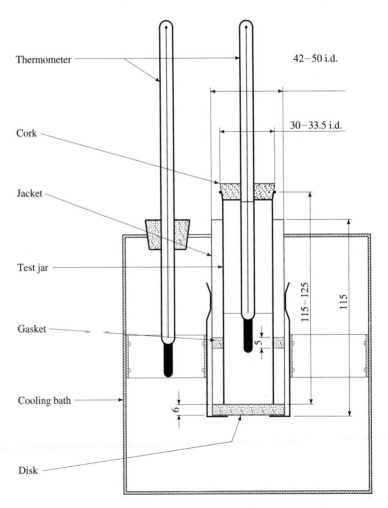

Figure 2–7   Apparatus for measuring cloud and pour points of petroleum oils.
Dimensions in millimeters. (*American Society for Testing and Materials, Philadelphia, Pa.*)

Another common thermal property for petroleum oils is the *cloud point*. At this temperature, impurities such as water and paraffin waxes crystallize and precipitate out of solution, forming a cloud of particles in the fluid. In some instances these precipitates can affect the liquid's lubricating and power transmission capabilities. Tests are done (D 2500) using the same apparatus as for pour point.

Two other significant temperatures are those of the *flash point* and the *fire point*. Flash point is the lowest temperature under atmospheric conditions at which a flame applied to the surface of the liquid will cause the liquid's vapor to ignite or flash. Fire point denotes

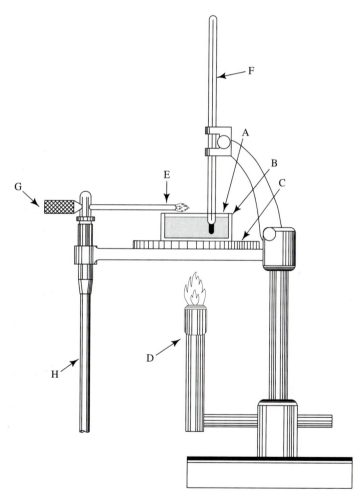

Figure 2–8 Cleveland open-cup apparatus for flash and fire points: A, oil sample; B, sample container; C, heating and support plate; D, heat source; E, movable open flame tip; F, thermometer; G, control handle for movable tip; and H, gas supply for flame tip. (*American Society for Testing and Materials, Philadelphia, Pa.*)

Table 2–5   Pour, Flash, and Fire Points for Various Liquids*

Liquid	Pour °C (°F)	Flash °C (°F)	Fire °C (°F)
Sun 2105	−42 (−44)	198 (388)	
Sunsafe F	−21 (−6)		
Royco 717	−55 (−67)	112 (234)	120 (248)
Royco 756	−62 (−80)	107 (225)	
Royco 782	−54 (−65)	232 (450)	254 (489)
Sunoco Ultra Super C Gold	−27 (−17)		
Sunoco Ultra	−21 (−6)		
Sunoco Type F	−45 (−49)	192 (378)	
Sunep 68	−15 (+ 5)	219 (426)	

*For descriptions of liquids and other properties, see Table 2–4.

the lowest temperature at which a specimen of liquid will sustain burning for 5 s. Both tests use the same apparatus—either the Pensky-Martens closed tester (D 93), the Tag open-cup apparatus (D 1310), or the Cleveland open cup (D 92)—selection of which is dependent upon the type of liquid being tested. Cleveland open cup apparatus, used for all petroleum products other than fuel oils or those having flash points below 175°F, is shown in Figure 2–8. These two ignition points, for vapor and liquid, are good indicators of the potential fire hazards involved with the use of any given liquid.

Table 2–5 presents typical values of pour, flash, and fire points for the same liquids listed in Table 2–4.

2.4   Other Notable Properties

Hydraulic fluids possess other measurable characteristics that can affect their selection and use in fluid power systems. Some of these properties (and their appropriate ASTM standards) include:

- *Corrosiveness* (D 130) and *oxidation stability* (D 4636), which can affect components or fittings, especially for petroleum oils containing high residual levels of sulfur compounds.
- *Vapor pressure,* which can be measured (D 323, D 2879) or estimated (D 2878) from other physical properties, is the pressure at which a liquid will begin to vaporize at a specified temperature; this property can affect the operation of hydraulic pumps, since inlet pressures are often low enough to produce *cavitation,* a process by which pockets or bubbles of vapor are formed and then carried into the pump where they collapse under increasing pressure and damage the pump impeller.

- *Foaming characteristics* (D 892) of oils, which can diminish their lubricating abilities, cause erratic system performance, and sometimes damage components.
- *Water content* (D 1744) and its effect on the stability, known as *hydrolytic stability* (D 2619), of hydraulic fluids. Hydrolytically unstable liquids may form acidic and insoluble contaminants that can alter the effective viscosity, promote corrosion, and cause erratic performance of system components.
- *Thermal stability* (D 2160), which indicates how well a liquid will maintain its integrity at elevated temperatures, and *thermal conductivity* (D 2717), which measures the ability of heat to flow within the fluid itself.
- *Wear characteristics* (D 2271), or the long-term effect of a fluid on system components, which is itself influenced by the degree of *insoluble contamination* (D 4898) of the liquid. Excessive wear of pump parts, for example, will decrease pump efficiency and lead to diminished system performance.
- Susceptibility to *microbial growth* in water-based hydraulic fluids. Growth of bacteria and fungi, which can clog filters and produce unwanted slime deposits, is generally controlled by additives whose effectiveness can be measured (E 979).

2.5   Types of Liquids

Four types of liquids are commonly used in hydraulic systems: petroleum oils, emulsions, synthetics, and water-glycols. *Petroleum oils* are exactly what their name implies—petroleum products of the same distillation process that yields gasoline, lubricating oils, and tars. The oils used in hydraulic systems frequently contain additives to enhance the natural properties of the liquid or inhibitors to control such undesirable characteristics as corrosion and foaming. Petroleum liquids have several advantages over other types of hydraulic fluids. Cost-efficient and readily available, they can operate over a wide range of temperatures, and because of their inherent lubricating abilities they tend to reduce both wear and corrosion during the life of a system.

Perhaps the single greatest disadvantage of petroleum oils is their flammability, a property that may, under certain operating conditions, present a severe threat to safety. This danger may be eliminated through the use of a *fire-resistant* liquid, defined by ANSI Standard B93.2 as "a fluid difficult to ignite which shows little tendency to propagate flame." (Such fluids are sometimes described as *nonflammable,* a designation disapproved both by ANSI and NFPA.) Although other major differences exist between the various categories of liquids, it is this property of fire resistance that best distinguishes the emulsions, synthetics, and water-glycol fluids from the petroleum oils.

An *emulsion* may be defined as a uniform dispersion of liquid in an immiscible "host" liquid. Unlike a solution, the two fluids do not mix but consist of tiny droplets of one liquid that form a homogeneous suspension in a continuous phase of the second liquid. The most common types of emulsions are oil-in-water and water-in-oil.

Liquids designated as high water content emulsions (often referred to as high water base fluids (HWBF) or high water content fluids (HWCF)) generally consist of oil-in-water fluids whose water content is 80% or higher. Such liquids are obviously fire resistant, and their viscosities, which are similar to that of water, are fairly independent of the actual water content. However, these fluids do possess several characteristics that can limit their use in a particular application. Evaporation of the working fluid must be considered during system operation, and operation near 0°F can affect the fluid chemistry or cause the fluid to freeze. Foaming can be greater than that of oils, and low viscosity of the liquid increases wear in certain components. These fluids also tend to hold particulate contamination in suspension more readily than oils and can promote galvanic action between different metals as well as an increase in the corrosion of aluminum. Finally, such liquids may affect seals made of butyl or ethylene-propylene, or seals impregnated with asbestos, leather, or cork. Certain of the above characteristics may be reduced significantly through the use of additives in the fluid.

Water-in-oil emulsions, often called *invert emulsions,* are made up of petroleum oils, water, emulsifiers, and selected additives. Both viscosity and fire resistance vary with water content, and although wear resistance is better than that of oil-in-water emulsions, separation of the water or a phase change of the fluid may greatly increase the wear rate. These liquids also exhibit similar effects in the areas of low-temperature operation, foaming, corrosion, and suspension of particulates. Water-in-oil emulsions are classified as *non-Newtonian* fluids, meaning that under conditions of high stress—such as those found in most pumps—the liquid undergoes a reduction in viscosity. For this reason, these liquids are normally manufactured to higher viscosity levels than those of petroleum oils used under similar conditions, and stated viscosity of the product is usually higher than effective viscosity.

Synthetic liquids may be formulated from chlorinated hydrocarbons, halogenated organic materials, esters, silicones, phosphate esters, or blends of phosphate esters with petroleum oils. Although they do not contain water, synthetics are generally fire resistant. In fact, the presence of water in a synthetic liquid can be detrimental, since it significantly increases the risk of corrosion. Water may be introduced into a system from such external sources as rain, condensation, heat exchangers, or metal-cutting fluids. As with emulsions, foaming and particulate suspension can be greater than that of petroleum oils; because of the wide range of blended synthetics available, corrosion of seals and metal surfaces can vary widely. Operation at elevated temperatures usually contributes to the corrosion process, since the liquid may undergo partial decomposition to acidic products. On a positive note, viscosity of a synthetic liquid often remains constant even after extended use, and wear resistance is generally good.

Water-glycol liquids are made up of water, glycols, thickeners, and additives. At high temperatures, evaporation may become a significant factor in operation of a system, but at low temperatures, properly formulated water-glycol solutions can provide very satisfactory performance. Both fire resistance and viscosity vary with water content; foaming and retention of particles in suspension are greater than with oils. Although wear resistance of water-glycols is good, high speeds and heavy loads can drastically reduce the life of antifriction bearings. For this reason, careful scrutiny should be given to the use of these liquids in certain components such as pumps, unless this use is approved by the component manufacturer or the component is modified according to manufacturer's specifications. Galvanic action and corrosion of aluminum can occur with water-glycol liquids, and they generally attack materials containing, or plated with, zinc, cadmium, and magnesium, forming sticky or gummy residues.

Selection of a liquid for use in any hydraulic system, then, is influenced not only by anticipated operating conditions but also by the materials used in components, piping, and seals. It is a task that should not be taken lightly, since the working fluid is the medium by which power is transmitted and converted in a fluid power system. Select only components that have been specified by their manufacturers to be compatible with the liquid being used, and maintain that liquid, with periodic testing, as recommended by the fluid supplier.

Occasionally the need arises to change from one type of liquid to another within a given system. This always involves a draining, cleaning, and flushing sequence followed by modifications to lines, components, and seals as necessary. Recommended practices for cleaning, flushing, and purifying petroleum fluid hydraulic systems are outlined in ASTM Standard D4174; practices for changing a hydraulic system over from one liquid to any other are presented in ANSI Standard B93.5.

2.6   Contamination of Fluid Power Systems

The working liquid in a fluid power system may become contaminated in three different ways, all of which contribute to a progressive deterioration in system performance.

Thermal contamination occurs when heat generated during the work cycle is not removed from the liquid, resulting in elevated temperatures of that liquid. Although this condition may be caused by a poorly designed circuit, improperly selected components, or inadequate air circulation in the environment around the system, most thermal contamination is caused by an improperly sized reservoir. If a system contains too little liquid, that liquid must pass through the system more frequently, picking up heat with each cycle. In a fluid power system having a 20-gal reservoir and driven by a 20-gal/min pump, for example, all the fluid must pass through the system once each minute; if the reservoir contained 100 gal, each particle of fluid would, on average, pass through the system once every 5 min. Those extra 4 min spent in the reservoir allow the liquid to "rest" and dissipate heat to the surroundings, thus reducing the overall operating temperature of the system.

Elevated temperatures decrease liquid viscosity, leading to higher wear rates, increased leakage, and erratic system performance. They also promote the formation of varnish deposits or sludge in petroleum liquids while increasing evaporation rates and microbial growth in water-based fluids.

Soluble contamination exists whenever dissolved solids or soluble liquids mix with the working fluid to form a solution. Examples of the latter include cleaning solvents, gasket sealants, and rust preventatives. These soluble contaminants can change the fluid viscosity, reduce its lubricating and power transmission capabilities, alter the flash point, attack rubber or synthetic seals, increase foaming, and lead to inconsistent system performance.

In petroleum liquids, soluble contaminants can also include by-products of the distillation process, such as waxes, as well as water that has entered the system through gaskets and seals or condensed from vapor in the air. Fluid power systems that are exposed to the elements or operate between extremes of temperature are especially prone to this type of contamination; most agricultural and construction equipment falls into this category. Water

Table 2–6 Relative Particle	Size		
Sizes of Some Common Objects (Parker Filtration, Metamora, Ohio)	Substance	(μm)	(in.)
	Grain of table salt	100	0.0039
	Human hair	70	0.0027
	Lower limit of visibility	40	0.00158
	White blood cells	25	0.001
	Talcum powder	10	0.00039
	Red blood cells	8	0.0003
	Bacteria (average)	2	0.000078

in solution can have an adverse effect on lubricating properties and can increase corrosion appreciably; a drop in temperature often causes water to separate from the working fluid, thus producing a variety of undesirable effects, such as an increase in oxidation, sludge formation, and foaming.

Prevention is the key to success in reducing soluble contamination. Care must be taken during system maintenance not to introduce foreign substances that can go into solution. Gaskets and seals must also be maintained to prevent leakage into the system during operation, while an adequately vented reservoir will reduce the likelihood of condensation. Should the working fluid become sufficiently contaminated, the system must be flushed and cleaned and a fresh charge of liquid added.

Perhaps the most significant degradation of a hydraulic liquid is caused by *insoluble contamination*. Solid particles of fiber, metal, and dirt can remain suspended in the liquid, lodge at various points in the system, or settle in areas such as the bottom of the reservoir. Although insoluble contaminants can cause components such as valves and cylinders to stick or leak and can result in clogged lines or filters, their most detrimental effect is the long-term abrasion and wear that takes place inside fluid power components, many of which depend upon very close tolerances for successful operation.

These solid contaminants need not be large; most, in fact, are invisible to the naked eye. To help put the sizes of these particles into perspective, Table 2–6 lists some common objects whose average dimensions are measured in *micrometers*. (One micrometer—abbreviated as μm and sometimes called a *micron*—is equal to one millionth of a meter, or 0.0000394 in.) Table 2–7 gives typical clearances for some fluid power components. Comparison of the tables reveals that even minute particles in suspension can seriously affect the performance and life of these precise (and therefore expensive!) hydraulic products.

As shown in Chapter 7, insoluble contaminants are removed from a liquid through the use of strainers and filters. To determine the degree of insoluble contamination, liquid samples may be extracted from the lines of an operating hydraulic system in accordance with ANSI Standard B93.19 or from the reservoir of an operating system according

Table 2–7   Average Clearances for Some Fluid Power Components
(*Parker Filtration, Metamora, Ohio*)

Component	Clearance (μm)	Component	Clearance (μm)
Slide bearings (vane pump)	0.5	Gear pump (tooth tip to case)	0.5–5
Tip of vane (control valve)	0.5	Vane pump (vane tip)	0.5–1
Roller element bearings	0.1–1	Piston pump (piston to bore)	5–40
Hydrostatic bearings	1–25	Servo valves (spool sleeve)	1–4
Gears	0.1–1		

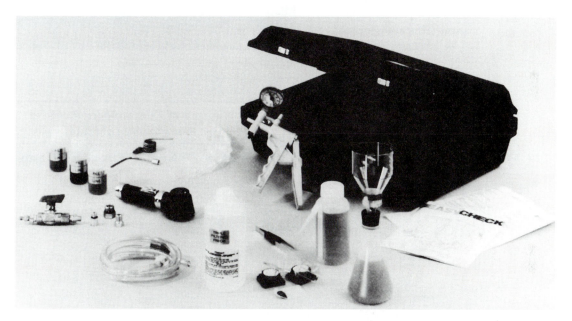

Figure 2–9   Typical patch kit for on-site hydraulic fluid contamination testing. (*Schroeder Industries, Inc., McKees Rocks, Pa.*)

to ANSI Standard B93.44. Samples may then be evaluated in one of several ways: particulate size distribution may be determined *optically* using a microscope; the samples may be sent to a laboratory where they are analyzed by one or more sophisticated *automatic counting* machines; on-site tests may be done by passing the sample through a patch of filter material and then comparing its color or reflectance with calibrated charts (see Figure 2–9) or through the use of a portable, fully automated fluid analyzer (see Figure 2–10).

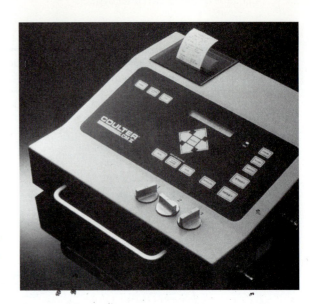

Figure 2–10 A portable, fully automated liquid contamination analyzer. (*Schroeder Industries, Inc., McKees Rocks, Pa.*)

Table 2–8 Cleanliness Level Correlated to ISO Code (*Parker Filtration, Metamora, Ohio*)		
ISO Code	**Number of Contaminant Particles per Milliliter of Fluid**	
	Particles ≥ 5 μm	**Particles ≥ 15 μm**
26/23	640,000	80,000
25/23	320,000	80,000
23/20	80,000	10,000
21/18	20,000	2,500
20/18	10,000	2,500
20/17	10,000	1,300
20/16	10,000	640
19/16	5,000	640
18/15	2,500	320
17/14	1,300	160
16/13	640	80
15/12	320	40
14/12	160	40
14/11	160	20
13/10	80	10
12/9	40	5
11/8	20	2.5
10/8	10	2.5
10/7	10	1.3
10/6	10	0.64

Table 2–9 Cleanliness Levels Required for Various Components
(*Parker Filtration, Metamora, Ohio*)

Component	ISO Code Cleanliness
Servo control valves	14/11
Vane and piston pumps/motors	16/13
Directional and pressure control valves	16/13
Gear pumps/motors	17/14
Flow control valves, cylinders	18/15

Table 2–10 Typical Ingression Rates
(*Parker Filtration, Metamora, Ohio*)

System	Ingression Rate*
Earthmoving and off-highway (extreme conditions)	10^9–10^{10}
Farm and other mobile equipment	10^8–10^9
Typical manufacturing plant	10^7–10^8
"Clean" plant environment	10^5–10^6

*Number of particles >10 μm ingressed into the system per minute from all sources.

Although fluid cleanliness may be reported several different ways, the two-number ISO Code system is perhaps the most widely used method. Table 2–8 gives typical ISO Code ratings and their corresponding contaminant distributions. The first number in the code indicates the relative number of particles greater than 5 μm, or silt size, that are present in the fluid, and the second number represents the relative number of large particles over 15 μm. Cleanliness levels required for various fluid power components are given in Table 2–9.

Contamination testing and maintenance of hydraulic fluids should be done at periodic intervals. Table 2–10 shows the average rate at which particles enter typical fluid power systems in a variety of operating environments. Even under relatively clean conditions, the ingression rate can be incredibly high. Progressive contamination can be controlled by proper filtration within the system, supplemented by the use of portable filtration units of the type shown in Figure 2–11. These units, most of which are designed to remove both water and particulates, can be used to service a number of individual systems. Wheeled on site, they are attached to an existing system whose working fluid is then circulated through the portable filtration system for a designated length of time. Such units are meant to supplement, not replace, a properly designed and maintained filtration system that is subject to continuous, and possibly extreme, operating conditions.

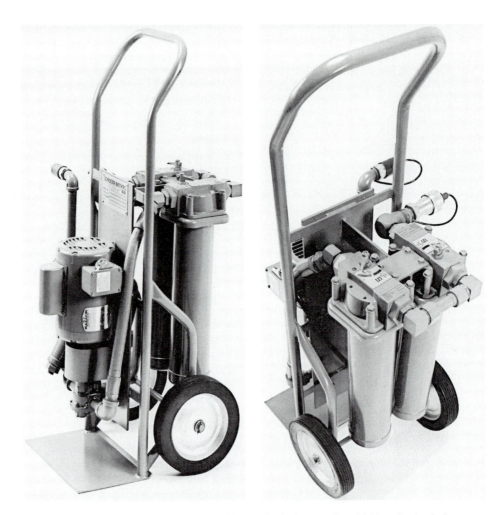

Figure 2–11   Portable filtration unit. (*Schroeder Industries, Inc., McKees Rocks, Pa.*)

Always remember that hydraulic fluid is the element that actually transmits power in a fluid power system. This simple fact is often ignored by engineers and technologists alike. It cannot be overly stressed that proper fluid maintenance and attention to possible sources of contamination are essential to satisfactory and reliable system operation.

Questions

1. A solid object is placed in water. If the specific gravity of the object is exactly 1.0, will the object float? Sink? Describe its behavior.

2. One property of a substance is its *specific heat.* This is defined as the amount of energy required to raise the temperature of a fixed amount of the substance by 1°. The most common SI and U.S.

SAFETY SIDEBAR

Use Hydraulic or Lubricating Liquids Safely

When working with any type of hydraulic or lubricating liquid, follow the manufacturer's recommendations for safe handling and use, and avoid ingestion by mouth, prolonged or repeated skin contact, and inhalation of vapors. Motor oils, for example, can cause mild gastrointestinal irritation if ingested, and continued contact can produce smarting and reddening of the skin. Ethylene glycol, the primary ingredient in many antifreeze solutions, is mildly toxic by ingestion and skin contact, but large doses or prolonged contact can result in permanent kidney damage or death.

Table 2–11 Specific Heats for Several Liquids	Liquid	**Specific Heat (BTU/lb−°F or cal/g°C)**
	Ethyl alcohol	0.60
	Gasoline	0.50
	Mercury	0.033
	Turpentine	0.42
	Water	1.0

customary units for specific heat are given here; because of the related definitions for BTUs and calories, numerical values of this property are identical in either system for any given material.

$$1.0 \, \frac{BTU}{lb-°F} = 1.0 \, \frac{cal}{g-°C}$$

Typical values of specific heat for several liquids are shown in Table 2–11. Of what use is this particular property? In a hydraulic system, would a high- or low-specific-heat liquid be preferable? How might the specific heat of a liquid be modified?

3. The most common automotive lubricating oils used today are *multigrade* oils, each with a dual-number designation, such as 5W-30. The first number (5W) indicates the oil's effective weight at 0°F, and the second number (30) represents the effective weight at 210°F. Why are such oils useful? What multigrade combinations (designations) are available in petroleum lubricants? Are synthetic oils rated in a similar manner?

4. What is the approximate range of specific gravities for carbon-based liquids?

5. Name the four types of liquids commonly used in hydraulic systems.

6. List several advantages of petroleum liquids over other types of hydraulic fluids. What is the greatest disadvantage of these petroleum oils?

7. What does HWBF designate?

8. What is an invert emulsion?

9. List several disadvantages of water-based hydraulic liquids.

10. Name the three types of liquid contamination that can occur in a hydraulic system.

11. How is the level of insoluble contamination measured?

12. What does an ISO Code rating of 18/15 indicate about the cleanliness of a hydraulic liquid?

Suggested Activities

1. (a) Using a graduated beaker and a weighing scale, measure the density or specific weight of any common liquid. Compute the specific gravity of the liquid. How accurate do you believe your measured and computed values to be?

   (b) Repeat this experiment using water. How well do your results compare with published values? What possible sources of error exist in this test?

   (c) Repeat (b) after heating or cooling the water sample. Is there a measurable change in properties?

2. It is a relatively easy process to determine the specific weight of any rectangular, cubical, or cylindrical solid, because the volume may be computed from a few linear measurements. To find the volume of an irregular object that is heavier than water, tie a light string around the object, lower the object into a graduated container, and fill the container with water (Figure 2–12). When the object is lifted out of the container, the water level drops an amount equal to the (solid) volume that has been removed. Use this procedure to determine specific gravities for several objects, such as stones, pieces of concrete or brick, and small metal parts.

3. Hydrometers used at automotive service stations estimate the concentration of antifreeze in the cooling system fluid by qualitatively determining specific gravity of the fluid at a given temperature, usually that of the hot engine. Using a commercial ethylene glycol antifreeze at room temperature, fill five small containers with: (a) pure water, (b) pure antifreeze, (c) a 25% water and 75% antifreeze mix, (d) a 50% water and 50% antifreeze mix, and (e) a 75% water and 25% antifreeze mix. Measure the specific gravity of liquid in each container, then slowly bring each liquid to a boil and measure the boiling temperature. Graph your results in the form of boiling point versus specific gravity of the mixture. Is there a measurable correlation between the two?

4. Using a test tube, cork, lead shot (or sand), and a glass rod (or wooden dowel), construct a float tube hydrometer to measure specific gravities from 0.6 to 0.9. Try changing parameters (overall weight, dimensions of bulb and stem) to determine their effects on stability and scale magnification of the instrument. If possible, compare these readings with those of a commercial hydrometer.

5. Place a sample of used crankcase oil between two glass slides and examine it under a microscope. Compare this with a similar specimen of fresh lubricating oil. Estimate the size of any visible contamination by using Table 2–6.

6. Although accurate numerical values of viscosity can be obtained only by using precision instruments, the *relative* viscosities of various liquids can be compared using just a funnel and a watch. Select several common household liquids such as ethyl rubbing alcohol, vegetable (cooking) or mineral oil, liquid car polish, dishwashing detergent, and automobile transmission fluid. Cap the bottom of a funnel, fill the funnel with water, then uncap the funnel and measure the time required for all the liquid to drain out. If the time is too short for reasonable measurement, decrease the size of the funnel opening by inserting a cork or rubber stopper that contains a small drilled hole. (This becomes your "test" orifice.) Repeat the experiment for each liquid, and create a table of "specific" viscosities by dividing all the measured times by the elapsed time for water. You may also want to examine the effects of temperature by cooling or heating the liquids slightly and testing each liquid

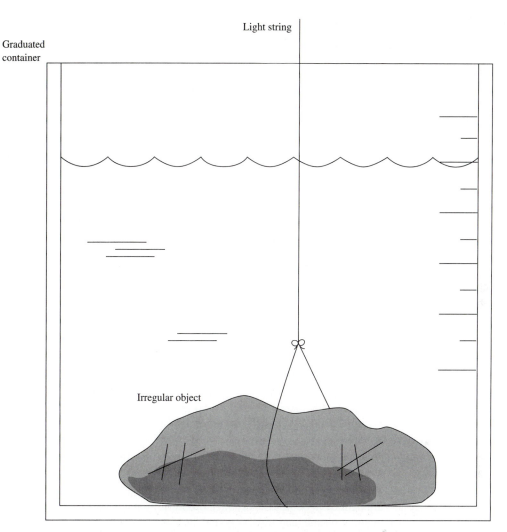

Light string

Graduated
container

Irregular object

Figure 2–12   To find the volume of an irregular object that is heavier than water, simply lower the object into a graduated container by using a light string. Next, fill the container with water to any convenient scale division. When the object is lifted from the container, the water level drops by an amount equal to the object's volume. (See suggested activity 2.)

once again. If so, plot a graph of elapsed time versus temperature for water, as well as for the liquid that exhibits the greatest change in viscosity. Can you now create your own "viscosity index"?

7. One way to determine if a hydraulic oil is contaminated by water is to conduct a "spatter" test. Heat an electric frypan or hot plate to a temperature above the boiling point of water but less than 300°F. Place a few drops of oil on the heated surface; if water is present, the oil will sizzle, spatter, or bubble. Prepare several samples with known amounts of water, and use this test to compare their behaviors with that of pure oil. (Be sure to wear safety glasses when conducting or observing this test.)

3

Fluids at Rest: Pascal's Principle

Fluid power systems are noted for their ability to provide smooth, controllable motion in linear, rotary, and reciprocating forms. Equally as important is their ability to magnify forces to the incredibly high levels routinely required in many construction and manufacturing applications. Both of these system characteristics are made possible by the behavior of a confined fluid, a unique phenomenon that is the subject of this chapter.

Most industrial fluid power systems are required to exert specified forces while moving through fixed distances in given periods of time. Such behavior is often described in terms of the work, energy, torque, or power produced by the system. A knowledge of the basic definitions and physical principles involved will allow us to correlate desired system output with the fluid conditions of pressure and flow rate necessary to obtain satisfactory operation.

The concepts studied in this chapter will be used to analyze different types of fluid power systems presented in Chapter 4 and Chapter 6.

3.1 The Nature of a Fluid

All matter consists of small particles of material known as *molecules* that are attached, or bonded, to each other in ways that give the material its distinctive properties. Depending upon the strength of these bonds, a material may be classified as either a *solid* or a *fluid*. The molecules of a solid, for example, are tightly bonded in a rigid lattice (Figure 3–1) that permits little movement of the molecules, thus allowing a solid object to maintain its specific shape.

In contrast, materials that do not maintain a fixed shape but conform to the shapes of their containers are defined to be fluids. Although this definition includes substances such as sand, molten metal or plastic, and fresh concrete, its use relative to fluid power systems is restricted to materials that are either liquids or gases at room temperature. A *liquid* is a fluid whose volume is essentially constant; 5 gal of water, for instance, will occupy the same amount of space whether it is placed in a 5-gal container or a 20-gal container. A *gas*, on the other hand, is a fluid that expands to fill the volume in which it is placed.

Molecular bonds in a liquid are weak enough to permit free movement of the molecules, yet strong enough to allow the liquid to maintain a discrete, cohesive mass. As shown in Figure 3–2, this organized but flexible arrangement of molecules is what gives the material its ability to flow and change shape. Gas molecules, however, are widely spaced, as evidenced by the fact that gases are invisible to the naked eye. For this reason, little or no bonding exists between molecules, thus allowing them to move randomly throughout the complete volume within which they are confined, as shown in Figure 3–3. This motion, which is due to the energy present in the molecules, keeps a balloon uniformly inflated

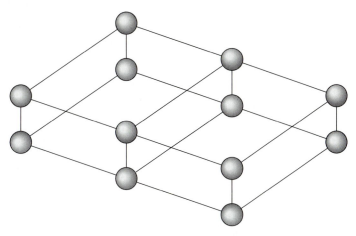

Figure 3–1   Molecules of a solid are tightly bonded in a rigid lattice arrangement.

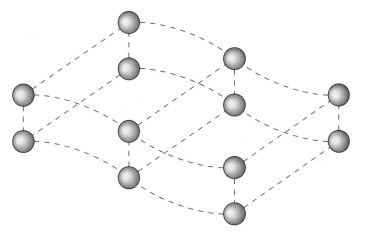

Figure 3–2   Level of bonding in a liquid permits free movement of molecules while maintaining a discrete, cohesive mass.

rather than allowing molecules to collect in one portion of the total volume. Some random molecular movement also occurs within liquids, but to a lesser degree because of the higher density and stronger bonding that exists.

The nature of a fluid allows it to transmit forces in a way that is uniquely different from that of a solid. If a force is applied to a solid object (Figure 3–4), the rigidly arrayed molecules can transmit this force only in the direction of the force itself. For this reason, if we touch the surface of a basement wall we do not experience the force carried by the wall and caused by the weight of the structure above.

Since the molecules of a liquid are free to move, however, any force applied to a confined fluid (Figure 3–5) will be transmitted in *all* directions. This occurs because those

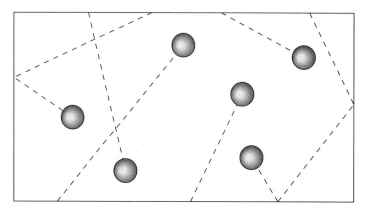

Figure 3–3 Molecules of a gas are virtually unbonded and follow random paths throughout the volume in which they are confined.

Figure 3–4 Transmission of an applied force by a solid object generally takes place only in the direction of the applied force.

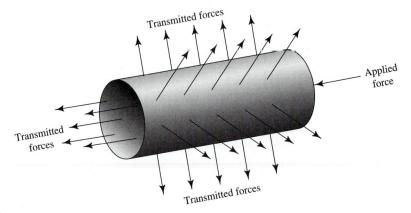

Figure 3–5 Because fluid molecules can move and adjust to the shapes of their containers, transmission of an applied force by a confined fluid occurs in all directions.

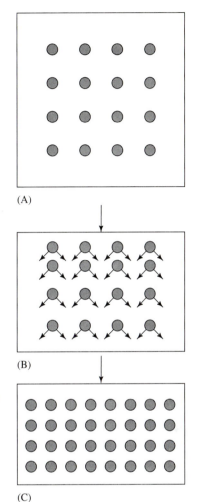

Figure 3–6   Liquid molecules adjust to an applied load. (A) Molecules in liquid at rest maintain a regular spacing. (B) Molecules under the applied load move to interior of liquid, sharing load with other molecules. (C) Molecules adjust so that each feels equal forces from all directions and is in *equilibrium.*

molecules in close proximity to the applied force will move away from the force until they encounter other molecules that, in turn, will move away, until finally the forces felt by each molecule are the same in all directions. These forces may be caused by adjacent molecules or the walls of the container. This process of adjustment, shown graphically in Figure 3–6, is similar to the one that takes place on any busy elevator. If the total space in that elevator is occupied by only two people, they tend to share the space equally. As more people board the elevator, the riders invariably arrange themselves so that each passenger has approximately the same amount of standing room.

In a confined gas, applied forces are also transmitted uniformly in all directions. Since the molecules are already impinging continually on the walls of their container, any external force will tend to reduce the available volume, resulting in more frequent collisions between molecules and walls. Forces used to compress an inflated balloon, for example, are transmitted equally to all areas on the balloon surface.

In summary, random molecular motion within a gas, and the ability of a liquid to re-arrange its molecules internally, allow fluids to change shape and cause them to transmit an applied load uniformly in all directions. These basic fluid phenomena provide for the wide variety of motions obtainable with simple fluid power systems, as well as for the smooth, continuous transmission of power that such systems offer.

3.2 Pressure

The degree of loading that exists on molecules in a confined liquid is indicated by the *pressure* of the fluid. Pressure is simply the *average force* applied to a fluid, and is computed as the *total force* applied to the liquid, divided by that *area* of fluid which supports the total load. Mathematically, pressure is defined as:

$$p = \frac{F}{A} \qquad\qquad (3–1)$$

where p is the fluid pressure, F is an external force, or load, applied to the fluid, and A is the surface area of fluid that supports the external load.

Consider the two fluids shown in Figure 3–7. Each is confined within a rigid cylin-der and sliding piston, with external forces applied by means of the movable piston. At first glance, the fluid in (B) appears to be most heavily loaded since it supports the larger of the two applied forces. Because the area of this piston is so great, however, the total load is distributed over a larger number of fluid molecules, so that the average load on each molecule is less than it is for the fluid in cylinder (A). From equation (3–1), the pres-sure, or unit load, in cylinder (A) may be computed as 1200 lb distributed on 3 in.^{2} of area, or 400 lb *per square inch,* while for the cylinder (B) it is 1800 lb distributed on 6 in.^{2}, or 300 lb *per square inch.*

Using pressure to describe the internal loads developed in a fluid is much like the method of unit pricing found today in most supermarkets and grocery stores. If a 12-oz package of soap, for example, costs $3.24 and a 15-oz package of the same product costs $4.35, then which package offers the lowest *average* cost to the consumer? To help the cost-conscious shopper evaluate these choices, the *unit* cost of each package must also be given: 27 cents *per ounce* for the small package, and 29 cents *per ounce* for the large. In other words, the prices are unitized on a per-ounce basis for easy comparison.

Similarly, pressure represents the average or unit loading that exists in a fluid com-puted on a per-area basis. Area of contact rather than the actual number of molecules ex-posed in that area is used as a basis for unitizing an applied force. It provides for easier computation, and yields numerical values of reasonable magnitude. It also provides a valid representation of molecular loading since fluids are assumed to be homogeneous, and the number of molecules contained in a particular volume or exposed to a given area is directly proportional to that volume or area. Later in this book we shall see that fluids may be pres-surized in ways other than applying a single concentrated load; static fluids are often pres-surized by their own weights, while pumps and compressors can pressurize continuous

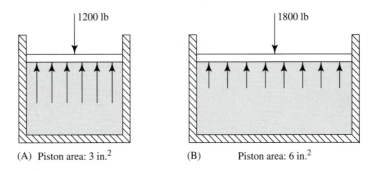

(A) Piston area: 3 in.^{2} (B)   Piston area: 6 in.^{2}

Figure 3–7   External forces applied to confined fluids: (A) piston area: 3 in.^{2}, (B) piston area: 6 in.^{2}.

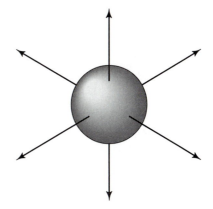

Figure 3–8   Because forces exerted by any liquid molecule are the same in all directions, pressure is *omnidirectional*.

streams of moving fluid. In all cases, however, pressure is used to describe and measure the degree of a loading experienced by the fluid.

Pressure is also *omnidirectional*. Any molecule in a liquid, for example, is surrounded on all sides by other molecules or the sides of its container. As these particles are squeezed together or pressurized, the forces that each molecule must exert on its neighbors in order to maintain its position in the liquid are the same in all directions (Figure 3–8). In a gas, random molecular movement ensures that moving particles continuously impinge on any given point from many different directions to produce a uniform state of pressure at that point (Figure 3–9).

Equation (3–1) shows that pressure is measured in force per unit area. The example of Figure 3–7 utilizes common U.S. customary pressure units of pounds per square inch, or *psi;* another conventional representation in this system is pounds per square foot (*psf*). The most common SI pressure unit is newtons per square meter, but this designation is seldom used. By definition, 1.00 N/m^2 is equivalent to 1 pascal (Pa), which is the preferred SI pressure unit and is named for an early student of fluid phenomena whose work we will

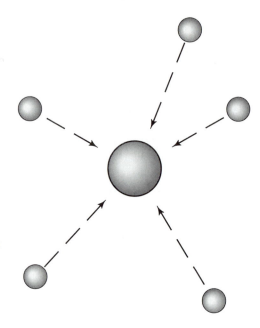

Figure 3–9   Gas molecules continuously bombard any point in the fluid from all directions, thus producing a uniform, omnidirectional state of pressure at that point.

examine shortly. Since 1 psi is equal to 6895 Pa, the two systems of measurement are orders of magnitude apart numerically; for this reason, the following prefixes are often used to maintain reasonable levels of computation:

$$1 \text{ kPa} = 1 \text{ kilopascal} = 1 \times 10^3 \text{ Pa}$$

$$1 \text{ MPa} = 1 \text{ megapascal} = 1 \times 10^6 \text{ Pa}$$

$$1 \text{ GPa} = 1 \text{ gigapascal} = 1 \times 10^9 \text{ Pa}$$

Another increasingly prevalent SI pressure unit is the millibar (mb), which by definition equals 100 Pa. And, as we shall see in Chapter 5, pressures may also be specified in terms of the height of a column of liquid that produces the same pressure at its base. Typical units are inches or mm of mercury (Hg) and inches or mm of water (H_2O). Human blood pressure, for example, is normally recorded in millimeters of mercury, with typical systolic/diastolic values of 120/80 (see Chapter 5, Example 9). As a technologist, you should possess a working knowledge of the various U.S. customary and SI pressure units, as well as a proficiency in converting from one system to another. Toward that end, some common pressure equivalents are listed in Table 3–1.

EXAMPLE I

A tank is pressurized to 4500 psf. Find the equivalent pressure in (a) psi, (b) lb/yd^2, (c) Pa, and (d) mb.

Table 3–1 Common U.S. Customary and SI Pressure Equivalents*	psi	psf	Pa	mb
	1	144	6895	68.95
	0.006944	1	47.88	0.4788
	0.0001450	0.02088	1	0.01
	0.01450	2.088	100	1

*All entries in a given horizontal row are equal.

Solution   Equivalent pressures may be computed by the mechanical application of conversion factors, as demonstrated in section 1.4. You should, however, be able to identify the appropriate areas involved in the computation and conversion of pressures.

(a) A pressure of 4500 psf indicates that every 1-ft^2 area is supporting 4500 lb, as shown in Figure 3–10(A). Since 1 ft^2 contains 144 in.^{2}, then this 4500 lb is actually being carried by 144 1-square-inch areas. Thus,

$$p = \frac{4500 \text{ lb}}{144 \text{ in.}^2} = \textbf{31.3 psi}$$

(b) Since one square yard contains 9 ft^2 (Figure 3–10(B)) and each foot-square area supports 4500 lb, the pressure is:

$$p = (4500 \text{ lb/ft}^2) \times (9 \text{ ft}^2/\text{yd}^2) = \textbf{40,500 lb/yd}^2$$

(c) Using a conversion factor of (6895 Pa/1 psi),

$$p = 31.3 \text{ psi} \times \left(\frac{6895 \text{ Pa}}{1 \text{ psi}}\right) = \textbf{216,000 Pa} = 216 \text{ kPa}$$

(d) Since 1 mb = 100 Pa,

$$p = 216,000 \text{ Pa} \times \left(\frac{1 \text{ mb}}{100 \text{ Pa}}\right) = \textbf{2,160 mb}$$

As we shall see in Chapter 7, pressure may be measured using a variety of instruments, including those whose sensing elements consist of a coiled tube, flexible diaphragm, or piezoelectric material. Table 3–2 gives typical values of pressure in both U.S. customary and SI units for some common pressurized fluids.

Pressure, then, is actually the reaction of confined fluid molecules to some type of applied load. If a pressurized fluid is stored in a rigid container, the pressure may be reduced by removing fluid particles from the container. Compressed air, for example, may be used to operate one or more pneumatic tools, but as this fluid is drawn from the tank, system pressure drops until eventually the air supply must be replenished. This is usually accomplished by a compressor that is activated automatically when system pressure drops below a preset value.

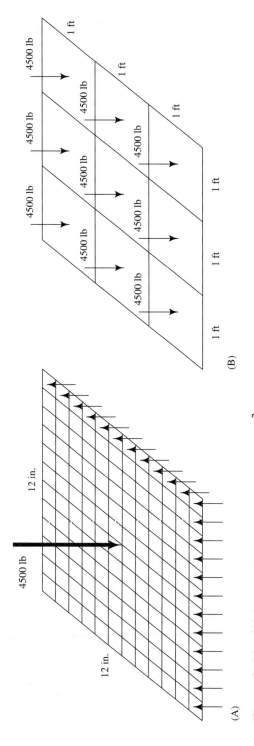

Figure 3–10 (A) A single 4500-lb force applied to 1 ft² is actually supported by forces on 144 one-square-inch areas. (B) One square yard contains 9 one-square-foot areas, each supporting a 4500-lb force.

Table 3–2   Typical Pressures of Some Common Pressurized Fluids

Fluid	psi	kPa
Human blood (systolic-diastolic)	2.3–1.5	15.9–10.3
Fresh water at bottom of 12-ft-deep swimming pool	5.2	35.9
Inflated automobile tire	30	207
Closed water faucet supplied by well	40–60	276–414
Municipal fire hydrants	70–200	483–1,380
Air supply—pneumatic tools	70–150	483–1,030
Industrial hydraulic systems	2000	13,800
Liquid jet cutter (see Figure 1–3)	50,000	345,000

If all of the fluid molecules are removed, or *evacu*ated, from a given volume, it seems logical that the pressure within that volume must be zero, since nothing is left in the volume to exert any pressure. This condition is known as a *perfect vacuum* and represents the absolutely lowest possible pressure that can be attained in nature. Still, students often ask, Is it possible to produce a negative pressure? The answer is—yes and no!

This apparent contradiction results from the existence of two separate pressure scales that are used to measure *absolute* and *gauge* pressures. Although we consider air to be weightless, when stacked to a depth of many miles it produces a pressure at the surface of the earth known as *atmospheric pressure,* average values of which are 14.7 psi or 101 kPa. The existence of atmospheric pressure was first theorized by Blaise Pascal, a French philosopher and mathematician, who tested his hypothesis atop a mountain in 1648. Although we frequently take this pressure for granted, its effects can be both powerful and unwanted as shown in Figure 3–11 (also see Figure 5–10).

Because virtually every object on earth is subject to atmospheric pressure, however, only pressures above or below this value are meaningful in most fluid power systems. Thus, pressure gauges that are not connected to a system are set to read zero pressure even though the actual or absolute pressure that exists is the atmospheric value. Pressures higher than atmospheric are read as positive pressures by most gauges, while pressures below atmospheric register as negative pressures and are referred to as *partial vacuums.* Absolute and gauge pressures are often indicated by the suffixes "a" or "g," respectively, such as 528 psia or 12 kPag; if no suffix is given, the pressure is assumed to be in gauge units.

In summary, then, the "zero reference point" of the absolute pressure system is a perfect vacuum, with all other pressures being measured as positive values. In the gauge pressure system, atmospheric pressure is used as the zero reference point, with pressures above this value measured as positive and pressures below this value measured as negative. The two systems are compared graphically in Figure 3–12.

Figure 3–11 This 55-gal ribbed steel drum collapsed under the effects of atmospheric pressure. Waste motor oil contained in the drum was being pumped into another tank for processing; although the suction hose inserted through the hole atop the drum was not sealed to the drum itself, liquid was removed so quickly by the pump that a partial vacuum was produced within.

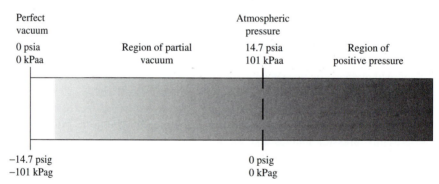

Figure 3–12 Comparison of absolute and gauge pressure scales.

EXAMPLE 2

    (a) Convert 56.8 psig to absolute pressure units.

    (b) Convert 720 psfa to psig.

    (c) Convert 3920 Pag to kPaa.

Solution   (a) Formalizing the relationship between absolute and gauge pressures yields:

$$p_{abs} = p_{gauge} + p_{atm}$$

Then,

$$p = 56.8 \text{ psig} + 14.7 \text{ psi} = \textbf{71.5 psia}$$

(b) This is a two-step conversion. First convert psf to psi, an operation that is the same whether gauge or absolute units are involved:

$$720 \text{ psfa} \times (1 \text{ ft}^2/144 \text{ in.}^2) = 5.00 \text{ psia}$$

Using the method from part (a), convert from psia to psig:

$$5.00 \text{ psia} - 14.7 \text{ psi} = \textbf{−9.70 psig}$$

(Notice that what initially appeared to be quite a large pressure was actually a partial vacuum.)

(c) Since 3920 Pag is equal to 3.920 kPag,

$$3.920 \text{ kPag} + 101 \text{ kPa} = \textbf{105 kPaa}$$

Equation (3–1) may also be rewritten as follows:

$$F = p \times A$$

This indicates that if a fluid at pressure p is in contact with some area A, then the *collective effect* of these unit forces may be represented as a single concentrated force, F. In other words, while a single applied force may be used to create pressure in a fluid, a pressurized fluid may in turn be used to produce a given concentrated force.

EXAMPLE 3

(a) Find the net force available if compressed air at 150 psi is applied to a piston whose diameter is 2.0 in.

(b) What piston diameter is required in order to double the force of part (a) if the pressure remains at 150 psi?

Solution   (a) The area of a 2.0-in.-diameter piston is 3.14 in.^{2}. Then available force may be computed as:

$$F = p \times A = \left(150 \text{ lb/in.}^2\right) \times (3.14 \text{ in.}^2) = \textbf{471 lb}$$

(b) Rearranging equation (3–1) yet again yields:

$$A = \frac{F}{p} = \frac{2 \times 471 \text{ lb}}{150 \text{ lb/in.}^2} = 6.28 \text{ in.}^2$$

and since a circular area is equal to $0.7854D^2$, then:

$$0.7854D^2 = 6.28 \text{ in.}^2$$

$$D^2 = 7.996 \text{ in.}^2$$

$$D = \mathbf{2.83 \text{ in.}}$$

Notice from Example 3 that it is not necessary to double the diameter of a circular piston in order to double the force available at a given pressure. This is because the area of a circle is not a linear function of diameter but involves the *square* of the diameter. Doubling the diameter increases force by a factor of four; tripling diameter increases the factor to nine; quadrupling diameter produces a sixteenfold increase. This effect is shown graphically in Figure 3–13, which indicates the net forces (in lb) available when various pressures (in psi) are applied to circular pistons whose diameters are specified in inches. Values for each point on these graphs were computed using the method of Example 3(a), i.e., by multiplying pressure times piston area. In many real systems, however, changes in total force, *F*, are accomplished by varying *both* system pressure and piston diameter.

EXAMPLE 4

The hydraulic machine shown in Figure 3–14 is a type that is widely used to measure the compressive strength of various engineering materials. The student at left is positioning a cylindrical concrete specimen in the test fixture. Her right hand rests on a movable platform, beneath which lies a large-diameter piston protected by the rubber bellows shown near her left hand. At the console, her lab partner will regulate oil flow to the piston, controlling the rate at which this piston raises the lower platform to crush the specimen against a stationary upper platform. (Note that tremendous forces can be developed during these tests, causing dramatic— and sometimes dangerous—failure of the specimens. For this reason, a large transparent safety shield is placed around the test fixture during actual testing.)

(a) The machine shown here contains a 14-in.-diameter piston and is capable of producing a maximum compressive force of $F = 120{,}000$ lb. What operating pressure is required in the hydraulic system to accomplish this?

(b) Larger models of the same machine can generate forces as high as 600,000 lb. This is accomplished by increasing both hydraulic pressure and piston diameter. If such a machine operates at a system pressure of 2000 psi, what is the required piston diameter?

Solution   (a) From equation (3-1), required system pressure is:

$$P = \frac{F}{A} = \frac{120{,}000 \text{ lb}}{0.7854 \times (14 \text{ in.})^2} = 780 \text{ psi}$$

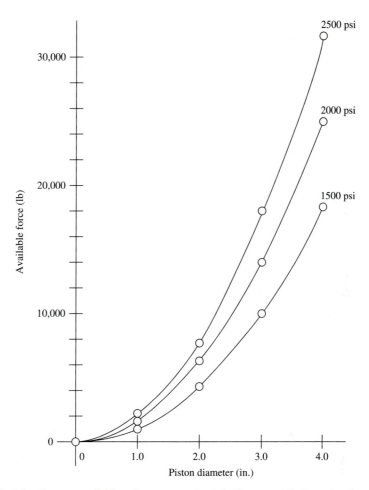

Figure 3–13   Forces available when pressurized fluids are applied to circular pistons of various diameters.

(b) Solving equation (3-1) for A gives:

$$A = \frac{600,000 \text{ lb}}{2000 \text{ lb/in.}^2} = 300 \text{ in.}^2$$

$$D = \sqrt{\frac{300 \text{ in.}^2}{0.7854}} = 19.5 \text{ in.}$$

Several points deserve mention in regard to Examples 3 and 4. First, hydraulic and pneumatic linear actuators are always referred to in fluid power terminology as cylinders.

Figure 3–14   Many engineering materials fail at extremely high loads. The machine shown here uses a large-diameter hydraulic piston to crush specimens (such as the cylindrical core sample of concrete located in the test fixture at left) by creating total forces up to a maximum of 120,000 lb. (See Example 4.)

Strictly speaking, only the round, rigid tubes or shells are actually cylinders; the movable elements that slide within these tubes to produce linear motion are pistons (see Figure 7–36). Throughout this text an attempt has been made to distinguish between pistons and cylinders when describing the operation of various components and systems.

Second, in discussing the use of compressors and pumps to produce a supply of fluid, virtually all fluid power publications stress the following points:

1. Pumps (and compressors) do *not* pressurize fluids.
2. Pressure is caused by a resistance to flow of the moving fluid.

Although these statements are essentially correct, they can be very misleading if not examined more closely.

Pumps and compressors are machines used to *move* fluids, often imparting significant *velocities* to these fluids as they travel through the machine. If a fluid is then discharged to the atmosphere, it is assumed to be at zero pressure (0 psig). Therefore, the pump or compressor, *by itself,* is deemed incapable of creating pressure in a fluid, thus forming the basis of statement 1. (Unfortunately, this viewpoint is technically incorrect, for as we shall see in Chapter 6, the velocity of a fluid is actually a form of pressure.) However, restricting the output of a pump or compressor by attaching hoses, fittings, tools, or cylinders creates a resistance to the flow of fluid, causing pressure to develop in the system. It is this behavior upon which statement 2 is based. To clarify these concepts, pumps and compressors are often defined as machines that have the ability to move a volume of fluid against a given pressure. In order to accomplish this task, of course, the machine must be capable of exerting those forces necessary to push fluid through the flow restrictions. For most students and technologists, this represents the ability of a pump or compressor to "pressurize" a fluid.

Problem Set 1

1. A pneumatic system operates at a pressure of 145 psi. Find this pressure in mbg and mba.
2. Which is the larger pressure in each case?
   (a) 30 psi or 3000 psf
   (b) 30 psia or 3000 psf
   (c) 30 kPa or 3000 psf
   (d) 15 kPa or 15 mb
   (e) 15 mb or 12 psi
3. A concentrated force of 2500 N is applied to a circular piston that is in turn supported by a confined fluid. Find the pressure created directly beneath this piston if the piston diameter is 8.00 cm.
4. Find the maximum force available in each case.
   (a) A pressure of 1000 psf is applied to a 3-in.-diameter piston.
   (b) A pressure of 20 kPa is applied to a 40-mm-diameter piston.
   (c) Of the two forces computed in (a) and (b), which is smaller?
5. A hydraulic pump is to be used with a 3.5-in.-diameter piston to produce a concentrated force of 16,000 lb.
   (a) Estimate the required pressure using Figure 1–13.
   (b) Compute the exact pressure using equation (3–1).
6. Find the diameter of a circular piston used to pressurize a hydraulic system to 400 psi using an applied load of 20 lb.

7. Using spreadsheets or a BASIC computer program, create a two-column table that lists the following.
   (a) The pressures required, in psi, on 1-in.-diameter to 10-in.-diameter cylinders (at increments of 0.5 in.) to produce a total force of 1000 lb.
   (b) The pressures required, in kPa, on 20-mm-diameter to 300-mm-diameter cylinders (at increments of 10 mm) to produce a total force of 1000 N.

3.3 Pascal's Principle: The Hydraulic Lever

The unique behavior of a fluid, including its ability to transmit an omnidirectional force, was summarized by Blaise Pascal in a theory known as *Pascal's principle*. This work was part of a book that he completed in 1653 but that did not appear until 1663, one year after his death. Although the concepts are now more than three centuries old, their validity has stood the test of time; they still form the basis of operation for most fluid power systems in use today.

Pascal's principle may generally be stated as follows:

The pressure applied to a confined fluid is transmitted
undiminished in all directions to every portion of the fluid.

In other words, if an applied force develops a specific level of loading (pressure) at one point in a fluid, the molecules in the fluid will adjust themselves until all molecules are at the same level of loading. Following the discussion of sections 3.1 and 3.2, these are not startling concepts and are based simply upon the ability of fluid particles to move and adjust to the shape of their container. When applied to a simple two-piston system, however, these basic ideas explain the operation and potential power capabilities of what can best be described as hydraulic or pneumatic levers.

Certain restrictions apply to Pascal's principle. It was originally intended to describe the behavior of *static* systems in which no fluid motion occurred, since fluid particles moving at different speeds are often at different pressures as well. We shall see in later chapters that although most useful fluid power systems involve some motion of the working fluid, the fluid velocity is frequently small enough that it may be neglected.

In addition, Pascal's principle neglects any pressure variations within a fluid that may be caused by the weights of standing columns of fluid. This also is not a serious restriction, since the vertical distances between high and low points in many hydraulic and pneumatic installations are usually insignificant.

Finally, although Pascal's principle is valid for both gases and liquids, the operational behavior of systems containing these fluids may differ somewhat due to the compressibility and temperature effects unique to a gas. To simplify the following examples, then, we shall restrict our analysis to systems that contain incompressible liquids as the working fluid.

To better understand the operation of a hydraulic lever, we first review the operation of a simple mechanical lever such as the one shown in Figure 3–15. A rigid member *A–C* is supported at *B* and is able to pivot about that point. Forces F_i and F_o applied at *C* and *A* balance the rigid member *A–C* on point *B* in the same way that children balance on a seesaw, or that weights are used on a balance beam scale.

In order for *A–C* to balance, a condition known as *equilibrium* must occur; i.e., the twisting or turning effects of F_i and F_o about point *B* must cancel. If F_o, for example, is removed quickly, member *A–C* pivots in a clockwise direction about *B;* likewise, F_i acting alone can cause a counterclockwise movement of *A–C* about *B*. As we saw in Chapter 1 (problem 12 in problem set 1), the ability of a force to cause rotation about a particular point is known as the *moment,* or *torque,* of the force about that point. Torque produced by some force *F* about a specific point *B* may be represented as T_B and is quantitatively defined as:

$$T_B = F \times d$$

where *d* is the *perpendicular* distance from point *B* to force *F* (Figure 3–16(A)) and is called the *moment arm.* If the force is not applied at right angles to the actual moment arm, the

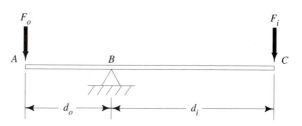

Figure 3–15   For a simple mechanical lever, equilibrium exists if the *clockwise* moment of F_i about point B is equal to the *counterclockwise* moment of F_o about the same point.

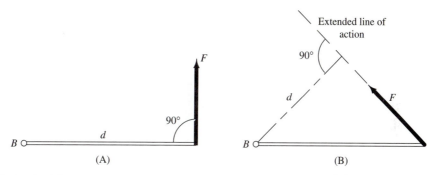

(A)   (B)

Figure 3–16   The moment, or *torque*, of a force about a particular point is defined as the product of that force times the perpendicular distance from the point to the line of action of the force. **(A)** If the force is applied at a right angle to the actual moment arm, then the length of this arm, *d*, is the distance used to compute torque. **(B)** If the force is *not* applied at a right angle to the actual moment arm, then the *effective* moment arm is the imaginary perpendicular distance, *d*, shown here.

effective moment arm is that distance from point B to the extended line of action of the force (Figure 3–16(B)).

Torque units may be any product of force times distance. In the U.S. customary system, in.-lb and ft-lb (or lb-in. and lb-ft) are common; preferred SI units are N·m. Numerical values of torque are useful in both mechanical and fluid power systems. The amount of twist, for example, applied to a threaded fastener is often specified in terms of the allowable torque; virtually all bolted connections (as well as spark plugs!) in an internal combustion engine are tightened to prescribed torques as measured by special wrenches known as *torque wrenches*. Special hydraulic and pneumatic tools known as *torque multipliers* are used to tighten nuts and bolts when high torques are required or available space is limited (Figure 3–17). Torque may also be used to designate the amount of twist available at the output shaft of an internal combustion engine, or an electric, hydraulic, or pneumatic motor. It is also common practice to indicate the amount of torque required to rotate the input shaft of a compressor or pump working against the load of a pressurized fluid.

Figure 3–17 This hydraulically driven torque multiplier can exert torques as high as 175,000 ft-lb and is able to function in restricted spaces. (Sweeney Co., a Dover Diversified Company, Englewood, Colo.)

Referring to the lever of Figure 3–15, equilibrium will be achieved only when the clockwise and counterclockwise torques about pivot point B are in balance. Stated in equation form,

$$F_i \times d_i = F_o \times d_o \qquad (3\text{–}2)$$

Equation (3-2) may be rearranged as follows:

$$\left(\frac{F_o}{F_i}\right) = \left(\frac{d_i}{d_o}\right)$$

The ratio (F_o/F_i) is defined as the *mechanical advantage* (M.A.) of the lever and represents the factor by which an input force, F_i, is magnified by this simple machine to produce some output force, F_o. In this case, M.A. equals the ratio of moment arms (d_i/d_o). Any change in this ratio produces a proportional change in M.A.—doubling the ratio doubles M.A.; tripling the ratio triples M.A. Consider the following example.

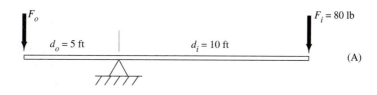

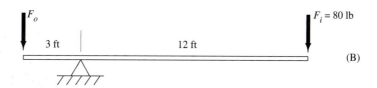

Figure 3–18   M.A. is proportional to ratio of lever arms.

EXAMPLE 5

(a) A 15-ft-long pipe is used as a lever (Figure 3–18(A)). For an input force, F_i, of 80 lb, what weight, W, can be raised? What M.A. does this represent?

(b) If the lever's *fulcrum*, or pivot point, is moved as shown in Figure 3–18(B), what is the new M.A.? What weight can now be raised by an input force of 80 lb?

Solution (a) Here, the weight being raised is represented by F_o. From equation (3- 2):

$$F_o \times d_o = F_i \times d_i$$

or:

$$F_o \times 5\ \text{ft} = 80\ \text{lb} \times 10\ \text{ft}$$

which yields:

$$F_o = \textbf{160 lb}$$

The mechanical advantage equals the ratio (F_o/F_i), or:

$$\text{M.A.} = \frac{160\ \text{lb}}{80\ \text{lb}} = \textbf{2.0}$$

Notice that this value for M.A. exactly equals the ratio of moment arms. In other words,

$$\text{M.A.} = \frac{d_i}{d_o} = \frac{10\ \text{ft}}{5\ \text{ft}} = 2.0$$

(b) For the new fulcrum position,

$$\text{M.A.} = \frac{d_i}{d_o} = \frac{12\ \text{ft}}{3\ \text{ft}} = \textbf{4.0}$$

Then:

$$F_o = \text{M.A.} \times F_i = 4.0 \times 80 \text{ lb} = \textbf{320 lb.}$$

What a wonderful device the simple lever appears to be! It takes any force applied to it and magnifies that force by a factor that we can control, virtually without limit, by moving the position of our fulcrum. Sound too good to be true? It is, because any magnification in force is matched by a corresponding reduction in the distance traveled by that force.

Consider the lever shown in Figure 3–19. When force F_i moves down some distance h_i, it raises force F_o some other distance h_o. These distances can be important since most real machines are called upon to *transmit* or move the applied forces through specified distances. Since triangles *ABC* and *DBE* are similar, we may write:

$$\frac{h_i}{d_i} = \frac{h_o}{d_o}$$

Rearranging gives:

$$\frac{h_o}{h_i} = \frac{d_i}{d_o}$$

or:

$$h_o = \frac{h_i}{\text{M.A.}}$$

Therefore, while input *force* F_i is *magnified* by factor M.A., the *distance* traveled by F_i is *reduced* by the same M.A. factor.

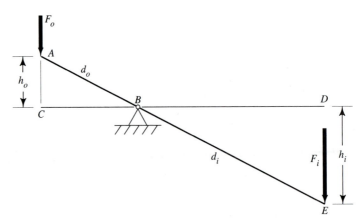

Figure 3–19   Lever movements are inversely proportional to M.A.

EXAMPLE 6

In Example 5, if the 80-lb weight moves down 2 ft, how high is the weight raised for each position of the fulcrum?

Solution   For part (a), from Example 5, M.A. = 2.0. Therefore, the weight is raised a distance h_o given by:

$$h_o = \frac{h_i}{\text{M.A.}} = \frac{2\text{ ft}}{2.0} = \textbf{1.0 ft}$$

Similarly for part (a) from Example 5 where M.A. = 4.0,

$$h_o = \frac{h_i}{\text{M.A.}} = \frac{2\text{ ft}}{4.0} = \textbf{0.50 ft}$$

Now consider the two-piston hydraulic system shown in Figure 3–20. The system consists of two pistons that are free to slide in their respective cylinders and that are connected by a continuous path of incompressible liquid. As we shall see, this system is equivalent to a hydraulic lever.

If an input force, F_i, is applied to a piston of area A_i, what force, F_o, can be supported on a piston whose area is A_o? What is the mechanical advantage of the system? The condition for equilibrium here is not one of balanced torques, but of equal pressures. Input force F_i creates a pressure directly beneath piston 1. This pressure, p_i, is simply:

$$p_i = \frac{F_i}{A_i}$$

But Pascal's principle states that because of the characteristics of a fluid, the pressure created by F_i must be transmitted undiminished throughout the system.

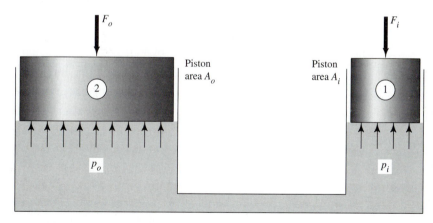

Figure 3–20   The two-piston system constitutes a hydraulic lever.

Therefore, the pressure, p_o, under piston 2 must be equal to p_i, or

$$p_o = p_i$$

Then:

$$\frac{F_o}{A_o} = \frac{F_i}{A_i}$$

and,

$$\frac{F_o}{F_i} = \frac{A_o}{A_i} \qquad (3-3)$$

For circular pistons, $A_o = 0.7854D_o{}^2$, $A_i = 0.7854D_i{}^2$, and equation (3–3) becomes:

$$\text{M.A.} = \frac{F_o}{F_i} = \frac{D_o{}^2}{D_i{}^2} = \left(\frac{D_o}{D_i}\right)^2 \qquad (3-4)$$

or:

$$F_o = F_i \times \frac{D_o{}^2}{D_i{}^2} = F_i \times \left(\frac{D_o}{D_i}\right)^2$$

Notice that there are several important differences between the mechanical and hydraulic levers. In order to obtain a magnification of force with the former, input force F_i is applied to the *larger* of the two moment arms; in a hydraulic lever, that input force is applied to the piston having the *smallest* cross-sectional area. Also, the M.A. for a mechanical lever is proportional to the simple ratio of lever arm lengths—doubling this ratio doubles the M.A. For circular pistons, however, the M.A. of a hydraulic lever is proportional to the *square* of the ratio of diameters—doubling this ratio causes a quadruple increase in M.A. This behavior was demonstrated graphically in Figure 3–13.

EXAMPLE 7

Find the piston diameters required in a two-piston hydraulic system that produces an output force of 800 lb for an input force of 50 lb and operates at a pressure of 600 psi. How far must the input force travel in order to move the output force through a distance of 2.0 in.?

Solution Referring to Figure 3–20, the 50-lb input force "activates" this system by creating a pressure of 600 psi directly beneath piston 1. The diameter may be found as follows:

$$A_i = \frac{F_i}{p_i} = \frac{50 \text{ lb}}{600 \text{ lb/in.}^2}$$

or:

$$A_i = 0.0833 \text{ in.}^2$$

Since $A_i = 0.7854D_i^2$,

$$0.7854D_i^2 = 0.0833 \text{ in.}^2$$

or:

$$D_i = \textbf{0.326 in.}$$

From Pascal's principle, the pressure under piston 2 must also equal 600 psi, so that:

$$A_o = \frac{F_o}{P_o} = \frac{800 \text{ lb}}{600 \text{ lb/in.}^2}$$

or:

$$A_o = 1.33 \text{ in.}^2$$

and:

$$D_o = \textbf{1.30 in.}^2$$

Notice that D_o could also have been computed using the M.A.

$$\text{M.A.} = \frac{F_o}{F_i} = \frac{800 \text{ lb}}{50 \text{ lb}} = 16.0$$

But,

$$\text{M.A.} = \frac{D_o^2}{D_i^2}$$

or:

$$16.0 = \frac{D_o^2}{(0.326 \text{ in.})^2}$$

so:

$$D_o = 1.30 \text{ in.}^2$$

To determine the relationship between piston movements, consider Figure 3–21. Part (A) of that figure shows pistons 1 and 2 in their initial positions, aligned here for clarity on the dotted reference line. In part (B), piston 1 has moved a distance h_i into its cylinder, producing a corresponding movement, h_o, of piston 2 out of its cylinder. Since the liquid is essentially incompressible and cannot escape from or accumulate in this system, space must be made available to accommodate the volume of liquid displaced *into* the system ahead of piston 1. This volume flows into the space left behind by piston 2 as it moves *out* of its cylinder. (The process is, of course, a continuous one, so that total volume between the pistons is always completely filled with liquid.) These volumes, V_i and V_o respectively, are represented by the shaded areas of Figure 3–22(A) and are shown pictorially in part (B) of that figure. Since $V_i = V_o$, it follows that:

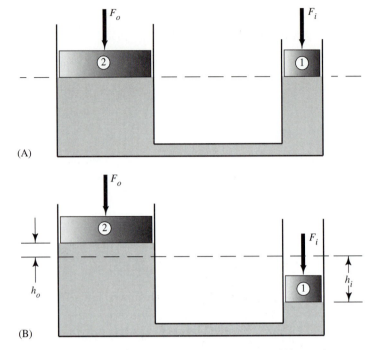

Figure 3–21 Piston movements in the two-piston system: (A) initial positions, (B) after movement.

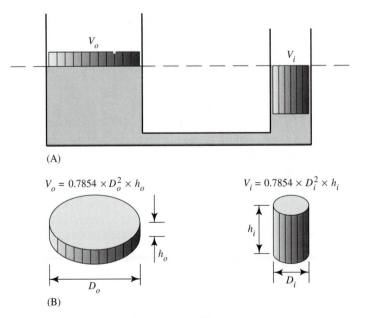

Figure 3–22 Displaced volumes caused by piston movements.

$$0.7854 \times (D_i)^2 \times h_i = 0.7854 \times (D_o)^2 \times h_o$$

or

$$h_o = \frac{h_i}{\left(\dfrac{D_o^2}{D_i^2}\right)} = \frac{h_i}{\text{M.A.}}$$

In other words, as with the mechanical lever, force magnification is accompanied by reduced movement of the output piston, this reduction being inversely proportional to M.A. For the system described here, whose M.A. is 16.0, a 2.0-in. motion of the output piston requires an input piston movement of

$$h_i = h_o \times \text{M.A.} = 2.0 \text{ in.} \times 16.0 = \textbf{32.0 in.}$$

Anyone who has used a hydraulic jack or other lifting device (see Figure 3–23) knows that such a large movement of the input piston is seldom accomplished in a single stroke, but rather by a series of short, repetitive strokes. Systems of this type typically contain check valves that lock the output piston in position between strokes and allow additional liquid to enter the system as the input piston is withdrawn. Operation of such a "pumping chamber" is shown in Figure 3–24. The check valves at A and B

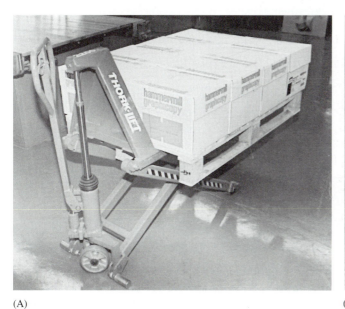

(A)   (B)

Figure 3–23   (A) The hand-operated hydraulic jack pictured here uses a two-piston system to lift pallets. (B) Through a lever handle, force is applied to the small piston at left; the large piston at right magnifies this force and provides lift through the telescoping cylinder rod visible in (A).

generally consist of smooth balls or tapered plugs that are held in contact with a mating surface, or seat, by springs, thus forming a leak-proof seal. If the pressure difference across the valve is sufficient to lift the ball or plug off its seat, flow will occur through the valve. A pressure difference in the reverse direction simply reinforces the spring action, seating the element more firmly. The check valve, then, allows flow in one direction only. As the input piston in Figure 3–24(A) moves into its

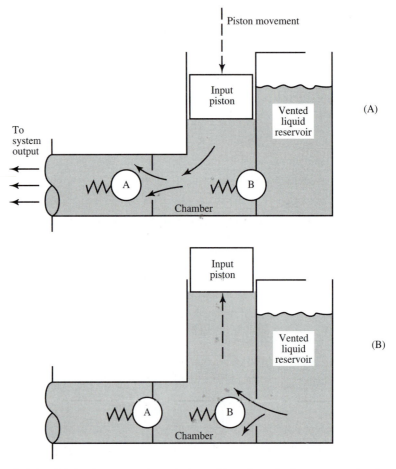

Figure 3–24   (A) As the input piston is forced into its cylinder, *pressure* developed in the pumping chamber lifts check valve A off its seat and allows liquid to flow to the system output. This same pressure aids the spring on check valve B in keeping that valve sealed so that no backflow occurs into the reservoir. (B) When the input piston is withdrawn from its cylinder, a *partial vacuum* is created in the chamber. This low pressure helps seal check valve A and opens check valve B so that liquid flows from the reservoir to the chamber, replacing the fluid that was pumped to the system output.

cylinder, fluid in the chamber is pressurized and check valve B closes, while the valve at A opens and liquid flows into the rest of the system. As we shall see in Chapter 5, if a solid, liquid, or gas is removed from the space it once occupied, then a partial vacuum may be created in that space. Therefore, when the piston is withdrawn as shown in Figure 3–24(B), pressure in the chamber drops and valve A closes, while the valve at B opens and liquid flows from the reservoir into the chamber. The entire sequence is then repeated.

In summary, the hydraulic lever is a powerful machine whose operation is described by Pascal's principle. Although the previous discussion centered upon systems that were pressurized by the application of single concentrated forces, the same principles of operation may be applied to other, more versatile systems in which the working liquid is pressurized either by a pump or by a compressed gas maintained in contact with the liquid.

Problem Set 2

1. The lever of Figure 3–25 has its pivot point, or *fulcrum,* at one end (point A).
   (a) Find the force, F, which must be exerted at end C in order to support the 180 lb force located at B as shown. (A wheelbarrow is equivalent to a lever of this type.)
   (b) What is the M.A. for this lever?

2. A lever may be used as a simple hydrometer to determine the specific gravity of an unknown liquid. In Figure 3–26, equal volumes of water and the unknown liquid are placed on opposite ends of a lever. The fulcrum is then adjusted horizontally until the lever is exactly balanced.
   (a) Show that $S_{liq} = \dfrac{L}{R}$.
   (b) What is S_{liq} if $L = 16$ cm and $R = 22.5$ cm?

3. A two-piston hydraulic system has input and output piston diameters of 6.0 cm and 15.0 cm, respectively. A force of 5.64 kN is applied to the smaller piston.
   (a) Compute the pressure developed in the system.
   (b) Find the force produced at the larger piston.
   (c) If the smaller piston moves into its cylinder a distance of 8.5 cm, how far does the larger piston move out of its cylinder?
   (d) If the 5.64-kN force is applied to the larger piston, moving it 8.5 cm into its cylinder, find the pressure in the system, the force produced at the smaller piston, and the distance traveled by this piston.

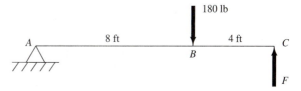

Figure 3–25   See problem 1
in problem set 2.

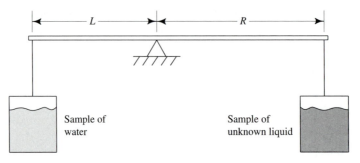

Figure 3–26   See problem 2 in problem set 2.

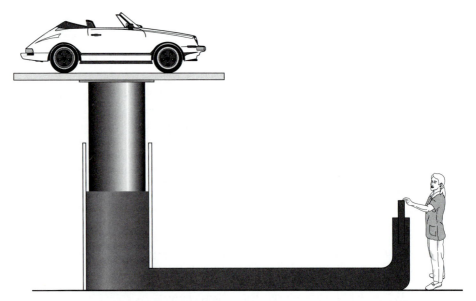

Figure 3–27   Theoretically, the mechanic can raise an automobile into the air with a push of her finger. (See problem 5 in problem set 2.)

4. A two-piston hydraulic system must raise a weight of 6000 lb on a 2.5-in.- diameter piston.
   (a) What is the working pressure of the system?
   (b) If the force applied to the input piston is restricted to 80 lb, what is the required input piston diameter? Is this a maximum or minimum allowable diameter?
   (c) What is the M.A. of this system?

5. The hydraulic hoist shown in Figure 3–27 is used to raise automobiles above the ground for repairs. A vehicle weighing 3500 lb rests on a cylinder whose diameter is 14 in., and a mechanic can easily exert a force of 1 lb with a push of her finger.
   (a) What is the required diameter of the piston upon which the mechanic pushes with her finger?
   (b) How far must the mechanic move this piston in order to raise the automobile 6 ft into the air?
   (c) What is the operating pressure of this system?
   (d) Is this a practical system? (Neglect any effects caused by weights of the liquid columns.)

Table 3–3   Common U.S. Customary and SI Work/Energy Equivalents*

ft-lb	kWh	J	Btu	cal
3.09	0.000116	4.19	0.00397	1
778	0.000293	1055	1	252
0.738	2.77×10^{-7}	1	0.000948	0.239
2,653,000	1	3,608,000	3410	8596
1	3.76×10^{-7}	1.36	0.00129	0.324

*All entries in a given horizontal row are equal.

3.4   Work, Energy, and Power

In order to analyze the transmission of power in fluid power systems, we must first define several terms used to describe the behavior of both the systems and their working fluids.

Work may generally be defined as the movement of a force through some specified distance, while *energy* is the ability of a substance or system to perform useful work. If the force, *F*, and the distance traveled, *d*, are in the same direction, then work, *w*, is computed as:

$$w = F \times d \qquad\qquad (3\text{–}5)$$

In the U.S. customary system, both work and energy are measured in ft-lb or lb-ft; comparable SI units are N·m or m·N, which are also known as *joules* (abbreviated as **J**) after English physicist James Prescott Joule. For systems that involve the *thermal* energy of a heated fluid or the stored *chemical* energy of a combustible fluid, work and energy are commonly measured in British thermal units (Btu) or calories (cal). (It was Joule who, in 1843, measured the equivalence between mechanical [ft-lb] and thermal [Btu] energy forms.) Food, for example, represents a source of energy that allows the human body to perform useful mechanical (physical) work, and we all know at least one person who is counting calories to limit the amount of excess energy stored chemically in the form of body mass. Finally, *electrical* work and energy are most often measured in watt-hours or kilowatt-hours (kWh). Table 3–3 lists the equivalencies between these common work/energy units.

Perhaps the simplest example of work involves the lifting of a weight through some vertical distance. In order to raise the 20-N block of Figure 3–28, a 20-N upward force must be exerted on the block, as shown. If the block is lifted through a vertical distance of 3 m, its motion is in the same direction as the force, so the work required can be computed from equation (3–5):

$$w = F \times d = 20 \text{ N} \times 3 \text{ m} = 60 \text{ N·m} = 60 \text{ J}$$

The ability of an object, material, or system to raise a weight through a vertical distance is often used as a test for the presence of energy. In Chapter 6 we will examine the three important forms of energy in a fluid—pressure, velocity, and elevation. As shown in Figure 3–29, a fluid possessing any of these three characteristics is capable of raising a weight; consequently, fluids exhibiting these traits must, by definition, contain energy. (Can

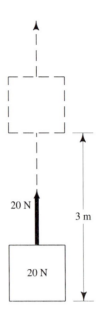

Figure 3–28 Force required to raise block is in same direction as block movement.

20 N

20 N

3 m

you think of a way by which the thermal energy (heat) contained in a fluid could be used to lift a weight through some vertical distance?)

The work required of a fluid power system may be specified in several ways. Systems that utilize a linear actuator, such as the hydraulic cylinder in the lifting device of Figure 3–23, are usually designed to provide a specified total force through a selected distance or *stroke*. These parameters represent quantities F and d in equation (3–5) and can be used to determine the work that must be performed by the system in a given series of operations called its *work cycle*. Since the performance of a system is actually accomplished by its working fluid, desired linear output can be used to compute the level of energy required in the fluid.

Rotary motion may also be used to produce useful work. In Figure 3–30, for example, weight F is attached to a cable that is being wound onto a rotating drum or pulley; for each revolution, the weight is raised a distance equal to the circumference of the drum. Although this drum could be driven by an electric motor or a gasoline engine, it could also be activated either by a hydraulic motor or a pneumatic rotary actuator. For such fluid power devices, desired work output of the system can again be used to determine the necessary energy of the fluid.

Notice that weight F is applied to the drum at radius r and that the cable is perpendicular to radius AB at point B. The product of F times r actually represents a torque created about point A, so that the suitability of a rotary actuator for use in any given system might be limited by the torque that the device is capable of producing.

Finally, some fluid power systems have outputs that are not as easily analyzed as simple linear or rotary motion. A dishwasher, for example, operates on the energy of a pressurized liquid, but the work it performs does not easily fit our definition of a force times a distance. For systems or devices such as these, required energy levels are usually specified by the manufacturer, who often determines, by experimental rather than analytical means, the levels necessary for satisfactory performance.

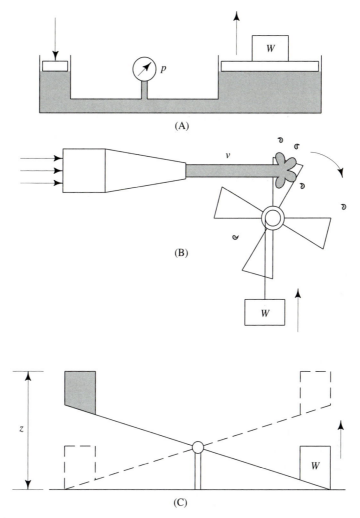

(A)

(B)

(C)

Figure 3–29   Forms of energy in a fluid: (A) pressure as energy; (B) velocity as energy; (C) elevation as energy. Each form has the capability of raising a weight W through some vertical distance.

Power is defined as the rate at which work is done, or the rate at which energy is utilized to produce that work. In equation form, power, *P*, is given as:

$$P = \frac{w}{t} = \frac{F \times d}{t} \tag{3–6}$$

or:

$$P = F \times \left(\frac{d}{t}\right)$$

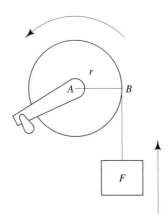

Figure 3–30 Rotary motion used to raise a weight.

Since distance traveled in a given time, (d/t), represents the velocity, v, at which force F is moving, equation (3–6) may also be written as

$$P = F \times v \qquad\qquad (3\text{–}7)$$

In the U.S. customary system, power is measured in *horsepower* (hp), a unit whose value of 550 ft-lb/s or 33,000 ft-lb/min resulted from tests done by James Watt during the late 1700s. It was the SI power unit, however, that was named after this Scottish engineer and inventor. One watt is defined as 1 J/s and is equal to 0.00134 hp. Mechanical power is customarily described in horsepower; electrical power, in watts or kilowatts. Although it is technically correct to ask the hardware clerk for a 0.10-hp lightbulb or to tell our friends about the "powerful" 187-kw V-8 engine in our new automobile, convention prohibits us from doing so.

EXAMPLE 8

(a) A hydraulic cylinder must provide a continuous force of 24,000 lb along its entire 30-in. stroke in a time of 6 s. What hp does this represent?

(b) If the allowable time is increased to 8 s, find the hp produced by this system.

Solution

(a) The 30-in. stroke is equivalent to a distance of 2.5 ft. From equation (3-6):

$$P = \frac{F \times d}{t} = \frac{24{,}000 \text{ lb} \times 2.5 \text{ ft}}{6 \text{ s}}$$

$$P = 10{,}000 \text{ ft-lb/s}$$

Since 1 hp = 550 ft-lb/s, the hp value is

$$P = \frac{10{,}000 \text{ ft-lb/s}}{550 \text{ ft-lb/s hp}} = \mathbf{18.2 \ hp}$$

(b) Similarly for $t = 8$ s, $P = 7{,}500$ ft-lb/s = **13.6 hp.**

Equation (3–6) can also be used to compute total work or energy:

$$w = P \times t$$

EXAMPLE 9

A 5-hp gasoline engine drives a water pump that moves liquid at the rate of 180 gal/min.

(a) How many ft-lb of work are performed by the engine for each minute of operation?

(b) How much energy does each gallon of water possess after passing through the pump?

Solution

(a) Each horsepower represents a work rate of 550 ft-lb/s. Then the total work, w, done in 1 min (60 s) is

$$w = P \times t = 5 \times 550 \; \frac{\text{ft-lb}}{\text{s}} \times 60 \text{ s} = \mathbf{165{,}000 \; ft\text{-}lb}$$

(b) During a 1-min period, 180 gal of liquid pass through the pump. If the pump is 100% efficient, it can utilize all this work to increase the energy of the liquid. Then the amount of energy received by each gallon of liquid may be computed as

$$\frac{(165{,}000 \text{ ft-lb})}{180 \text{ gal}} = \mathbf{917 \; ft\text{-}lb/gal}$$

(Since 1 gal of water weighs 8.34 lb, the pump could raise this liquid to a height of approximately 110 ft.)

Power transmission is often accomplished through rotating shafts. The power available from such a shaft is directly related to the rotational speed, N, of the shaft, as well as the torque, T, that the shaft can produce. From Figure 3–30, T controls the weight that can be raised by a drum of radius r, while N determines how fast cable is wound onto the drum and, therefore, the speed at which this weight can be raised. In equation form:

$$P = \frac{(N \times T)}{63{,}000} \tag{3–8}$$

where P is in hp, N is in rpm, and T is in units of in.-lb. The equivalent SI form of this equation is

$$P = 6.28 \times T \times \omega \tag{3–9}$$

where P is in watts, T is in N·m, and ω is the *frequency* of shaft rotation in revolutions per second (commonly designated as Hz).

In Chapter 4 we will examine a typical hydraulic system and determine the input power required to drive this system based solely upon the mechanical output for a given work cycle. We will also compute the liquid pressure required to achieve the desired performance. In Chapter 6, we will discuss another method of analyzing the power of a liquid based upon its individual energy forms—pressure, velocity, elevation—and its rate of flow. This general method allows us to determine either the required input power or available output power when a liquid is used as the transmission medium.

Problem Set 3

1. A 118-lb college student climbs a 1200-ft mountain in 2 h.
   (a) Compute the total work done during this activity.
   (b) Find the average rate, in hp, at which she is expending energy to produce this work.
   (c) If this same student can sprint up a 10-ft high flight of stairs in 3 sec, what average hp does she generate?

2. A countertop water heater heats 2 cups (16 oz) of liquid for use in tea, instant coffee, or soup. The device is rated at 1450 W and can heat this amount of liquid in 2 min 10 s.
   (a) How many kwh of energy are used?
   (b) How many ft-lb of energy does this represent?
   (c) How high on a vertical ladder would a 150-lb person have to climb in order to expend the same amount of energy as computed in part (b)? Could this reasonably be accomplished in the heating time specified?

3. A water pump moves 200 gal of water per minute to a height of 65 ft. If water weighs 8.34 lb/gal, find the hp required for this activity. Assume the pump is 100% efficient and neglect any energy contained in the velocity of the liquid.

4. If a hydraulic cylinder moves a force of 50 kN through a distance of 0.6 m in 4 s, find the total energy required and the power, in kw, that this activity represents.

5. An automated hydraulic system lifts crates weighing 145 lb from a conveyor belt through a vertical distance of 42 in. to a storage platform. If each crate is lifted in 2.5 s, how many hp is the hydraulic system producing during this operation?

6. Many electric motors are designed to run at a constant speed of 1750 rpm and are often used as power supplies for both hydraulic and pneumatic systems. How much torque would be available from a 1/2-hp electric motor operating at 1750 rpm?

Questions

1. The top of a 30-in. × 30-in. folding table experiences a total force of 13,230 lb due to atmospheric pressure. Why is it so easy to lift such a table despite this enormous force applied to its top?

2. In 1654, Otto von Guericke (inventor of the air pump) gave an amazing demonstration. Assembling two brass hemispheres, as shown in Figure 3–31, he evacuated the air from inside the sphere and hitched a team of horses to each of the two halves. Even with their best efforts, the horses could not pull the hemispheres apart.
   (a) What holds the assembled sphere together?
   (b) Could the hemispheres ever be pulled apart?

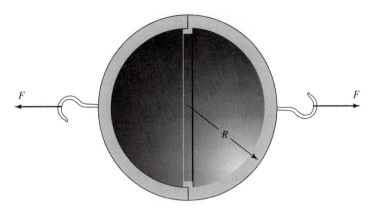

Figure 3–31   When air is removed from inside this sealed, two-hemisphere assembly, the sphere is not easily pulled apart. (See question 2.)

(c) Would the size (radius R) of the sphere have any effect on the force required to pull the hemispheres apart?

(d) What might we do to make the sphere simply fall apart?

3. Why is pressure omnidirectional?

4. Is it possible to achieve a pressure of -20 psig? -20 psia? Explain.

5. Define the mechanical advantage of a lever and of a two-piston hydraulic system.

6. List several common dimensional units, other than ft-lb, used to describe work and energy.

Review Problems

1. The reservoir for a hydraulic system has a rectangular top whose dimensions are 2 ft × 3 ft. Find the total force experienced by this member due to atmospheric pressure.

2. Write a computer program that produces, in spreadsheet format, the total force available when pressures are applied to pistons of various diameters. Typical output might be as shown in Table 3–4. Repeat this procedure using SI units.

3. Mechanical and hydraulic levers are often combined in a single system such as the one shown in Figure 3–32. Here a hand-operated lever activates the two-piston system whose output is used to clamp a cylindrical workpiece. If the machine operator applies a 50-lb force to the lever:
   (a) Calculate the force applied to the 2-in. diameter input piston.
   (b) Find the pressure developed in the system.
   (c) Compute the clamping force applied by the 3.5-in. diameter output piston.

4. A two-piston system can also be used to magnify displacements rather than force. In Figure 3–33, for example, how far will piston B rise in its cylinder if piston A moves down 1 mm? Can you think of a possible application for this phenomenon?

5. Compressed air at 70 psi is applied to the top of a 1.75-in.-diameter piston (Figure 3–34(A)). This piston acts as the input piston in the two-piston hydraulic system shown.
   (a) Find the working pressure of the hydraulic system.
   (b) Find the maximum weight, W, that the output piston can lift.

Table 3–4 Available Force (lb) for Various Combinations of Pressure (psi) and Piston Diameters (in.). (See Review Problem 2)

Piston Diameter (in.)	Pressure (psi)				
	1000	1500	2000	2500	3000
2.0	3142	4712	6283	—	—
2.5	4909	7363	—	—	—
3.0	7069	—	—	—	—

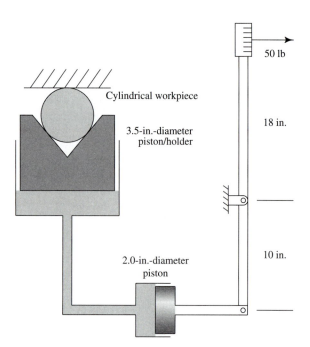

Figure 3–32 A mechanical and hydraulic lever combination for review problem 3.

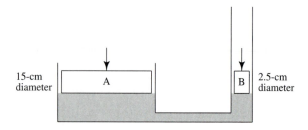

Figure 3–33 Two-piston system for review problem 4.

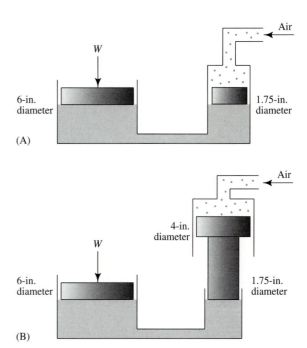

Figure 3–34 Two-piston hydraulic system for review problem 5.

(c) Repeat (a) and (b) if the compressed air is instead applied to the 4-in. diameter side of a *stepped* input piston as shown in Figure 3–34(B).

Suggested Activities

1. Using a T-handle tire pump of the type shown in Figure 3–35, measure the barrel diameter, D (or piston diameter if the unit can be disassembled), and install a pressure gauge on the outlet hose. Now apply a known force, F, to the handle (by balancing a concrete block, for example), and measure the pressure recorded by the gauge. How well does this pressure agree with the computed value obtained from equation (3-1)?

2. Apply a suction cup to a smooth horizontal surface, as shown in Figure 3–36. What holds the suction cup in place? Now measure the cup diameter, and hang a container from the assembly. Slowly add weight to the container (water or sand works well), and measure the total weight at which the cup begins to separate from the horizontal surface. How does this force compare to that computed from equation (3-1) if p is assumed to be atmospheric pressure?

3. Measure the specific gravities of several unknown liquids using the method of problem 2 in problem set 2, and compare these results to those obtained in suggested activity 1 in Chapter 2. A meterstick makes a convenient lever, and measured liquid samples are easily contained in paper or styrofoam cups or small plastic bags.

4. Have the members of your class weigh themselves. Next, measure the vertical distance spanned by a single straight stairway. Using a stopwatch, find the time required for each member of the class to sprint up the stairway. Compute the energy expended and horsepower developed by each individual during this activity.

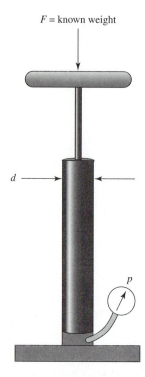

Figure 3–35 Using a simple tire inflation pump, it is possible to compare the measured pressure, p, with the pressure predicted by equation (3–1). (See suggested activity 1.)

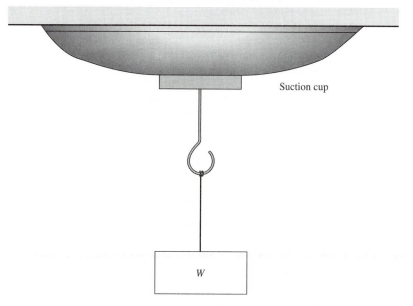

Figure 3–36 What holds this suction cup on a smooth horizontal surface despite the application of weight W? (See suggested activity 2.)

4

Sizing the Closed Hydraulic System

Some of the most common and useful fluid power circuits are those in which hydraulic cylinders extend and retract in a specified manner. Such linear actuators can be used to lift, clamp, or position objects and tools in a variety of manufacturing processes; they can also be used to activate the brakes in a truck or automobile, raise the blade on a snowplow, or compact trash in a vehicle designed for that purpose.

In this chapter, we examine a typical device that utilizes several hydraulic cylinders. Because the desired mechanical performance for such machines is usually specified, these criteria can often be used to determine the minimum power input required by the hydraulic system. In addition, we will compute the operational characteristics necessary for the actual power transmission components, namely the hydraulic pump and cylinder, in order to obtain the required outputs of force and motion.

By the end of this chapter, you should be able to select a suitable motor-pump-cylinder combination that satisfies the requirements for a given work cycle. You should also be able to correlate the hydraulic parameters of pressure and flow rate with the mechanical performance of a system and use this knowledge to predict the behavior of hydraulic circuits assembled from selected components.

4.1 Closed Versus Open Systems

Because the words closed and open are used in different ways to describe fluid power systems, a few basic definitions are necessary. In this chapter, a *closed* hydraulic system is simply one from which the working fluid does not escape. The liquid may move through the system continuously or intermittently, but it is used over and over again to perform its specific function. In other words, a fixed quantity of liquid is sealed into the closed system. Most fluid power systems that utilize an oil as the working fluid are closed systems since it would not be economically or environmentally feasible to discard large volumes of a fluid that has seen only light use. Automobile braking and engine lubrication systems are common examples of closed systems.

Similarly, an *open* system generally receives a constant supply of fresh working fluid that is discarded after a single use. Most water supply systems, for example, deliver liquid to a particular location at a certain pressure and velocity. Once the liquid has been used to wash a car, fight a fire, or generate hydroelectric power, it leaves the system and is not recycled. Pneumatic systems may also be classified as open systems, because the spent fluid (air) is usually discharged to the atmosphere.

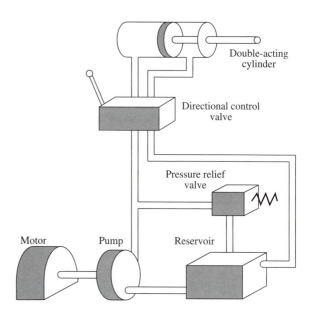

Figure 4–1 Major components of a typical fluid power system.

When applied specifically to hydraulic fluid power systems, however, a *closed* circuit often refers to one in which the liquid leaving a linear or rotary actuator is fed directly back to the inlet of the pump. Under this definition, the typical hydraulic system shown in Figure 4–1 is considered an *open* circuit since fluid leaving the actuator must first pass through a reservoir before reaching the pump inlet. On the other hand, this system satisfies our initial definition of a closed system because it utilizes a fixed amount of working fluid.

Finally, as seen in Chapter 11, fluid power systems may be classified as closed-loop or open-loop, depending upon the method by which they are controlled. The *closed-loop* system monitors output behavior and feeds appropriate command signals back to various control elements. These components then adjust the operating conditions to maintain desired system performance. Such closed-loop, or *feedback,* systems provide *automatic control* of many manufacturing and processing operations. Fluid power circuits subject to active human control are known as *open-loop* systems. Typical examples include most types of joystick-operated, or lever-operated, construction and agricultural equipment, as well as the sample system in Figure 4–1 whose directional control valve is activated using a hand lever.

Both open and closed systems utilize a change in the energy of the working fluid. Chapter 6 explains that energy and power contained in a liquid may be computed from the changes in pressure, velocity, and elevation that take place as the liquid moves through the system. As pointed out earlier, however, many closed systems use this energy to perform some mechanical operation such as extending and retracting a piston or driving a rotary actuator. By examining this mechanical output, it is frequently possible to determine the energy and power characteristics of a closed system without regard to the condition, or state, of the liquid itself.

4.2   Specifying System Performance

The major performance criteria for a closed system such as the one shown in Figure 4–1 include the following:

1. **Force** to be exerted by the piston
2. **Distance** through which the force is to operate
3. **Cycle time** for complete extension and retraction of piston

The sequence of events for which the piston extends and then retracts to its original position is called the *work cycle* of the system; *cycle time* is the time required for one work cycle.

An additional factor that may influence the selection of components is the total cost of the system. Usually several combinations of components will satisfy the given performance criteria; each of these combinations must be evaluated individually if total system cost is to be minimized.

4.3   Component Characteristics

The primary components that influence system behavior and the characteristics for which they are selected are as follows:

Hydraulic cylinder: Bore and stroke

Pump: Rated pressure and flow rate

Motor: Power available

These three pieces of hardware constitute the actual *power train* of a hydraulic system. Specifying the characteristics for a *matched* set of components that will achieve any given system performance is demonstrated by the following example.

EXAMPLE I

The hydraulic wheel crusher shown in Figure 4–2 is used in scrap yards to quickly separate tires from the steel wheels on which they are mounted. Once a deflated tire has been placed in the machine, three pistons extend simultaneously to crush the wheel, which is then simply lifted out of the tire. This process is much quicker than conventional methods of dismounting a tire. Each piston of the crusher must be capable of exerting 38,000 lb of force on a 13-in. stroke in a cycle time of 11.5 s.

(a) Select an appropriate piston for this device.
(b) Determine the necessary pump characteristics.
(c) Determine the minimum horsepower required to drive the system.

(A)   (B)

(C)

Figure 4–2   (A) This commercial wheel crusher utilizes three hydraulic cylinders that extend simultaneously (B) to crush and deform steel rims. Once the cylinders have retracted to their original positions, the bent wheel is simply lifted out of its tire (C) and sent to the scrap heap. Because this procedure is much faster than conventional dismounting methods, it is used extensively in salvage yards that process large quantities of mounted tires.

Solution   (a) *Piston Selection*

Virtually any piston having a stroke of 13 in. may be selected. Table 4–1 lists several possible choices found in one manufacturer's catalog, while Figure 4–3 shows the construction details of a typical commercial cylinder. A piston is attached to the long rod called an *output rod,* which extends through an opening in one *end cap* of the cylinder. This piston/rod assembly slides within the cylinder under the action of hydraulic fluid supplied through the threaded

Table 4–1   Typical Hydraulic Cylinder Dimensions

Cylinder*	Piston Diameter, D (in.)	Stroke, s (in.)	Output Rod Diameter, D_r, (in.)
A	4.0	13	2.50
B	5.0	13	3.50
C	6.0	13	4.00

*Cylinders A, B, and C are referred to frequently throughout Chapter 4.

Figure 4–3   Cutaway view showing construction of a typical hydraulic cylinder. *(Milwaukee Cylinder, Cudahy, Wis.)*

ports in each end cap. Seals are used on the piston and around the opening in the end cap to prevent leakage of the pressurized liquid. We will design a different system around each type of cylinder listed in the table.

(b) *Determining Pump Characteristics*

Once the piston has been selected, required pressure may be computed by applying equation (3–1) to a circular area:

$$p = \frac{F}{A} = \frac{F}{(0.7854 \times D^2)}$$

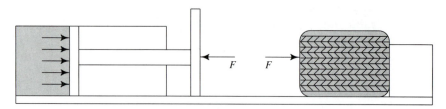

Figure 4–4   Forces (*F*) experienced by piston and tire assembly are equal in magnitude but opposite in direction.

| **Table 4–2**   Working | | **Cylinder** | | |
| Pressures Required to Produce | | **A** | **B** | **C** |
38,000 lb				
	Pressure (psi)	3024	1935	1344

where p is the required pump pressure in psi, F is the crushing force in lb, and D is the piston diameter in inches. Note in Figure 4–4 that as the piston exerts a force of 38,000 lb on the tire assembly, the tire exerts an equal but opposite force on the piston. This resisting force creates a pressure on the liquid inside the cylinder, and it is this pressure against which the pump must be able to move fluid. The computed values of required working pressures for cylinders A, B, and C appear in Table 4–2.

To calculate the pump flow rate necessary for the specified system performance, it is necessary only to determine the total volume that the pump must fill during one complete extension/retraction; cylindrical volumes required for each portion of the work cycle are shown in Figure 4–5. Since volume is computed as the product of cross-sectional area and length (see Chapter 1, Example 6), the volume V_{ext} required to extend each piston may be computed as:

$$\text{Volume} = \text{piston area} \times \text{stroke}$$

or:

$$V_{ext} = 0.7854 \times D^2 \times s$$

where V_{ext} is in cubic inches if piston diameter D and stroke s are both in inches. Notice that V_{ret}, the volume required to retract the piston, is equal to V_{ext} *minus* the volume V_{rod} occupied by the output rod:

$$V_{ret} = (0.7854 \times D^2 \times s) - (0.7854 \times D_r^2 \times s)$$

where D_r is the output rod diameter in inches. Combining the expressions for V_{ext} and V_{ret} yields the total volume V_{tot} to be filled during one complete work cycle. In simplified form:

$$V_{tot} = 0.7854 \times (2D^2 - D_r^2) \times s \tag{4–1}$$

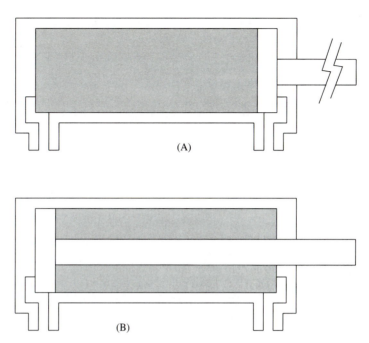

Figure 4–5   Cylinder volumes filled during one work cycle. (A) Volume filled during extension. (B) Volume filled during retraction.

Table 4–3 Volumes and Flow Rates for Cylinders A, B, and C from Table 4–1	Cylinder		
Results	**A**	**B**	**C**
V_{ext} (in.^{3})	163	255	368
V_{rod} (in.^{3})	64	125	163
V_{ret} (in.^{3})	100	130	204
V_{tot} (in.^{3})	263	385	572
V_{sys} (in.^{3})*	789	1156	1715
q (in.^{3}/s)	68.6	101	149
q (gal/min)	17.8	26.2	38.7

*V_{sys} = 3 (cylinders) $\times$ V_{tot}.

Equation (4–1) yields the volume of liquid required by a single cylinder during the work cycle; since the wheel crusher utilizes three cylinders, the required flow rate for the system is triple that for an individual cylinder. The pump must move this volume of liquid during the 11.5-s cycle time specified, so $V_{tot}/(11.5$ s) yields the total flow rate needed in cubic inches per second. Results for each cylinder, including flow rates in gallons per minute, appear in Table 4–3.

You should notice several points about the extend/retract characteristics of a system such as the one presented here.

1. Cylinder diameter affects both required pressure and flow rate. A large piston requires a smaller pressure to produce a given output force, but because the cylindrical volume is large, a high flow rate is necessary to achieve the desired cycle time. Obviously then, small pistons may be used with pumps that are able to operate at high pressures and low flow rates, while larger pistons may be used with low-pressure pumps operating at high flow rates. Because of the many piston/pump combinations possible, system performance criteria can generally be satisfied by any one of several matched sets of components.

2. Since the "retract" volume is smaller than the "extend" volume due to the presence of the output rod inside the cylinder, all pistons with single-output rods will retract faster than they will extend. To find the actual extend and retract times, it is necessary only to determine what percentage of the total volume the respective extend and retract volumes represent. Cylinder A, for example, has an extend volume of 163 in.^{3} and a total volume of 263 in.^{3}, so the extend volume represents $(163/263) = 0.620$, or 62.0%, of the liquid moved during one complete 11.5-s work cycle. If we assume that the pump delivers a uniform supply of liquid, the time required to extend the piston, t_{ext}, must be proportional to the percentage of total volume filled during the extend portion of the cycle. In other words:

$$t_{ext} = \frac{V_{ext}}{V_{tot}} \times \text{cycle time} \tag{4-2}$$

For cylinder A:

$$t_{ext} = (0.620) \times (11.5 \text{ s}) = 7.13 \text{ s}$$

Similarly for t_{ret}, the time to retract the piston:

$$t_{ret} = \frac{V_{ret}}{V_{tot}} \times (\text{cycle time}) \tag{4-3}$$

and for A:

$$t_{ret} = \frac{100}{263} \times (11.5 \text{ s}) = 4.37 \text{ s}$$

The extend and retract times for pistons A, B, and C appear in Table 4–4.

3. During retraction, the pressure that moves the piston axially may be applied only to that area of the piston not covered by the output rod, as shown in Figure 4–6. Since this *annular* area is less than the total area of the piston, the amount of force that the piston can exert during retraction may be considerably less than that exerted during extension. Piston C, for example, can exert only 16,800 lb of force while retracting, as opposed to the 38,000 lb available while extending. This may be an important consideration depending upon the specific performance requirements of a particular system.

Table 4–4 Extend and Retract Times of Cylinders A, B, and C for a Cycle Time of 11.5 s		Cylinder		
	Times	**A**	**B**	**C**
	t_{ext} (s)	7.13	7.62	7.40
	t_{ret} (s)	4.37	3.88	4.10

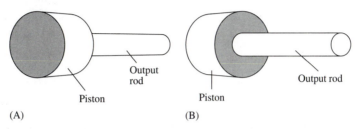

(A)   (B)

Figure 4–6   Piston areas (shaded) on which axial pressure acts during (A) extension and (B) retraction.

(c) *Calculating Required Horsepower*

In Chapter 6 we will examine the power contained in a liquid as evidenced by its pressure, velocity, elevation, and flow rate. This is the primary method used for determining fluid power in both closed and open systems. For closed systems, however, it is also possible to compute the power required to drive a system merely by using the mechanical power output indicated in the original performance criteria.

Mechanical power was discussed in Chapter 3 and was defined in equation (3–6) as:

$$\text{Power} = \frac{\text{force} \times \text{distance}}{\text{elapsed time}}$$

Notice that the system being designed here is capable of moving a force of 38,000 lb on the extend stroke through a distance of 13 in. (1.08 ft) in the elapsed times (t_{ext}) given in Table 4–4. Using cylinder A, for example, the power output of the system may be computed as:

$$\text{Power} = \frac{38,000 \text{ lb} \times 1.08 \text{ ft}}{7.13 \text{ s}} = 5760 \frac{\text{ft-lb}}{\text{s}}$$

Since 1 horsepower equals 550 ft-lb/s, the output of this system is 10.5 hp. Again, note that this is the power required to drive a *single* cylinder; for the three-cylinder wheel crusher, 31.5 hp are necessary. Similar results may be obtained for cylinders B and C; they are listed in Table 4–5. These figures represent the mechanical power output of each system, as well as the *minimum theoretical input horsepower* that must be provided to the system in order for it to satisfy the original performance criteria. This assumes that each system is 100% efficient,

Table 4–5   Summary of Component Characteristics

Cylinder	Piston Diameter (in.)	Pump Pressure (psi)	Pump Flow Rate (gpm)	Required Input Horsepower (hp)
A	4.0	3024	17.8	31.5
B	5.0	1935	26.2	29.4
C	6.0	1344	38.7	30.3

i.e. no frictional or hydraulic losses occur anywhere in the circuit during actual operation. Although available force, stroke, and cycle time are the same for each set of matched components, required input horsepower varies slightly with the ratio of output rod volume to total volume of the cylinder.

Notice that the pump and motor characteristics on each line of Table 4–5 are matched to the specified cylinder listed at the left. Any of these three cylinder-pump-motor combinations will, theoretically, deliver the specified system performance.

Problem Set I

1. A 3.5-in. diameter hydraulic cylinder has a stroke of 10 in. and an output rod diameter of 1.125 in. It is to be used in a system capable of producing 75 tons of force and having a cycle time of 8 s.
   (a) Compute the required pump pressure.
   (b) Compute the required flow rate.
   (c) Find the minimum horsepower needed to drive the system.

2. A hydraulic cylinder 16 cm in diameter has a stroke of 1 m and an output rod diameter of 5 cm. This cylinder must produce a force of 90 kN during extension and have a cycle time of 14 s.
   (a) Compute the required pump pressure.
   (b) Compute the required flow rate.
   (c) Determine the necessary power input for this system.

3. Referring to Example 4 in Chapter 3, find the pump flow rate (gpm) and ideal input horsepower for the compression testing machines described in parts (a) and (b) if both machines contain pistons that move upward at the rate of 2 in./min while delivering their maximum loads.

4. Write a computer program with the **performance** inputs of *stroke, available force,* and *cycle time,* and **cylinder characteristics** of *piston diameter* and *output rod diameter.* This program should compute and print the required **pump characteristics** of *working pressure* and *flow rate,* as well as the minimum *input horsepower* to drive the system.

5. Develop a spreadsheet that computes pump psi, pump gpm, required horsepower, and total cycle time for a single-cylinder hydraulic system whose extend motion moves 50,000 lb through a distance of 24 in. in 4 s. Compute these values for each of the following cylinders.

Cylinder	Piston Diameter (in.)	Stroke (in.)	Output Rod Diameter (in.)
A	3.0	24	1.00
B	3.5	24	1.25
C	4.0	24	1.50
D	4.5	24	1.50
E	5.5	24	1.75
F	6.5	24	2.25

4.4 Adjustments to Theory

As mentioned above, the preceding analysis applies to the performance of an ideal system, one in which the input energy or power exactly equals the output energy or power. Real life systems, however, are not ideal, and some adjustments to theory must be considered.

1. **Pump Efficiency.** Although frictional losses occur in the system as the liquid flows through fittings and hoses, and also as the piston extends/retracts in its cylinder, these losses are generally small compared to those that occur within the pump. Pump losses are directly related to the overall pump efficiency, which is itself composed of three parts. *Hydraulic efficiency* is a measure of how well input horsepower is actually transmitted to the liquid. It is controlled by the shape of the pump's moving parts and how closely the actual operating conditions match the design conditions of the pump. *Mechanical efficiency* is an indication of the losses that occur due to friction between moving parts, principally in the pump bearings. *Volumetric efficiency* accounts for losses due to the fluid slippage that invariably occurs between the rotor and housing of the pump. The *overall pump efficiency, η,* is a combination of hydraulic, mechanical, and volumetric efficiencies, and is defined as:

$$\eta = \frac{\text{power output}}{\text{power input}} \qquad (4\text{--}4)$$

To help evaluate pump efficiency, we make use of equation (4-5). This formula, which is derived in Chapter 6, allows calculation of pump output power, P, as:

$$P = 0.000583 \times q \times p \qquad (4\text{--}5)$$

where P is in horsepower if flow rate, q, is in gallons per minute and rated pump pressure, p, is in psi. Figure 4–7 represents the performance characteristics of a typical gear pump for various horsepower inputs. Point A on the graph, for example, indicates that the pump can move 4 gpm against a pressure of 1500 psi. From equation (4–5), this represents a power level in the liquid of 3.50 hp. At B and C the indicated performance levels are 9 gpm at 750 psi (3.94 hp) and 13 gpm at 500 psi (3.79 hp), respectively. Since input power to the pump is constant at 5 hp for these three points, the efficiencies computed using equation (4–4) are 0.70, 0.79, and 0.76. You should select several points along the 10-hp and 15-hp operating curves and compute the pump efficiencies at these points. Obviously, operating conditions

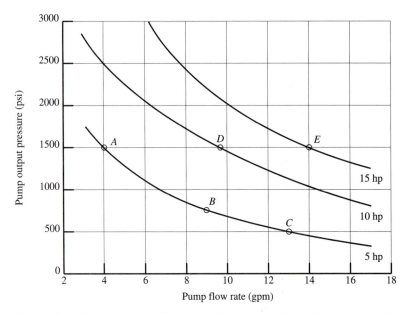

Figure 4–7   Typical gear pump performance for constant input horsepower. *(From data provided by Eaton Corp., Hydraulics Division, Eden Prairie, Minn.)*

do affect pump efficiency, and most gear pumps that operate at their optimum or design conditions normally exhibit efficiencies in the 85% to 93% range.

The effect of efficiency on system design, then, is as follows: In order to overcome losses in the system, input horsepower must be increased. Rearranging equation (4-4) yields:

$$\text{Power input} = \frac{\text{power output}}{\eta}$$

Most pump manufacturers provide recommended power inputs for their products, but if such information is not available, a conservative pump efficiency of approximately 0.70 should be assumed. In Example 1 of the preceding section, the values of required input horsepower listed in Table 4–5 should each be divided by 0.70.

2. **The Underpowered System.** If a fluid power system is driven by an inadequate power supply, how is the system's performance affected? Referring to Figure 4–7, assume that a system has been designed to move 14 gpm against a pressure of 1500 psi and requires a power input of 15 hp (point *E* on the diagram). If a 10-hp engine is used to drive the system, then the operating conditions become those shown at point *D*, namely, 9.5 gpm at 1500 psi. Similarly, a 5-hp engine could move 4 gpm at 1500 psi (point *A*).

In Chapter 3, the relationship between horsepower, torque, and engine speed was discussed. There it was shown that torque and engine speed are inversely proportional at constant horsepower (see equation 3-8). In an "ideal" underpowered system, the engine would operate continuously at maximum power, while a perfect transmission (consisting of gears, clutches, and/or pulleys) inserted between the engine and pump would provide the torque necessary to turn the pump against the required pressure. This would be accomplished with a corresponding decrease in speed of the transmission output shaft, thus reducing the flow rate provided by the pump. In most real-life systems, however, the pressure load is applied directly to the engine, decreasing engine speed until an adequate torque level is achieved. If the load is excessive or is applied too quickly, the engine will stall or burn out.

Comparing operating conditions *A, D,* and *E* on Figure 4–7, it is apparent that the underpowered system will continue to provide the required piston force (pressure) but at reduced flow rates, resulting in proportional increases in cycle times.

3. **The Two-Stage Pump.** In Chapter 7, the two-stage pump will be discussed. This device actually consists of two pumps, one with high-flow–low-pressure capabilities, the other having low-flow–high-pressure characteristics. The purpose of a two-stage pump is to take up slack in a system that is not exactly repetitive. For example, each cylinder of our wheel crusher of Example 1 is capable of exerting 38,000 lb of force continuously through its entire stroke of 13 in. If a small-diameter tire is placed in the crusher, the piston must travel through several inches of space just to reach the tire, moving at a constant and relatively slow speed even though the piston exerts no force along this distance. Had a two-stage pump been used, the high-flow–low-pressure section would have moved the piston quickly through the intervening distance until contact was made with the tire. As pressure in the system began to rise, flow from this section would bypass back to the reservoir and subsequent advancement of the piston would take place as a result of flow from the low-flow–high-pressure stage. The net effect, then, is that high-pressure capability is provided only when needed. Depending upon what portion of the work cycle occurs at near zero pressure, the two-stage pump offers a 25% to 70% reduction of required input horsepower from that of a single-stage pump that will accomplish the same task. Although not universally applicable, two-stage pumps should be considered as a viable option in the design of fluid power systems.

4. **Liquid Supply and Operating Pressure.** In Chapter 7 we will discuss several design considerations for the tanks, or *reservoirs,* used in most hydraulic systems to store an adequate supply of working fluid. We should at least note here, however, that a good rule of thumb for sizing such a reservoir is that its volume should be two to three times the pump flow rate. The system designed around cylinder A in Example 1 required a pump whose flow rate was 17.8 gpm. A good reservoir for such a pump might contain a liquid volume of 2.5 × 17.8 gal, or approximately 45 gal. A large supply volume reduces degradation and thermal contamination of the liquid and is particularly important in industrial machines that experience sustained, continuous operation. Also, the working pressures computed for the various

systems of Example 1 are not necessarily the maximum pressures that may occur in every work cycle of the machine. In the case of cylinder C, for example, it is highly unlikely that we could find a pump whose maximum pressure rating was the 1344-psi value required. A typical pump might be capable of working at pressures in the 2500-psi range, pressures that could occur if the system jams, if it is overloaded, or if a component fails to operate properly. For this reason, the circuit must be protected, usually through the use of a pressure relief valve located in the high-pressure output line of the pump. In addition, system components such as valves, cylinders, motors, hoses, and fittings are often selected whose bearings, seals, and housings can withstand the maximum rated pressure of the pump.

4.5 Analyzing Existing Systems

Frequently a fluid power system will be assembled from components that are readily available or already on hand. In other instances an existing system may be modified by selectively replacing components. Using the methods of the preceding sections, it is possible to predict the approximate behavior of the resulting system so that an evaluation may be made beforehand as to whether or not the required time and expense are justified.

EXAMPLE 2

A hydraulic system is to be assembled from available components having the following characteristics:

Cylinder: 4.0-in. diameter
            26-in. stroke
            1.75-in. diameter output rod

Pump: 8.6 gpm at 3600 psi

Motor: 7.5 hp

(a) Estimate the maximum available force and cycle time of the resulting system.

(b) How many horsepower are necessary to drive the pump to its specified capacity? If driven by such a power supply, what is the available force and cycle time of the system?

Solution

(a) Assume that the motor is capable of producing sufficient torque to operate the pump against a pressure of 3600 psi. (Otherwise it is not possible to create a functional system using these components.) Then the available force, F, may be computed as follows:

$$F = p \times A = (3600 \text{ psi}) \times 0.7854 \times (4.0 \text{ in.})^2$$

$$= \textbf{45,200 lb}$$

Assuming a pump efficiency of 0.7, the actual power transmitted to the liquid is $0.7 \times 7.5 \text{ hp} = 5.25 \text{ hp}$. Using equation (4-5), the approximate flow rate is:

$$q = \frac{(5.25 \text{ hp})}{(3600 \text{ psi} \times 0.000583)} = 2.5 \text{ gpm} = 9.63 \text{ in.}^3/\text{s}$$

From equation (4-1), the total cylinder volume filled in one complete cycle is 591 in.^{3}, so the cycle time for this system should be:

$$t = \frac{591 \text{ in.}^3}{9.63 \text{ in.}^3/\text{s}} = \textbf{61.4 s}$$

(b) The ideal horsepower required to drive this system to full capacity is:

$$P = 0.000583 \times (8.6 \text{ gpm}) \times (3600 \text{ psi}) = 18.1 \text{ hp}$$

Again assuming a pump efficiency of 0.7, the actual required input power becomes

$$P = \frac{(18.1 \text{ hp})}{0.7} = \textbf{25.8 hp}$$

Driving the system with such a power supply will not increase the pressure, since pressure is actually determined by the resistance of the load, assumed here to be its maximum of 45,200 lb. The increased horsepower should, however, allow attainment of the rated flow rate of 8.6 gpm (33.1 in.^{3}/s), dropping the estimated cycle time to:

$$t = \frac{591 \text{ in.}^3}{33.1 \text{ in.}^3/\text{s}} = \textbf{17.9 s}$$

Problem Set 2

1. One of your neighbors comes to you with a 2.5-in.-diameter hydraulic cylinder having a 20-in. stroke and an output rod 1.0 in. in diameter. He also has a used pump that moves 3.5 gpm against a pressure of 1800 psi. Using these components he wishes to build a log splitter such as the one shown in Figure 4–8.
   (a) What is the maximum splitting force he can expect with these components?
   (b) What would be the cycle time of his splitter?
   (c) What horsepower engine would be necessary to drive this system to its full capacity?
   (d) What is your recommendation regarding this project?

2. The following information is known concerning a fluid power system and its components. Compute the three unknown quantities.

   Motor: 6kW
   Piston: _____ cm diameter
                 45-cm stroke
                 5-cm diameter output rod
   Pump: _____ m^3/s flow rate
                 15.5 MPa rated pressure
   System: _____ kN extension force
   Cycle time: 17 s

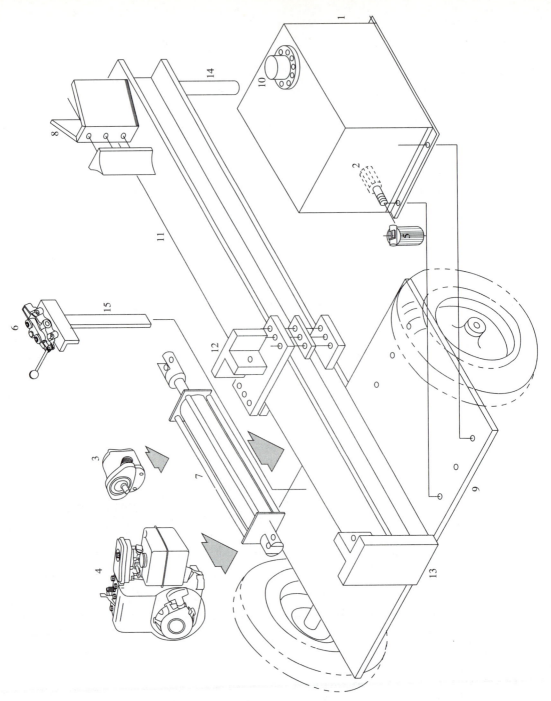

Figure 4–8 General configuration of hydraulic log splitter, showing its components: 1, reservoir; 2, strainer; 3, pump; 4, power supply; 5, filter; 6, flow control valve; 7, double-acting cylinder; 8, splitting wedge; 9, axle and mounting platform; 10, vented cap; 11, main frame member; 12, ram assembly; 13, cylinder thrust plate; 14, support leg; and 15, pedestal mount.

4.6   Rotary Actuators

Although a hydraulic pump converts the mechanical power of a *rotating* input shaft into the fluid power of a liquid, the term *rotary actuator* is generally used to describe any device that acts like a hydraulic *motor*. Such machines use either the pressure or velocity of an incoming liquid to produce rotary motion of an output shaft. (The simplest type of rotary actuator may well be the picturesque water wheel of colonial America. These wheels harnessed the power of moving water in brooks and streams to rotate grindstones used in the milling of flour or saw blades used to turn logs into boards.)

Performance and power requirement specifications for rotary actuators are often presented in differing forms. Many tools, such as the hydraulic earth drill in Figure 4–9, for example, simply list the pressure and flow rate required for "satisfactory" operation, although such operations may vary considerably. Input power may then be computed using equation (4-5) or its SI equivalent.

Performance data for other rotary devices, such as the pedestal-mounted hydraulic motor of Figure 4–10, relate mechancal output—torque and rotational speed—to given flow conditions of pressure and flow rate. Output power may then be computed using equations (3-8) or (3-9), while input power may again be determined from the U.S. customary or SI versions of equation (4-5). A typical set of performance data is presented in Table 4–6.

Figure 4–9   Hydraulically driven earth drill, with safety cages and shields removed for illustration purposes. *(Danuser Machine Co., Inc., Fulton, Mo.)*

Figure 4–10 Typical pedestal-mounted hydraulic motor.
(Lamina, Inc., Royal Oak, Mich.)

Table 4–6 Typical Performance Data for Hydraulic Motor
(Lamina, Inc., Royal Oak, Mich.)

gpm	rpm and Torque*	A-100 Motor (psi)														
		100	200	300	400	500	600	700	800	900	1000	1100	1200	1300	1400	1500
1	rpm	80	76	71	66	59	52	45	37	28	19	9				
	Torque	38	78	117	156	195	233	272	310	348	386	423				
2	rpm	151	147	142	136	129	122	115	106	98	88	78	67	56	44	31
	Torque	35	74	114	153	192	231	269	308	346	384	422	459	496	534	570
3	rpm	223	219	214	208	201	194	186	178	168	159	148	137	126	113	100
	Torque	30	70	109	148	187	226	265	304	342	380	418	456	494	531	568
4	rpm	297	293	287	281	275	267	259	250	241	231	220	209	197	185	171
	Torque	23	63	103	142	181	220	259	298	337	375	413	451	489	527	564
5	rpm	373	368	363	357	350	342	334	325	316	305	294	283	270	258	244
	Torque	15	55	94	134	173	213	252	291	330	368	407	445	483	521	559
6	rpm		446	440	434	427	419	411	402	392	381	370	358	346	332	319
	Torque		44	84	124	163	203	242	282	321	360	398	437	476	514	552
7	rpm		522	520	514	507	499	490	480	470	459	448	436	423	409	395
	Torque		32	72	112	151	191	231	270	310	349	388	427	466	504	543
8	rpm		595	590	582	576	570	564	558	551	540	528	515	502	488	474
	Torque		17	57	97	137	177	217	257	296	336	375	415	454	493	532

*Torque in in.-lb.

Problem Set 3

1. An earth drill such as that pictured in Figure 4–9 requires 10 gpm of liquid at 2500 psi.
   (a) What horsepower is contained in the incoming liquid?
   (b) If the auger produces 700 ft-lb of torque, find its operating speed in rpm. Assume that the machine is 100% efficient.

2. The hydraulic motor specified in Table 4–6 operates using 5 gpm of fluid at 1200 psi.
   (a) What hydraulic power input does this represent?
   (b) From the table, what are the output torque and shaft speed?
   (c) What mechanical power output does this represent?
   (d) Using the power input from part (a) and the power output from part (c), compute the efficiency of the motor at this operating condition.

Questions

1. Define "work cycle" as it applies to a hydraulic system.

2. List the three important parameters used to describe performance of a system that utilizes hydraulic cylinders.

3. Name the characteristics used to select these components:
   (a) Cylinder
   (b) Pump
   (c) Motor

4. Why does a standard double-acting cylinder retract more quickly than it extends?

5. A hydraulic cylinder is fed by a pump, which, in turn, is driven by a motor that is operating at its maximum power level. What would be the effects on cycle time and total force of the cylinder in each case?
   (a) The stroke of the cylinder is doubled, and piston and output-rod diameters remain the same.
   (b) The piston and output rod diameters are both doubled, and stroke remains the same.
   (c) The stroke, piston, and output-rod diameters are all doubled.

6. Consider a hand-held, single-speed electric drill. If the material being drilled is exceptionally hard or if the operator forces the drill bit into the workpiece, an increased torque is required to rotate the bit. What effect does this have on drill speed? If the drill cannot produce the required torque *at any speed,* what happens to the drill? Can you see any advantage to a variable-speed drill? What behavior might you expect from a hydraulic system that is powered by an electric motor if the hydraulic load is gradually increased?

7. Most automotive repair facilities use a variety of hand tools—such as wrenches, drills, and chisels—that operate on compressed air. Why are these *pneumatic* tools so popular? What advantages and disadvantages are involved with their use? Do you think that *hydraulic* hand tools would be feasible? What might be some of the potential advantages and drawbacks of such tools?

8. Define pump efficiency. Explain the factors that affect this quantity, and give typical numerical values.

9. What is the rule of thumb for determining reservoir capacity?

Review Problems

1. Modify equation (4-5) to yield fluid power in watts as a function of flow rate, q, in m^3/s, and rated pump pressure, p, in kPa.

2. A fluid power system provides 55,000 lb of force through a distance of 40 in. in a cycle time of 32 s. The hydraulic cylinder used is 6.0 in. in diameter and has an output rod diameter of 2.0 in.
   (a) Find the required pump pressure.
   (b) Find the required flow rate.
   (c) Calculate the minimum horsepower necessary to drive this system.

3. Write a computer program whose inputs are **component characteristics** of *pump pressure, pump flow rate, cylinder stroke, cylinder diameter, cylinder output rod diameter,* and *power supply.* This program should compute and print anticipated **performance data** of *extension force, extension time, retraction force, retraction time,* and *total cycle time.*

5

Columns of Liquid: Manometers, Barometers, and Hydrometers

Chapter 3 dealt with the analysis of systems in which pumps, concentrated loads, and compressed gases were used to develop pressures within a confined liquid. It is also possible, however, for pressure to exist in a liquid due to the weight of the fluid itself. For example, anyone diving to the bottom of a swimming pool experiences an increased pressure on his or her eardrums. At a depth of 12 ft, this pressure amounts to just over 5 psi and is entirely due to the weight of liquid in the pool pressing down on fluid particles below.

In this chapter we investigate the pressures caused by standing columns of liquid, examine the effects of this phenomenon on fluid power systems, and analyze the behavior of several common devices whose operation is based upon the physical principles presented here.

5.1 Pressures Caused by Columns of Liquid

The column of liquid shown in Figure 5–1 has constant cross-sectional area A and is of height h. If the specific weight of the liquid is γ_f, then the total weight of the liquid column may be computed as:

$$W = \gamma_f \times \text{volume} = \gamma_f \times A \times h$$

This weight is supported by the shaded area at the bottom of the column, so the pressure on that area becomes:

$$p = \frac{W}{A} = \frac{(\gamma_f \times A \times h)}{A}$$

$$p = \gamma_f \times h \tag{5–1}$$

EXAMPLE 1

Compute the pressure produced by a 60-m-high column of oil whose specific weight is 7360 N/m³.

Solution Using equation (5–1):

$$p = 7360 \, \frac{N}{m^3} \times 60 \text{ m} = 442{,}000 \text{ N/m}^2 = \textbf{442 kPa}$$

Note that the units of length in both γ_f and h must agree.

117

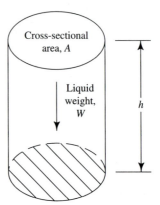

Figure 5–1   Standing liquid
column of uniform cross-
sectional area A.

EXAMPLE 2

Find the pressure, in psi, created by the weight of a 40-ft column of water.

Solution   From equation (5–1),

$$p = \left(62.4 \ \frac{\text{lb}}{\text{ft}^3}\right) \times 40 \ \text{ft} = 2500 \ \text{psf}$$

and

$$p = 2500 \ \text{psf} \times \frac{1 \ \text{ft}^2}{144 \ \text{in.}^2} = \textbf{17.4 psi}$$

Notice from example 2 that for columns of *water,* γ_f always has a value of 62.4 lb/ft³, and the conversion from psf to psi always involves the factor (1 ft²/144 in.²). Equation (5–1) may then be written as:

$$p \ (\text{psi}) = \left[\frac{62.4 \ \text{lb/ft}^3}{\left(\dfrac{144 \ \text{in.}^2}{1 \ \text{ft}^2}\right)}\right] \times h \ (\text{ft})$$

or

$$p \ (\text{psi}) = 0.433 \ \text{psi/ft} \ \times h \ (\text{ft}) \tag{5–2}$$

The numerical factor 0.433 psi/ft obtained by dividing the two constants (62.4/144) is often rounded to 0.5 and used to estimate pressures created by standing columns of water. Expressed as a rule of thumb: the pressure created by a column of water is *approximately* 0.5 psi per foot of column height.

EXAMPLE 3

(a) *Estimate* the pressure caused by a column of water whose height is 80 ft.

(b) *Approximately* how high must a column of water be in order to produce a pressure of 12 psi at its base?

Solution

(a) At 0.5 psi per foot, the 80-ft column produces a pressure of approximately 0.5 × 80, or **40 psi.** The exact pressure is obtained using equation (5–2):

$$p = 0.433 \text{ psi/ft} \times 80 \text{ ft} = 34.6 \text{ psi}$$

(b) Since a 2-ft water column produces approximately 1.0 psi of pressure, the estimated height needed for 12 psi is 12 × 2, or about **24 ft.** The exact height may be computed by rearranging equation (5–2):

$$h \text{ (ft)} = \frac{p \text{ (psi)}}{(0.433 \text{ psi/ft})} = 2.31 \text{ ft/psi} \times p \text{ (psi)}$$

$$h \text{ (ft)} = 2.31 \text{ ft/psi} \times 12 \text{ psi} = 27.7 \text{ ft}$$

The rule of thumb for standing water columns is a convenient method for estimating pressures in the field and for spotting gross errors in numerical pressure calculations. The metric equivalent to equation (5–2) may also be used for this purpose since:

$$p \text{ (kPa)} = 9.807 \text{ kPa/m} \times h \text{(m)} \tag{5–3}$$

or approximately 10 kPa of pressure per meter of column height. This method may be extended to other liquids by making use of the fact that for any liquid, $\gamma_f = S_f \times \gamma_{water}$, where S_f is the liquid's specific gravity discussed earlier in Chapter 2. Equation (5–1) may then be written as

$$p = S_f \times \gamma_{water} \times h \tag{5–4}$$

and equations (5–2) and (5–3) become:

$$p \text{ (psi)} = S_f \times 0.433 \text{ psi/ft} \times h \text{ (ft)} \tag{5–5}$$

and,

$$p \text{ (kPa)} = S_f \times 9.807 \text{ kPa/m} \times h \text{ (m)} \tag{5–6}$$

If the specific gravity of a liquid is known, equations (5–5) and (5–6) can be used to compute the column height necessary to produce any given pressure. The effect of S_f is shown graphically in Figure 5–2, where each column of liquid creates a pressure of 1.0 psi. Note that column height and S_f are inversely proportional; doubling S_f, for example, decreases required column height by half (compare the columns for $S = 0.6$ and $S = 1.2$).

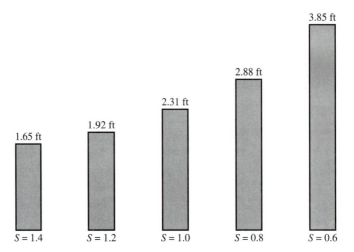

Figure 5–2   Column heights required to produce a pressure at their bases of 1.0 psi.

EXAMPLE 4

Find the approximate and exact pressures caused by:

(a) A column of oil ($S = 0.75$) that is 40 ft high.

(b) A column of mercury ($S = 13.6$) that is 2 m high.

Solution   (a) A 40-ft column of water would create an approximate pressure of 20 psi. If oil is only 75% as heavy as water, an equal column of oil would produce only 0.75 × 20, or about **15 psi,** and an exact pressure of:

$$p = 0.75 \times 0.433 \text{ psi/ft} \times 40 \text{ ft} = \textbf{13.0 psi}$$

(b) A column of water 2 m high would exert a pressure of about 20 kPa. Since mercury weighs 13.6 times as much as water, the estimated pressure caused by this column would be 13.6 × 20 = **272 kPa,** and an exact pressure of:

$$p = 13.6 \times 9.807 \text{ kPa/m} \times 2 \text{ m} = \textbf{267 kPa}$$

Distance h in equations (5–1) through (5–6) need not be the total height of the liquid column. Pressure may be computed at any depth in the liquid if h represents the vertical distance from the surface of the liquid down to the desired level. This is illustrated by the following example.

EXAMPLE 5

Compute the pressures at depths of 5 ft, 10 ft, and 15 ft in the open tank of seawater ($S = 1.03$) shown in Figure 5–3(A). Plot a graph of pressure versus depth for this tank.

Solution   At $h = 0$, equation (5–5) yields a pressure of $p = 0$ psi. This seems reasonable since no liquid column exists above the surface of the pool. But is the pressure at the surface actually 0 psi? Since the tank is open to atmosphere, pressure here may be specified as either 0 psig or 14.7 psia. Our computed values for the tank must reflect both the numerical difference and type of pressure units selected for $h = 0$. To allow for this choice, all of the preceding equations must be modified, beginning with (5–1), which becomes

$$p = p_o + (\gamma_f \times h) \qquad (5\text{–}7)$$

where p_o represents pressure at the liquid surface ($h = 0$).

Whenever equation (5–7) or any of its developed forms is used, however, p and p_o must have consistent units; both must be represented either in absolute or gauge units. On the other hand, the ($\gamma_f \times h$) term has "generic" pressure units and a numerical value that is the same in either measurement system. Since the choice of units is accounted for by the p_o term, attempting to convert the ($\gamma_f \times h$) term into gauge or absolute units has the effect of taking atmospheric pressure into account *twice*. The rules for consistent use of equation (5–7) are given next.

$$p \quad = \quad p_o \quad\quad + \quad\quad (\gamma_f \times h)$$

Must have consistent units, i.e., both in absolute or gauge pressure.	"Generic" pressure units; do not convert to absolute or gauge pressure.

Returning now to the tank of seawater, assume that at $h = 0$, $p_o = 0$ psig. Then the pressures at depths of 5 ft, 10 ft, and 15 ft become

$$p_5 = 0 \text{ psig} + \left(1.03 \times 0.433 \text{ psi/ft} \times 5 \text{ ft}\right) = \textbf{2.23 psig}$$

$$p_{10} = 0 \text{ psig} + \left(1.03 \times 0.433 \text{ psi/ft} \times 10 \text{ ft}\right) = \textbf{4.46 psig}$$

$$p_{15} = 0 \text{ psig} + \left(1.03 \times 0.433 \text{ psi/ft} \times 15 \text{ ft}\right) = \textbf{6.69 psig}$$

These pressures are represented by the solid line on the graph of Figure 5–3(B). If atmospheric pressure at $h = 0$ is instead selected as $p_o = 14.7$ psia, then the respective pressures become:

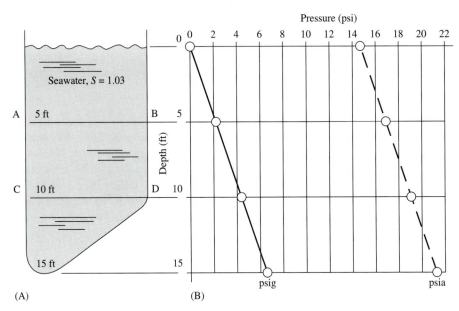

Figure 5–3   Pressure distribution in open tank of liquid.

$$p_5 = 14.7 \text{ psia} + \left(1.03 \times 0.433 \text{ psi/ft} \times 5 \text{ ft}\right) = \textbf{16.9 psia}$$

$$p_{10} = 14.7 \text{ psia} + \left(1.03 \times 0.433 \text{ psi/ft} \times 10 \text{ ft}\right) = \textbf{19.2 psia}$$

$$p_{15} = 14.7 \text{ psia} + \left(1.03 \times 0.433 \text{ psi/ft} \times 15 \text{ ft}\right) = \textbf{21.4 psia}$$

These pressures are indicated by the broken line plotted in Figure 5–3(B).

Surface pressure p_o is transmitted by the liquid itself to all points in the standing column and need not be restricted to atmospheric pressure. It may instead consist of a vacuum created at the inlet of a pump or a positive static pressure caused by a piston or compressed gas. Figure 5–4, for example, shows how a tank of carbon dioxide (CO_2) at 900 to 1000 psi is commonly used in a beverage delivery system. Applied to the liquid surface, this gas forces the beverage out of its storage container through tubing to individual dispensers at separate locations.

Because equations (5–1) through (5–7) are independent of cross-sectional area A, they imply that their results are valid for any column of liquid regardless of the shape of its container. Is this a reasonable assumption, or merely the outcome of having used a column of uniform cross section to derive equation (5–1) initially?

To examine this question, consider the static system shown in Figure 5–5. A reservoir of unlimited volume (R) feeds a pump (P), which in turn maintains constant pressure

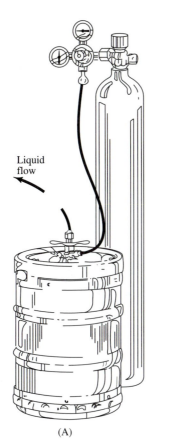
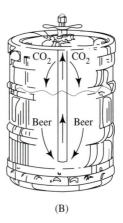

(A)   (B)

Figure 5–4   (A) Compressed CO_2 applied to the liquid surface is used to maintain natural carbonation in the beverage while pushing it through plastic tubing to dispensers at various locations. (B) Cross section illustrates CO_2 effect on liquid surface. *(Anheuser-Busch, Inc., Draught Marketing Services, St. Louis, Mo.)*

in a closed pipe (*N*) and is capable of supplying pressurized liquid to the pipe as needed. If a hole of cross-sectional area *A* is now created in the pipe at B and a tube of uniform cross section inserted into this hole, then liquid will rise in the tube to some height *h* until the back pressure exerted by the column exactly balances the pressure in the pipe. This condition is shown in Figure 5–5(A). The weight, *W*, of liquid in this vertical tube is supported on area, *A*, so that the pressure at the base of the column, and therefore pressure in the pipe, is $p = W/A$.

    If another hole of cross-sectional area 2*A* is now opened in the pipe at *C* and a tube inserted, then liquid will also rise to height *h* in this tube, as shown in Figure 5–5(B). This

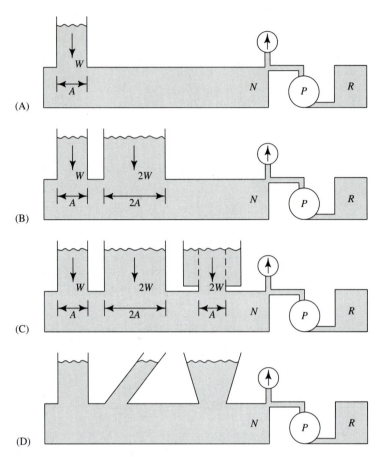

Figure 5–5
Liquid level is constant regardless of tube shape.

occurs because the weight of liquid in the vertical tube is now $2W$ and is supported on area $2A$, so that pressure at the base of this column is $p = 2W/2A = W/A$, which equals the constant pressure being maintained in the pipe.

Now consider a third hole whose area is A but that contains a stepped tube of area $2A$ (Figure 5–5(C)). If we assume that liquid also rises to some height in this irregular tube, then the weight of liquid in the tube is essentially $2W$ but is supported on area A, *apparently* creating a pressure of $p = 2W/A = 2(W/A)$, or twice the pressure being maintained in the pipe. How can this be? Closer inspection reveals that only the volume of liquid directly over the opening is supported on area A and that this volume has weight W, which produces the expected pressure of W/A. The remaining liquid, located between the dotted lines and the walls of the tube, is actually supported by the horizontal portions of the tube, which in turn transmit the weight of this liquid to the rigid walls of the pipe.

This example merely illustrates the old adage that liquid seeks its own level regardless of the size, shape, or orientation of its container, a phenomenon further confirmed by the liquid levels of the tapered and inclined tubes of Figure 5–5(D). Although this result establishes the validity of equations (5–1) through (5–7), it also indicates that any two points at the same depth in the same body of liquid are at equal pressures. In Figure 5–3(A), for example, the pressure at point A equals the pressure at point B and the pressures at points C and D equal each other as well. If this were not true, then a horizontal flow of liquid would occur from the point of higher pressure toward the point of lower pressure.

Although pressures created by standing columns of liquid are often significant in open systems, such as those to be analyzed in Chapter 6, they may frequently be neglected in closed systems of the type discussed in Chapter 4. Fluid power systems commonly found on construction, agricultural, and industrial machinery, for example, do not ordinarily contain large vertical separations (and therefore large standing columns of liquid) between the supply of pressurized liquid and the linear or rotary actuators. On a backhoe, bulldozer, or front-end loader, the hydraulic cylinders are usually not more than 6 to 10 ft above the pump used to supply them. Compared to the working pressures of such systems, static pressures caused by the vertical columns are often insignificant, as illustrated by the following example.

EXAMPLE 6

The aerial bucket shown in Figure 5–6 is typical of those used by telephone and electrical linepeople. Besides raising and lowering this device, the hydraulic system must also lift the oil ($S = 0.8$) and maintain a "residual" pressure of 2500 psig at a control panel located inside the bucket. The pressurized liquid may be used to operate a variety of hand tools needed for maintenance and repair procedures (also see Figure 6–30). If the pump supplying this control panel is located 3 ft above the ground, find its required working pressure when the bucket is located 35 ft above the ground. What percentage of the total pressure does this standing column represent?

Solution   For the column of liquid represented by the supply line between pump and bucket:

$h = 32$ ft
$S_f = 0.8$
$p_o = 2500$ psig
$p = ?$

Putting equation (5–5) into the form of equation (5–7):

$$p = p_o + \left(S_f \times 0.433 \text{ psi/ft} \times h(\text{ft})\right)$$

or

$$p = 2500 \text{ psig} + \left(0.8 \times 0.433 \text{ psi/ft} \times 32 \text{ ft}\right)$$

$$p = 2500 \text{ psig} + 11.1 \text{ psi} = \textbf{2511 psig}$$

Figure 5–6   A typical aerial bucket.

Note that the increase in required pressure (11.1 psig) due to the standing column is less than 0.5% of the working pressure of the system. In practice, such a slight increase is generally neglected.

Problem Set 1

1. (*Note:* Whenever possible, first estimate column pressures using the rules of thumb given in the preceding section.)
   (a) Find the pressure created by a column of water whose height is 42.9 ft.
   (b) What height must a column of gasoline ($S = 0.68$) have in order to produce the same pressure as the water column described in (a)?
   (c) Repeat (b) for carbon tetrachloride ($S = 1.59$).

2. The water tank shown in Figure 5–7 contains liquid to a height of 16.3 m. If the lower gauge reads 184 kPa, find the reading of the upper gauge. Give a possible explanation for this reading.

3. A vented storage tank contains industrial solvent to a depth of 24 ft. A gauge located on the side of this tank 5 ft above the base reads a pressure of 42.5 psi. Compute the specific gravity of the solvent.

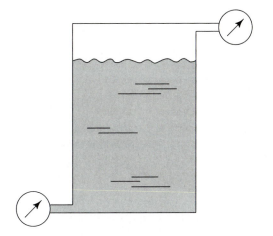

Figure 5–7   Depiction of
water tank used in problem 2.

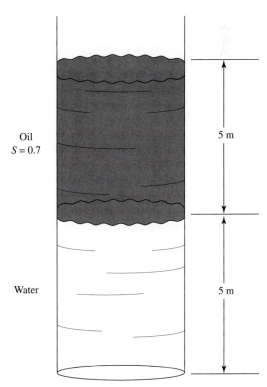

Oil
$S = 0.7$

5 m

Water

5 m

Figure 5–8   (See problem 4
in problem set 1.)

4. For a particular liquid column, 5 m of oil ($S = 0.7$) sits atop 5 m of water, as shown in Figure 5–8.
   Find the pressure at the base of this column.

5. An automobile skids off a roadway and plunges into a lake that is 15 ft deep. If the driver's door
   is 35 in. wide by 44 in. high, what is the total force keeping this door shut?

5.2 Vacuums

Another way of viewing columns of liquid may be obtained by rearranging equation (5–7), as follows:

$$(p - p_o) = \gamma_f \times h$$

This form of the equation suggests several important aspects of standing liquid columns:

(a) The difference in pressure between the base of the column and the surface of the column is what actually supports a column of liquid.

(b) By increasing the pressure at the base, p, or by reducing pressure at the surface, p_o, it is possible to raise a column of liquid.

One common method of implementing (b) is to create a partial vacuum at the surface of a liquid and allow atmospheric pressure to raise a column of the liquid. This is exactly the process involved in drinking through a straw; reducing pressure at the top of the straw causes a column of liquid to rise within the straw due to the "push" of atmospheric pressure. Suction pumps achieve the same results on a continuous basis as they move specified volumes of liquid through typical fluid power systems. The maximum height, called *lift*, to which a column of liquid may be raised by a vacuum depends on several factors, most important of which is the amount of vacuum that may be produced at the surface of the liquid.

In Chapter 3, pressure was ultimately attributed to the action of liquid or gas molecules as they impinged against the walls of their containers. The lowest pressure attainable, then, is achieved by removing all of the fluid particles from a given space. As you may remember, this condition is known as a *perfect vacuum,* is arbitrarily assigned a pressure of 0 psia, and is the starting point for the "absolute" pressure scale.

Although both absolute and gauge pressures were discussed in Chapter 3, it is worth noting here that atmospheric pressure, the connecting link between these two pressure scales, is itself caused by a standing column of fluid. Because they have little apparent physiological effect upon us, we generally neglect those gases which comprise the surrounding atmosphere. A column of these gases several miles high, however, is still able to create that pervasive effect known as atmospheric pressure which has an assigned value of 14.7 psia, or 101 kPa, on the absolute pressure scale. Since most useful work performed by fluid power systems takes place at pressures above or below atmospheric, this point was selected as a reference level on the gauge pressure scale, and has an assigned value of 0 psig. The relationship between these two pressure scales, reproduced from Chapter 3, is shown in Figure 5–9. From this comparison, the following conditions should be apparent.

(a) The term *vacuum* is used to describe any pressure that is less than atmospheric pressure.

(b) A perfect vacuum represents the total absence of pressure and is assigned a value of 0 psia (0 kPaa) on the absolute scale or -14.7 psig (-101 kPag) on the gauge scale.

Another common way of describing a vacuum is in terms of the height of a column of liquid that the vacuum could raise under the effect of atmospheric pressure. The

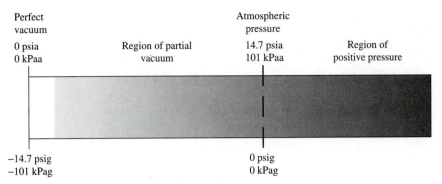

Figure 5–9 Comparison of absolute and gauge pressures.

units, which reflect this important application of vacuums, are often in inches or millimeters of mercury (usually written as inches Hg or millimeters Hg), inches of water, or w.c. (for water column). The lifting capability of vacuums will be discussed quantitatively in the following section, as will the conversion between units of psi and inches of a liquid.

How are vacuums created? In most fluid power systems, this condition is achieved by removing material from a given space. At the inlet of a pump, for example, liquid is swept away by the pump impeller, leaving a partial vacuum that draws additional liquid into the system to fill this void. Performed on a continuous basis, this action creates the desired flow through the pump.

The removed material need not be a liquid. Movement of a piston within a cylinder, for example, often results in the creation of a vacuum. Following the power stroke of a carbureted four-cycle engine, the piston moves up in its cylinder, pushing spent gases out through the exhaust valve. As the piston moves back down into the cylinder, it leaves behind a partial vacuum that, through the open intake valve, draws in a fresh quantity of the gasoline-air mixture provided by the carburetor. Similarly, the action of a plunger in a hypodermic syringe allows liquid solutions to be drawn into that device.

Vacuums are sometimes formed inadvertently. Consider the ordinary metal can shown in Figure 5–10. If the cap is removed from this empty container and a small quantity of water is added, placing the container on a stove or hot plate will cause some of the water to vaporize. In a short time, all of the air initially in the can will have been displaced by water vapor. If the can is now capped and allowed to cool, the vapor condenses back into a liquid, leaving a partial vacuum in the can. The effects of atmospheric pressure are apparent in Figure 5–10(B).

This example also illustrates what can happen when liquid is removed from an improperly vented tank or reservoir; failure to provide a path for atmospheric air to replace the removed liquid can produce results similar to those shown in the figure. It should also be apparent that any conductor of liquids, from a drinking straw to a pump inlet hose, that is subject to vacuums should possess sufficient strength and rigidity to resist collapse.

(A)   (B)

Figure 5–10   Effect of vacuum formed in unvented tank. (A) Water is heated in an
uncapped metal can. (B) Can is capped and allowed to cool.

5.3   Barometers and the Lifting Ability of Pumps

In 1643, an Italian philosopher named Evangelista Torricelli developed a simple instrument
that would measure atmospheric pressure. This device, called a *barometer*, produces values
of atmospheric, or "barometric," pressure directly in inches of mercury. Since it is greatly
affected by weather conditions, this pressure is a standard piece of meteorological data used
by professional and amateur weather forecasters to predict the future. More importantly,
however, barometric pressure readings provide limiting values to the amount of lift that
may be applied to a column of liquid at any given place and time.

Construction of a barometer requires a glass tube approximately 34 in. long, closed
at one end and open at the other. The tube is filled with mercury and its open end sealed
temporarily. Next the tube is inverted, its end inserted into an open container of mercury,
and the temporary seal is then removed. Final configuration of this device is as shown in
Figure 5–11. (*Note:* Although construction of a mercury barometer is an interesting and
relatively simple lab activity, mercury itself can represent a health hazard, and as such
should be handled in accordance with the guidelines presented in the Safety Sidebar on
page 135.)

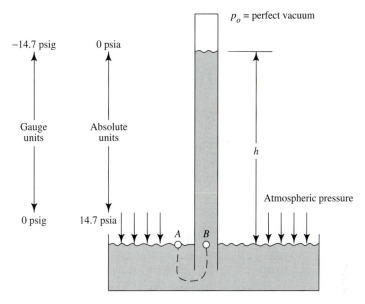

Figure 5–11   Construction of a simple mercury barometer. Points A and B, at the same level in the same liquid, are at equal pressures.

Referring to Figure 5–11, the obvious question is, Why doesn't the mercury in the glass tube merely run down into the open container? In fact, it does begin to slide down the tube. As it does so, however, the space it leaves behind in the closed end of the tube is, ideally, a perfect vacuum. Since the open end of the tube is below the liquid surface in the open container, air cannot enter the system to destroy the vacuum created in the closed end of the tube. As a result, the liquid column slides down the tube, decreasing in height until the pressure exerted at its base exactly matches the atmospheric pressure that acts on the liquid surface in the open container. The equilibrium height of this standing column of mercury, then, is used as a measure of the existing atmospheric or barometric pressure. This height can be computed as shown in Example 7.

EXAMPLE 7

Find the height of a mercury ($S = 13.6$) barometer on a day when atmospheric pressure is 14.7 psia.

Solution   Equation (5–7) is applied to the barometer shown in Figure 5–11. Height of the liquid column, h, is unknown, and the vacuum at the top of this column, p_o, may be designated as either 0 psia or -14.7 psig. Because point A on the open surface of the liquid and point B within the closed tube are both distance h below the top surface of the column and are part of the same body of liquid (as indicated by the dotted line), the pressures at these two points are equal. Since the pressure which exists at point A is atmospheric, pressure p at the base of the column may be designated either as 14.7 psia or 0 psig. The appropriate form of equation (5–7) to be used is:

$$p \text{ (psia)} = p_o \text{ (psia)} + [13.6 \times 0.433 \text{ (psi/ft)} \times h \text{ (ft)}]$$

or

$$p \text{ (psig)} = p_o \text{ (psig)} + [13.6 \times 0.433 \text{ (psi/ft)} \times h \text{ (ft)}]$$

Substituting numerical values into these equations yields

$$14.7 \text{ psia} = 0 \text{ psia} + [13.6 \times 0.433 \text{ psi/ft} \times h \text{ (ft)}]$$

or

$$0 \text{ psig} = -14.7 \text{ psig} + [13.6 \times 0.433 \text{ psi/ft} \times h \text{ (ft)}]$$

Solution of either equation results in a value of $h = 2.50$ ft, which converts to 30.0 in. Standard atmospheric pressure, then, may be represented as 30.0 in. Hg (762 mm). Remember that units for p and p_o must be consistent and that the two systems of units simply offer two ways of visualizing the lifting process. If absolute units are selected, no pressure exists at the top of the liquid column and a positive pressure of 14.7 psi at its base "pushes" liquid up into the tube. If gauge units are used, no pressure exists at the base and a negative pressure of -14.7 psi within the tube creates a suction that "pulls" liquid into the tube. Commercial mercury and digital barometers are shown in Figure 5–12.

From Example 7 it should be apparent that the amount of lift available for any liquid depends upon the amount of vacuum that may be applied to the liquid, as well as the existing value of barometric pressure. Since a perfect vacuum represents an ideal limiting value, the maximum amount of lift attainable for any liquid may readily be computed if barometric pressure is known.

EXAMPLE 8

The suction pump shown in Figure 5–13 produces a perfect vacuum at its inlet. On a day when atmospheric pressure is 14.7 psia, find the maximum height to which this pump can lift a column of water.

Solution   Except for the specific gravity of the liquids, this problem is identical to Example 7. A perfect vacuum exists at the top of the liquid column and atmospheric pressure at its base. Applying equation (5–7) to the column of water yields:

$$14.7 \text{ psia} = 0 \text{ psia} + [1.0 \times 0.433 \text{ psi/ft} \times h \text{ (ft)}]$$

or:

$$0 \text{ psig} = -14.7 \text{ psig} + [1.0 \times 0.433 \text{ psi/ft} \times h \text{ (ft)}]$$

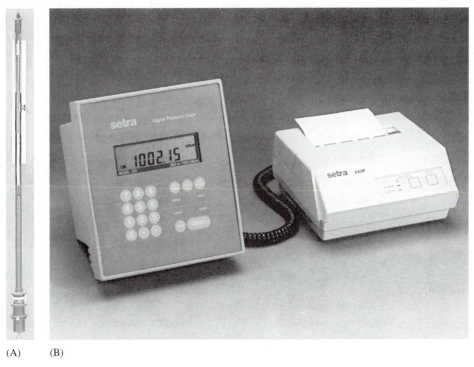

(A)   (B)

Figure 5–12   Commercial mercury and digital barometers. (A) Weather service mercurial barometer. *(Princo Instruments, Inc., Southampton, Pa.)* (B) Digital barometer with printer. *(Setra Systems, Inc., Acton, Ma.)*

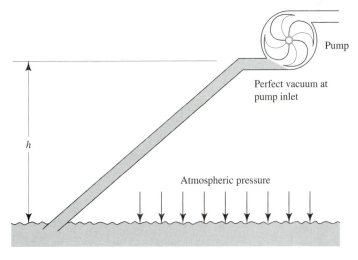

Figure 5–13   Ideal suction pump raising a column of water. (See Example 8.)

Solving either equation results in a value of $h = 33.9$ ft or 10.3 m. This figure represents the maximum height to which a column of water may be raised by suction under conditions of a perfect vacuum and standard atmospheric pressure. Notice that this maximum height is inversely proportional to the specific gravity of the liquid; since the mercury of Example 7 is 13.6 times as heavy as water, it may be lifted only (1/13.6) times as high, or (1/13.6) × 33.9 ft = 2.50 ft. The maximum lift, h_{max}, for any liquid of specific gravity S_f may be computed using

$$h_{max} = \frac{(33.9 \text{ ft})}{S_f}$$ (5–8)

In metric units:

$$h_{max} = \frac{(10.3 \text{ m})}{S_f}$$ (5–9)

The limiting value of lift for water obtained in Example 8 has significant implications for real life hydraulic systems. In rural areas, for example, firefighters must often rely on water available in ponds, lakes, or rivers, since hydrant systems are usually nonexistent at these locations. If water is to be obtained from such sources using suction pumps, the pump inlets must be considerably less than 33.9 ft above the surface of the liquid. Similarly, wells for residential or agricultural use must bring water to within 33.9 ft of ground level if the liquid is to be lifted the remaining distance by a suction pump. In wells deeper than 33.9 ft, water may be lifted in stages using several suction pumps at various levels, but this same task is generally accomplished through the use of a *submersible pump* that is placed (submerged) at the bottom of the well. Such a pump *pushes* water to the surface; compared to suction pumps, a submersible pump capable of working against a pressure of 2500 psi can raise a column of water well over a mile!

As a final note, several other factors affect the maximum lift available for any liquid. Imperfections in the suction tube or the use of strainers at the inlet cause hydraulic losses that reduce the maximum lift available. Liquid temperature also has an effect; increasing the temperature of a liquid while reducing the pressure at its surface encourages vaporization of the liquid. A pump designed to move liquids will have difficulty maintaining the desired inlet vacuum when significant amounts of vapor are present. This, in turn, also reduces the amount of lift available. Finally, reductions in atmospheric pressure will cause a loss of lift. Although variations in barometric pressure over time at any given location are generally small, atmospheric pressure decreases significantly with altitude. Since this is the driving force in any suction system, the amount of lift that can be obtained in Denver (elevation 5280 ft) is less than that available in Boston (elevation 0 ft at sea level). Table 5–1 summarizes the effect of altitude on lift. Taking all of these factors into account, New England firefighters, for example, generally anticipate a maximum possible lift of only 25 to 28 ft under actual service conditions.

SAFETY SIDEBAR

Handle Quicksilver with Care

Mercury is also known as *quicksilver* for its ability to disperse rapidly when spilled, often forming tiny droplets that can adhere unnoticed to skin or clothing. This liquid, which is commonly used in barometers and manometers, is corrosive to skin, eyes, and mucous membranes, and prolonged exposure to this liquid may cause psychic, kidney, and cardio-vascular disturbances. Take adequate steps to protect hands and eyes when working with this substance, do not ingest it, and avoid exposure over an extended time period.

Table 5–1   Effect of Altitude on Maximum Lift

Elevation Above Sea Level		Loss of Lift	
ft*	m	ft of Water*	m of Water
1000	305	1.22	0.372
2000	610	2.38	0.725
3000	914	3.50	1.07
4000	1220	4.75	1.45
5000	1524	5.80	1.77
6000	1829	6.80	2.07
7000	2134	7.70	2.35

*From *Fire Service Hydraulics*, 1974, Fire Engineering Books & Videos, a Division of Pennwell Publishing Company, Saddle Brook, N.J.

Problem Set 2

1. A suction pump develops an inlet pressure of -10 psig. On a day when the barometric pressure is 14.2 psia, how high can this pump lift a column of water? A column of oil whose specific gravity is 0.7?

2. If a pump similar to that of problem 1 develops an inlet pressure of -82.8 kPag under standard atmospheric conditions, how high can it lift a column of hydraulic oil of specific gravity 0.74?

3. Several common devices use vacuums to provide liquids "on demand," such as poultry and livestock waterers and the hummingbird feeder shown in Figure 5–14. Most of these devices consist of an upper housing where liquid is stored and a lower housing where liquid is dispensed. The feeder shown has an upper bulb whose narrow stem threads into a lower bulb from which the birds feed. Typically, the upper bulb contains a partial vacuum above a standing column of liquid. Openings on the bottom of the stem are normally covered by liquid in the lower bulb. As fluid is withdrawn from this bulb through consumption, leakage, or evaporation, these openings become uncovered, allowing air to enter the stem and rise to the surface of the liquid column. This reduces the vacuum above the column, releasing liquid through the openings and allowing it to flow down into the lower bulb. As fluid level in the lower bulb rises, the openings become covered, thus stopping the flow of liquid. In this manner, the liquid level in the lower bulb becomes

Figure 5–14 Liquid stored in this standing column automatically maintains a constant liquid level in the lower bulb.

self-regulating and automatically remains constant until all of the fluid stored in the upper bulb and stem has been depleted. Bottled water dispensers work in a similar manner, except that an opening is manually created at the base via a simple valve. Assuming standard atmospheric conditions, find the amount of vacuum required above the water surface to support a water column 40 cm high.

5.4 Net Positive Suction Head

It is frequently desirable to quantify the amount of lift available to a suction pump, particularly those pumps that are part of a permanent installation. The theoretical values derived in the preceding section often must be modified for the operation of a specific pump produced by a particular manufacturer.

As mentioned in section 2.4, when the pressure anywhere within a pump falls below the vapor pressure of the liquid, bubbles of vapor form in the liquid. This condition can exist at the pump inlet, on the inside walls of the pump, or, most commonly, at points along the pump impeller. As these bubbles move with the flow into regions of higher pressure, the bubbles collapse, generally with detrimental effects to the pump. This process of bubble formation and collapse is known as *cavitation*. It can reduce the efficiency of the pump dramatically and in extreme cases can cause the pump to lose its prime (liquid in the suction tube drops back into the reservoir, leaving the tube filled with air or vapor). Over extended time it can cause severe damage to the pump impeller, an example of which is shown in Figure 5–15. A pump that is experiencing cavitation

Figure 5–15 Effect of cavitation on a 17-in.-diameter pump impeller made of cast stainless steel. (*Ingersoll-Rand Co., Pump Group, Phillipsburg, N.J.*)

will usually emit sounds ranging from a steady high-pitched whine similar to that produced by a defective bearing to a loud and irregular rapping or knocking that leaves no doubt as to its effect on the internal workings of the pump.

To avoid cavitation, total pressure at the inlet of a pump must be greater than the vapor pressure of the liquid by some amount called the *net positive suction head* (NPSH). Numerical values of the minimum NPSH for any given pump are generally provided by the manufacturer; as long as the pump inlet pressure is maintained above the minimum NPSH, satisfactory pump operation will occur. It should now be obvious, therefore, that Example 8 of the preceding section was unrealistic in its assumption that a perfect vacuum could exist at the pump inlet during normal flow of a real liquid; besides lifting liquid to the pump inlet, a portion of atmospheric pressure in the form of NPSH accelerates fluid into the pump and fills the pump cavity. That assumption, however, did yield a theoretical limit to the amount of lift available with a suction pump.

As we shall see in Chapter 6, the total pressure of a liquid consists of two parts, static pressure caused by compression of the fluid and dynamic pressure due to the motion or velocity of the fluid. NPSH is therefore defined as:

$$\text{NPSH} = [(v^2/2g) + (p/\gamma_f)]_{\text{inlet}} - (p_v/\gamma_f) \qquad (5\text{–}10)$$

where v and p represent velocity and static pressure at the pump inlet, g is the gravitational acceleration, γ_f is specific weight of the liquid, and p_v is the vapor pressure of the liquid. Note that each term in equation (5–10) has units of length.

Using energy considerations, the maximum lift, h_{max}, obtainable with a given pump becomes

$$h_{\text{max}} = \left[\frac{p_a - p_v}{\gamma_f} \right] - \text{NPSH}_{\text{min}} - h_L \qquad (5\text{–}11)$$

where p_a is atmospheric pressure, and h_L represents the decrease in lift due to hydraulic losses between the reservoir and pump inlet. If h_{max} is negative, the pump inlet must be lower than the liquid surface in the reservoir, and thus ceases to be classified as a suction pump.

5.5   The Manometer

Columns of liquid may be used to measure low pressures or small differences in pressure. The device shown in Figure 5–16, a *manometer*, consists of a U-tube containing some liquid such as water, oil, or mercury that is immiscible with the fluid whose pressure is being measured. To measure gauge pressure, one end of the tube is left open to atmosphere while the other end is connected to the pressurized liquid or gas. When both ends of the U-tube are connected to fluids at different pressures, the manometer reads the differential pressure between the two sources.

To magnify displacement of the manometer liquid, several variations of the basic U-tube have been developed, including the well-type manometer (Figure 5–17(A)) and the inclined manometer (Figure 5–17(B)). Commercial examples of the basic device and its variations are shown in Figure 5–18 and Figure 5–19.

EXAMPLE 9

Blood circulation through the human body is accomplished by the pumping action of the heart. (In fact, this circulatory system represents our own personal hydraulic circuit!) Between beats, the heart relaxes. During these periods, known as *diastoles*, valves in the heart open to allow a supply of blood to flow into the heart. This process reduces resistance in the system and causes blood pressure to decrease. Contraction of the heart, called the *systole*, then sends a pulse of increased pressure into the system, delivering blood through arteries and capillaries to the major organs and muscles throughout the body. The device used to measure blood pressure is called a *sphygmomanometer*, from the Greek words for pulse (sphygmos) and measure (metron); a typical example is pictured in Figure 5–20. A pneumatic cuff wrapped around the upper arm of the patient is pressurized using a small, hand-operated bulb and shutoff valve, while a stethoscope placed on a vein monitors the flow of blood. Pressure in the pneumatic cuff is also applied to a small reservoir and tube containing mercury, while the other end of the tube is open to atmospheric pressure. To actually measure blood pressure, a physician pressurizes the cuff until no pulse is heard through the stethoscope. At this point, pressure in the cuff is greater than systolic pressure of the heart, and blood flow under the cuff has essentially stopped. This pressure is reflected by a change in the mercury level of the sphygmomanometer. Using the shutoff valve, the physician then slowly "bleeds off" pressure in the cuff until a pulse is heard through the stethoscope. At this point, usually near 120 mm of mercury, pressure in the cuff equals systolic pressure, and blood flow has resumed. Cuff pressure is slowly reduced until no pulse is heard, generally near 80 mm of mercury, indicating that cuff pressure equals the heart's diastolic pressure. It is conventional to represent blood pressure as systolic/diastolic with both readings understood to be in millimeters of mercury. (The pressure caused by 1 mm of mercury is also known as a *torr*.)

(a) A patient's blood pressure is measured as 120/80. Find the actual pressures both in kPa and psi.

(b) Atmospheric pressure can easily vary between 29.5 in. (749.3 mm) and 31.0 in. (787.4 mm) of mercury. Since this pressure is applied to the top of the

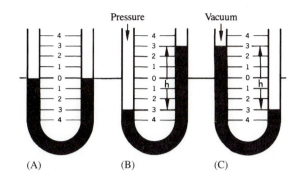

Figure 5–16 Operation of a simple manometer. (A) In its simplest form the manometer is a **U**-tube about half filled with liquid. With both ends of the tube open, the liquid is at the same height in each leg. (B) When positive pressure is applied to one leg, the liquid is forced down in that leg and up in the other. The difference in height, *h,* which is the sum of the readings above and below zero, indicates the pressure. (C) When a vacuum is applied to one leg, the liquid rises in that leg and falls in the other. The difference in height, *h,* which is the sum of the readings above and below zero, indicates the amount of vacuum. *(Dwyer Instruments, Inc., Michigan City, Ind.)*

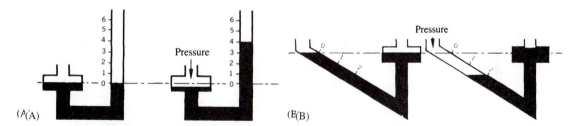

Figure 5–17 Manometer variations. (A) Well-type manometer. At left, equal pressure is imposed on the fluid in the well and in the indicating tube. Reading is zero. At the right, a positive pressure has been imposed on the liquid in the well causing the level to go down very slightly. Liquid level in the indicating tube has risen substantially. Reading is taken directly from scale at liquid level in the indicating tube. The scale has been compensated for the drop in level in the well. (B) Inclined manometer. At left, equal pressure is imposed on the liquid in the well and the indicating tube. Reading is zero. At the right, a positive pressure has been imposed on the liquid in the indicating tube, pushing it down to a point on the scale equal to the pressure. Liquid level in the well rises proportionately. Inclining the indicating tube has opened up the scale to permit more precise reading of the pressure. *(Dwyer Instruments, Inc., Michigan City, Ind.)*

(A)

(B)

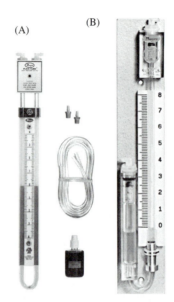

Figure 5–18 Commercial manometers: (A) conventional **U**-tube; (B) well-type. *(Dwyer Instruments, Inc., Michigan City, Ind.)*

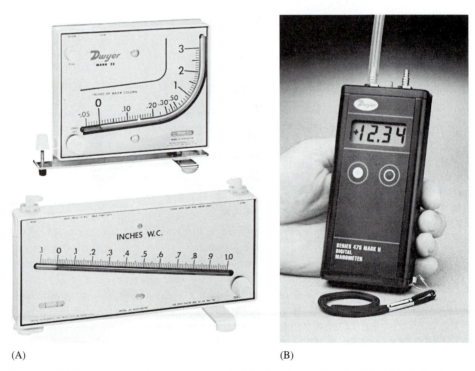

(A)

(B)

Figure 5–19 Commercial manometers: (A) inclined type; (B) hand-held digital. *(Dwyer Instruments, Inc., Michigan City, Ind.)*

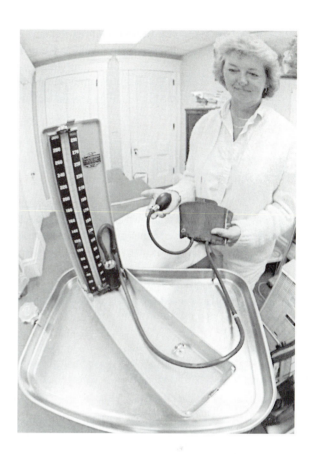

Figure 5–20   The sphygmomanometer used in hospitals and laboratories consists of a pneumatic tube and cuff, rubber air-bulb and shutoff valve, and a mercury column.

mercury column, could the 38.1 mm variation result in different readings for blood pressures measured on separate days?

Solution   (a)   Blood pressure may be computed directly from equation (5–1):

$$p_{\text{systolic}} = \gamma_f \times h = 133.4\,\frac{\text{kN}}{\text{m}^3} \times 0.120\ \text{m} = \textbf{16.0 kPa}$$

$$= \frac{16.0\ \text{kPa}}{6.895\ \text{kPa/psi}} = \textbf{2.32 psi}$$

Similarly,

$$p_{\text{diastolic}} = 133.4\,\frac{\text{kN}}{\text{m}^3} \times 0.080\ \text{m} = \textbf{10.7 kPa}$$

$$= \frac{10.7\ \text{kPa}}{6.895\ \text{kPa/psi}} = \textbf{1.55 psi}$$

(b) No. Bodily fluids experience the same variations in atmospheric pressure as those seen by liquid in the instrument. An increase in barometric pressure, for example, will not only increase overall pressure of the blood, but will also increase back pressure on the sphygmomanometer's mercury column, so that the net readings remain unchanged.

EXAMPLE 10

The operating principle of a sphygmomanometer may be used to measure specific gravity of an unknown liquid. The simple hydrometer shown in Figure 5–21 may easily be constructed using glass tubes, rubber stoppers, and plastic tubing. One jar contains a liquid whose specific gravity is known (usually water), while the other jar contains the unknown liquid. If air pressure above both samples is increased using any hand-operated air pump (such as a bicycle pump or air bulb and shutoff valve), a column of liquid will rise in each tube as shown. Since the pressure at the liquid surface is the same in both jars, it follows from equation (5–1) that:

$$\gamma_L \times h_L = \gamma_R \times h_R$$

or,

$$S_L \times \gamma_{\text{water}} \times h_L = S_R \times \gamma_{\text{water}} \times h_R$$

This reduces to:

$$S_L = \frac{h_R}{h_L} \times S_R$$

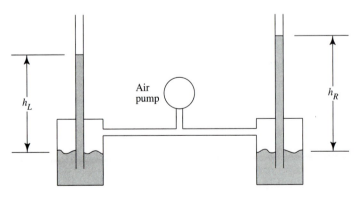

Figure 5–21   A simple liquid column hydrometer.

If the applied pressure raises a column of water 12.8 in. high while raising 17.6 in. of the unknown liquid, find the specific gravity of the unknown liquid.

Solution

$$S_{unknown} = \frac{h_{water}}{h_{unknown}} \times S_{water}$$

$$= \frac{12.8 \text{ in.}}{17.6 \text{ in.}} \times 1.00$$

$$= \mathbf{0.727}$$

Problem Set 3

1. A pipe contains water at some unknown static pressure p. When a U-tube containing mercury (S = 13.6) is attached to the pipe, pressure p causes a deflection of the mercury as shown in Figure 5–22. If the free end of this U-tube is open to atmospheric pressure (14.7 psia), find pressure p in psig. (*Hint:*   Because points A and B are at the same elevation and are connected by a continuous path of the same liquid, $p_A = p_B$. This equality may be written using expressions for the left and right legs of the U-tube as:

$$p + (\gamma_{water} \times h_{water}) = p_{atm} + (\gamma_{merc} \times h_{merc})$$

The only unknown in this equation is p.)

2. Tanks A and B in Figure 5–23 each contain a pressurized gas. If a manometer whose working fluid is water is connected between the two tanks and deflects as shown find the differential pressure $p_A - p_B$.

3. A U-tube is open to the atmosphere at both ends and is partially filled with water. If oil is added to one leg of this tube and produces the conditions shown in Figure 5–24, find the specific gravity of the oil. (How does this method for measuring specific gravity compare with that presented in Example 10?)

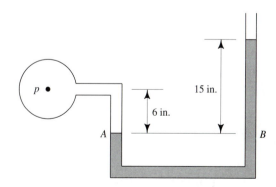

Figure 5–22   Liquid columns used to measure a small pressure for problem 1.

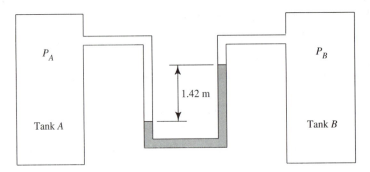

Figure 5–23 A manometer used to measure differential pressure for problem 2.

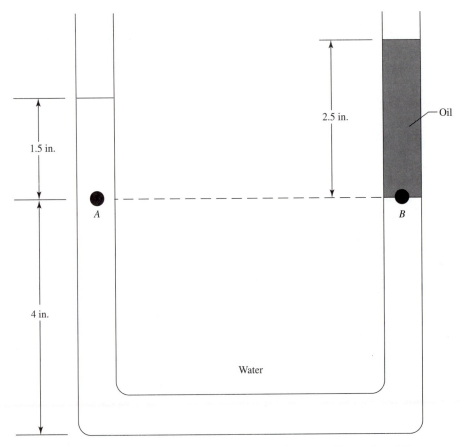

Figure 5–24 A U-tube may be used as a simple hydrometer. Points A and B are at equal pressures. (See problem 3 in problem set 3.)

5.6   The Float Tube Hydrometer

As pointed out in Chapter 2, the float tube hydrometer is often used to measure specific gravity of a liquid. Since these devices are supported by pressures that exist at various depths in a fluid, it is appropriate to examine their operation here.

Consider the solid cylindrical rod of length L and uniform cross-sectional area A shown in Figure 5–25. If the rod is of weight W, then the bearing stress developed between the rod and any solid surface used to support it is merely rod weight divided by contact area, or

$$\text{Bearing stress} = \frac{W}{A}$$

(Bearing stress is simply the mechanical pressure between two solid objects that are in surface contact.) As far as the rod is concerned, any material—solid, gas, or liquid—that is capable of sustaining a pressure of W/A is also able to support the rod as shown.

If the rod is placed in a liquid and released, it will sink into the liquid as it seeks a depth, h, at which the static pressure equals W/A. Finding such a level, the rod will float in the liquid as shown in Figure 5–26. For this condition, equating the pressure at depth h with the required bearing stress yields:

$$S_{\text{liq}} \times \gamma_{\text{water}} \times h = \frac{W}{A}$$

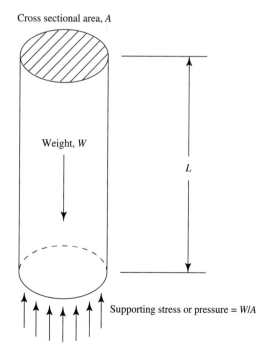

Figure 5–25   Dimensions of a simple float tube hydrometer.

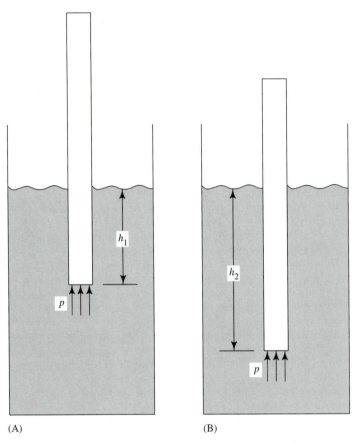

Figure 5–26   Float tube hydrometer in different liquids: (A) liquid with high specific gravity; (B) liquid with low specific gravity.

or

$$S_{liq} = \frac{\left[\dfrac{W}{A \times \gamma_{water}}\right]}{(h)}$$

Since γ_{water} is a constant, and both W and A are fixed for a particular rod, the depth of flotation, h, is inversely related to the specific gravity of the liquid, S_{liq}; in heavy liquids, where pressure increases rapidly with depth, the rod floats "high," while in light liquids the rod must float "low" to reach a depth of equivalent pressure. These two conditions are illustrated in Figure 5–26(A) and (B). If the rod is graduated to simplify measurement of float depth, it becomes the simplest type of *float tube hydrometer* and is the most commonly used instrument to determine the specific gravity of a liquid.

As seen from the instruments of Figure 2–2, commercial hydrometers are rarely of uniform cross-sectional area. For stability, weight is generally concentrated at the bottom of a large-diameter hollow tube often referred to as the bulb. To increase scale magnification, the graduated upper end, or stem, of the device is normally much smaller in diameter than the bulb end. By controlling overall weight of the hydrometer, as well as the shape and size of bulb and stem, it is possible to produce an instrument capable of measuring any range of specific gravities.

The cumulative effect of pressure on the bottom of a float tube hydrometer is known as the *buoyant force* of the liquid. For a flat-ended tube or rod of uniform cross-sectional area, such as the one used in our example, computation of the pressure required for a buoyant force equal to rod weight is a relatively simple task. If the submerged end of any floating object is irregular, however, it is not easy to predict the net buoyant force by analyzing pressure effects on areas at varying depths within the liquid. To eliminate the need for such an analysis, the Greek philosopher Archimedes showed that buoyant force on an object is equal to the weight of liquid displaced by the object. In other words, the buoyant force on any object partially or totally immersed in a liquid may be computed as the weight of liquid whose volume equals the submerged volume of the object. This theorem, known as *Archimedes' principle,* is the method most often used to compute buoyant forces on an object. It has also been very effective at confusing students through the ages by obscuring the fact that these forces are indeed caused by pressures that exist at various depths in a liquid due to the weight of the liquid itself. Example 11 illustrates both methods of computation.

EXAMPLE 11

A flat-bottomed float tube hydrometer 14 in. long has a uniform diameter of 7/8 in. and a total weight of 4 oz. Find its depth of flotation in water:

   (a) By computing the level at which sufficient pressure exists to support the tube (Figure 5–27(A)).

   (b) Using Archimedes, principle (Figure 5–27(B)).

Solution   (a)   The cross-sectional area, A, of the tube is:

$$A = 0.7854 \times (7/8 \text{ in.})^2 = 0.6013 \text{ in.}^2$$

If the hydrometer's weight (4 oz or 0.25 lb) is to be supported on its base, then the required pressure is:

$$p = \frac{W}{A} = \frac{(0.25 \text{ lb})}{0.6013 \text{ in.}^2} = 0.4158 \text{ psi}$$

At what depth in water is the pressure equal to this value? Using equation (5–7) for U.S. customary units:

$$p \text{ (psig)} = p_o \text{ (psig)} + (S_f \times 0.433 \text{ psi/ft} \times h \text{ (ft)})$$

$$0.4158 \text{ psig} = 0 \text{ psig} + (1.0 \times 0.433 \text{ psi/ft} \times h \text{ (ft)})$$

or:

$$h = 0.9602 \text{ ft} = \textbf{11.5 in.}$$

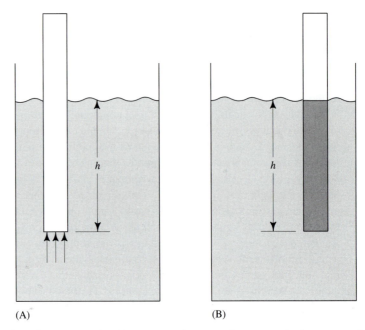

Figure 5–27   Two methods of computing buoyant force on a hydrometer of constant cross-sectional area: (A) pressure at depth h; (B) weight of displaced liquid . (See Example 11.)

(b) According to Archimedes' principle, buoyant force equals the weight of liquid displaced by the submerged volume of the hydrometer. Since the tube weighs 0.25 lb, then 0.25 lb of water must be displaced. What volume of water weighs 0.25 lb? Since total weight, W, equals specific weight, γ_f, times volume, V, then:

$$W = \gamma_f \times V$$
$$0.25 \text{ lb} = 62.4 \text{ lb/ft}^3 \times V$$

or

$$V = 0.004006 \text{ ft}^3 = 6.922 \text{ in.}^3$$

Then 6.922 in.^{3} of water must be pushed aside by the submerged portion of the hydrometer, or:

$$6.922 \text{ in.}^3 = A \times h = 0.6013 \text{ in.}^2 \times h$$

thus,

$$h = \textbf{11.5 in.}$$

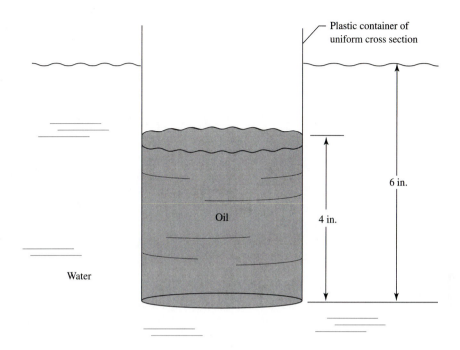

Figure 5–28   This simple hydrometer of problem 1 works best if weight of the plastic container is small compared to weight of the oil sample.

Problem Set 4

1. A weightless plastic container of uniform cross section is filled to a depth of 4 in. with an unknown liquid. If the container is placed in water and floats at the level shown in Figure 5–28, what is the specific gravity of the unknown liquid?

2. A solid block of composite material weighs 214 lb in air. When completely submerged in water, the block has an apparent weight of 139 lb.
   (a) What is the specific gravity of this composite material?
   (b) If the same block has an apparent weight of 165 lb in an unknown liquid, what is the specific gravity of the liquid?

3. When a solid object has the same specific gravity as the liquid in which it is placed, that object is said to possess *neutral buoyancy*. With this property, if the object is moved to any point in the liquid, it will neither rise nor sink but will remain in equilibrium at that point. (Scuba divers strive to achieve this condition in order to maintain their positions at any given depth with a minimum of effort.)
   (a) When empty, an average plastic 2-liter soda bottle, including cap, has a mass of 54.1 g. What is the apparent specific gravity of this bottle?
   (b) How much mass must be placed inside the bottle so that the sealed container will have neutral buoyancy in water?

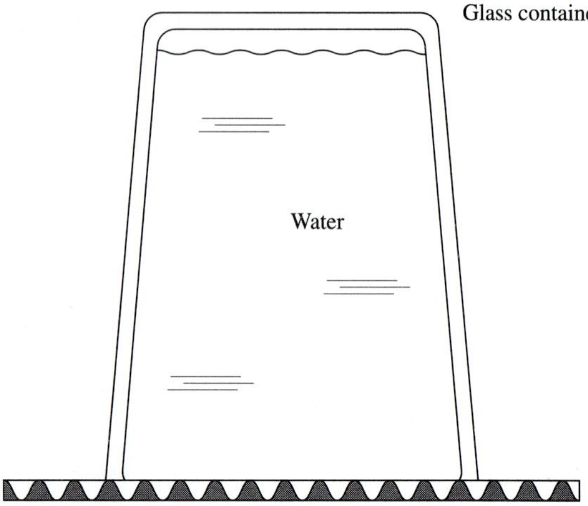

Figure 5–29 Assembled with care, a piece of unsupported cardboard will hold water in an inverted glass. (See question 4.)

(c) Measure out an amount of sand or pebbles whose mass equals that computed in part (b). Place this mass inside the bottle, and apply the screw-on cap. When placed at any depth in water, does the container exhibit neutral buoyancy?

Questions

1. Water tanks are often placed atop towers, and reservoirs are generally located on hills above the communities which they serve. Why is this a common practice?

2. Dams are always thicker at their bases than at their tops. Why? If a reservoir is of uniform depth, does its total volume have any effect on the design of its dam?

3. A typical gasoline can used for fueling lawn mowers, chain saws, or outboard motors invariably has a small vent cap located on its top surface. What is the purpose of such a vent? If closed, how does the vent affect gasoline flow?

4. A glass is filled with water and an index card or other piece of cardboard is placed across the rim. If the card is held in place by hand while the glass is inverted, it is possible to achieve the condition shown in Figure 5–29, where the water remains held in place by the unsupported card. How is this possible? (Try this feat yourself; what is the greatest difficulty in achieving the desired result?)

5. Why does the lifting ability of a suction pump decrease with altitude? What is the approximate loss of lift for each 1000-ft increase in elevation?

6. Define cavitation. What effect can this phenomenon have on the components in a hydraulic system?

7. What does NPSH represent?

8. Explain how buoyant forces are created. What is Archimedes' principle?

9. What is the difference between a hydrometer and a manometer?

10. Why are inclined and well-type manometers used?

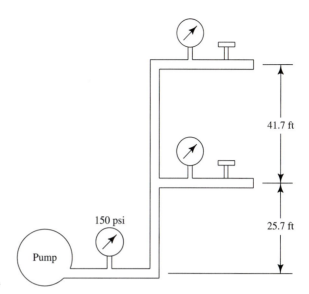

Figure 5–30 Manufacturing
facility plan for review problem 1.

11. If a small cube of solid steel is dropped into a very deep part of the ocean, will the cube eventually sink to a particular depth and stop, remaining in equilibrium at that level?

12. A block of ice is floating in a bucket of water. When the ice melts, will the water level in the
bucket rise, fall, or stay the same?

Review Problems

1. A hydraulic pump feeds two outlets at different elevations in the manufacturing facility of Figure
5–30. If the pump can work against a maximum pressure of 150 psi, find the static (no-flow) pressures at the two outlets when the liquid is hydraulic oil ($S = 0.72$)—(a) top outlet, (b) bottom outlet.

2. The average vacuum created in the intake manifold of a gasoline automobile engine at idle is 18
in. Hg. Such vacuums are normally measured using dial gauges.
   (a) Convert this vacuum into units of psig and psia.
   (b) The same gauge also reads in mm of Hg. Find the gauge reading in these units, and convert
   to kPa.

3. On August 25, 1992, one locality hit by Hurricane Andrew recorded a barometric reading of 27.52
in. Find the loss of lift, compared to standard atmospheric pressure, for a water column on that day.

Suggested Activities

1. Could a device such as the feeder of Figure 5–14 be used as a simple qualitative barometer to indicate changes in atmospheric pressure? Construct a replica by filling any clear glass bottle and a
shallow tray with water. While slowly pouring some water out of the bottle, submerge the mouth
of the bottle below the surface of the liquid in this tray. Clamp the bottle in position, and observe
any changes in liquid level within the bottle for a week. To minimize evaporation during this period, be sure to cover the tray loosely.

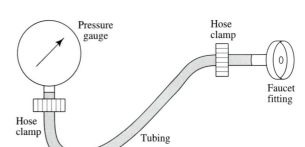

Figure 5–31   Apparatus for suggested activity 2.

2. Using a pressure gauge, rubber hose, or plastic tubing, two hose clamps, and a faucet fitting, construct the simple apparatus shown in Figure 5–31. On different floors in the same building, thread this gauge arrangement onto a faucet, open the tap, and measure the static pressure at that location. (Rest rooms or janitors' water closets are convenient for this activity.) Using these measurements, compute the actual difference in elevation between the faucets. Do your numbers seem realistic? Could the pressure gauge be used as an altimeter within the building? Can you see how a pressure gauge is able to function as a depth gauge for skindivers as well as submarines?

3. Using a pitcher of water and a length of clear plastic tubing, see how high the suction of your lungs can raise a water column. Does the diameter of the tubing affect your results? The length of tubing? Can you devise a method for raising a measurable column of water by blowing?

Liquids in Motion: The Open System

In Chapter 4, the power requirements for a closed system were determined by examining the mechanical output of a linear or rotary actuator during its work cycle. Operation of an open system, however, generally does not involve motion of an actuator, so a different method of analysis must be used, a method based upon the energy content and flow rate of the liquid.

Although open systems are not common in industry, and are often excluded from the study of fluid power, the flow characteristics that they exhibit apply to any system employing a moving liquid. Techniques used to examine the dynamic behavior of such liquids allow us to predict velocity and pressure variations, frictional losses occurring in conductors or fittings, and overall fluid power requirements for both open and closed systems.

This chapter begins, then, by examining several phenomena associated with the flow of a liquid and defining the three most significant energy forms that can exist within that liquid. Two types of flow measurement devices—pitot gauges and venturi meters—are used to illustrate these flow principles. Fluid power is then defined quantitatively in terms of the energy content and flow rate of a liquid and this definition is used to compute the power contained in a moving liquid. Finally, frictional losses are discussed and the results used to modify the predicted behavior of an ideal system.

6.1 Continuity and Types of Flow

The behavior of a two-piston system such as that shown in Figure 6–1 was discussed in Chapter 3. It was assumed that the system was completely filled with an incompressible liquid and that the liquid could neither escape from nor accumulate anywhere in the system. Based upon these assumptions, the relationship between piston displacements became a matter of solid geometry. In order for piston 1 to move into its cylinder a distance d_1, the shaded volume of liquid shown in Figure 6–1(A) had to move into the system ahead of piston 1. To accommodate this liquid, piston 2 was required to move out of its cylinder a distance d_2, leaving an empty space represented by the shaded volume of Figure 6–1(B). Since these volumes must be equal, it followed that:

$$A_1 \times d_1 = A_2 \times d_2$$

where A_1 and A_2 represent the cross-sectional areas of piston 1 and piston 2 respectively. If these pistons are circular, then the relationship can be stated as

$$(D_1)^2 \times d_1 = (D_2)^2 \times d_2$$

where D_1 and D_2 are the appropriate piston diameters.

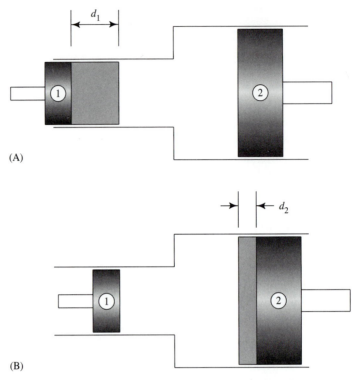

Figure 6–1   Piston displacements in a multipiston system: (A) motion of input piston; (B) motion of output piston.

For an incompressible fluid, displacements d_1 and d_2 must occur simultaneously during some time interval, t. Dividing both sides of the equation on page 153 by t results in:

$$A_1 \times \frac{d_1}{t} = A_2 \times \frac{d_2}{t}$$

Since distance divided by time is velocity, this result becomes:

$$A_1 \times v_1 = A_2 \times v_2 \tag{6–1}$$

which for circular areas can be written as:

$$(D_1)^2 \times v_1 = (D_2)^2 \times v_2 \tag{6–2}$$

Equations (6–1) and (6–2) describe the *conservation of mass* for a flowing liquid and apply not only to closed, multipiston systems but to open, continuous-flow systems as well. Any mathematical statement of this condition is known as an *equation of continuity*.

Consider the flow of a liquid through the stepped conductor shown in Figure 6–2(A). Fluid is forced into this system not by a piston but by the push of a continuous stream

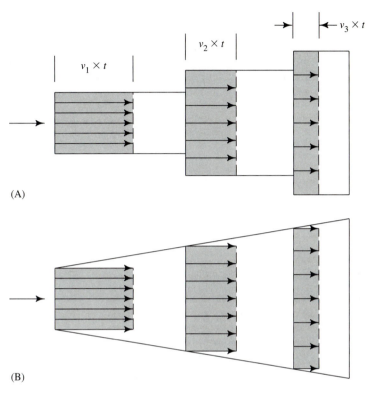

(A)

(B)

Figure 6–2   Continuity of flow: (A) volume flow in a stepped conductor during time interval t; (B) similar flow in a smoothly tapering conductor.

of liquid moving in the direction of flow. (Anyone who has ever exited a stadium or auditorium after a large musical performance or sporting event has experienced this same effect of being swept along with the crowd, moved forward by the press of those spectators following behind.) During a given time interval t, the volume of fluid passing through any cross section of the conductor must be the same. But volume equals cross-sectional area times length, and the volume of liquid moving across any section of pipe during the time interval has a length equal to the product of liquid velocity times the duration of the interval. In other words:

$$A_1 \times (v_1 \times t) = A_2 \times (v_2 \times t) = A_3 \times (v_3 \times t)$$

Division by t results again in equation (6–1). Notice that the continuity equation is valid for any shaped conductor, including the smoothly tapering one shown in Figure 6–2(B). Here the concept is illustrated clearly:

As cross-sectional *area increases,* fluid *velocity decreases.*
As cross-sectional *area decreases,* fluid *velocity increases.*

Continuity, then, has two major applications:

1. It provides a method of predicting variations in liquid velocity based solely upon changes in the cross-sectional area of the conductor.
2. It allows calculation of either the volume flow rate, q, or weight flow rate, Q, through the conductor, since:

$$q = A \times v \qquad (6\text{--}3)$$

and

$$Q = q \times \gamma_f = A \times v \times \gamma_f \qquad (6\text{--}4)$$

where γ_f represents the specific weight of the flowing liquid.

EXAMPLE I

Water flows in the pipe shown in Figure 6–3.

(a) Find the liquid velocities at the 3-in.-diameter and 6-in.-diameter sections.

(b) Compute the volume flow rate in the pipe in cubic feet per second and gallons per minute.

(c) Compute the weight flow rate in the pipe in pounds per second.

Solution

(a) From equation (6–2), the velocities at the 4-in. and 3-in. sections are related by:

$$v_3 = v_4 \times \left(\frac{D_4}{D_3}\right)^2$$

or:

$$v_3 = (36 \text{ ft/s}) \times \left(\frac{4 \text{ in.}}{3 \text{ in.}}\right)^2$$

$$= \mathbf{64\ ft/s}$$

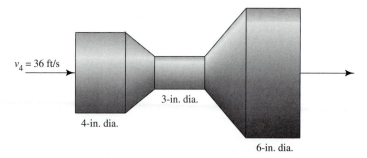

Figure 6–3 Water pipe for Example I.

Similarly for the velocity at the 6-in. section:

$$v_6 = v_4 \times \left(\frac{D_4}{D_6}\right)^2$$

or:

$$v_6 = (36 \text{ ft/s}) \times \left(\frac{4 \text{ in.}}{6 \text{ in.}}\right)^2$$

$$= \textbf{16 ft/s}$$

(b) Volume flow rate is constant and can be computed at any section in the pipe using equation (6–3). For the 4-in.-diameter section:

$$A_4 = 12.6 \text{ in.}^2 = 0.0873 \text{ ft}^2$$

Then,

$$q = A_4 \times v_4 = 0.0873 \text{ ft}^2 \times 36 \text{ ft/s} = \textbf{3.14 ft}^3\textbf{/s}$$

Because flow rates are usually computed in cfs but are more usefully applied in gpm, the following conversion factors are helpful:

$$1\frac{\text{ft}^3}{\text{s}} \times \frac{1728 \text{ in.}^3}{1 \text{ ft}^3} \times \frac{1 \text{ gal}}{231 \text{ in.}^3} \times \frac{60 \text{ s}}{1 \text{ min}} = 449 \text{ gal/min}$$

or:

$$1 \text{ cfs} = 449 \text{ gpm}$$

Similarly,

$$1 \text{ gpm} = 0.00223 \text{ cfs}$$

Therefore,

$$q = 3.14 \text{ cfs} \times 449 \frac{\text{gpm}}{\text{cfs}} = \textbf{1410 gpm}$$

(c) For water, $\gamma_f = 62.4 \text{ lb/ft}^3$, so from equation (6–4):

$$Q = q \times \gamma_f = 3.14 \frac{\text{ft}^3}{\text{s}} \times 62.4 \frac{\text{lb}}{\text{ft}^3} = \textbf{196 lb/s}$$

EXAMPLE 2

Consider flow in the smoothly tapering pipe of Figure 6–4. Compute the liquid velocities at the 45-cm-, 60-cm-, and 75-cm-diameter sections and plot a graph of liquid velocity along the pipe.

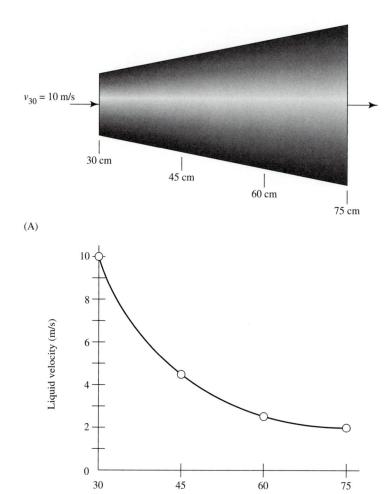

(A)

(B)

Figure 6–4   Flow characteristics for pipe of Example 2: (A) flow through smoothly tapering pipe; (B) velocity distribution along pipe.

Solution   Using the method of Example 1 results in the following values, which are represented graphically in Figure 6–4(B).

D (in cm)	30	45	60	75
v (in m/s)	10.0	4.44	2.50	1.60

Note that although the pipe profile varies linearly, our graph of liquid velocity along the pipe is *not* a straight line.

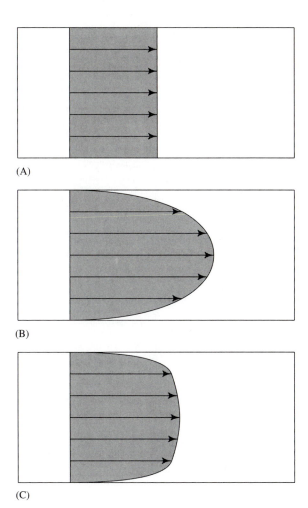

Figure 6–5   Velocity distributions for various types of flow: (A) flow of an ideal liquid; (B) laminar flow of a real liquid; (C) turbulent flow of a real liquid.

Each equation and figure given previously assumes that the velocity distribution across any section of pipe is uniform, as shown in Figure 6–5(A). This condition would exist if the flowing liquid were an *ideal fluid,* one that has no viscosity or internal resistance to flow. As shown in Chapter 2, most liquids possess some viscosity and are therefore able to "feel" the effects of friction between the moving liquid and its surroundings or even between layers of liquid that are moving at different velocities. The net effect, then, is that liquid particles near the pipe wall are slowed down by friction. (In fact, particles along the wall in a thin layer called the *boundary layer* have velocities approaching zero.) These slower particles, in turn, tend to reduce the velocities of other particles flowing past and this effect is passed along through adjacent layers of fluid.

Such frictional effects in a real liquid diminish rapidly with distance from the pipe wall but result in nonuniform flow velocities. Figure 6–5(B) shows a typical velocity profile for *laminar flow,* while Figure 6–5(C) shows the velocity distribution for *turbulent flow.* Notice

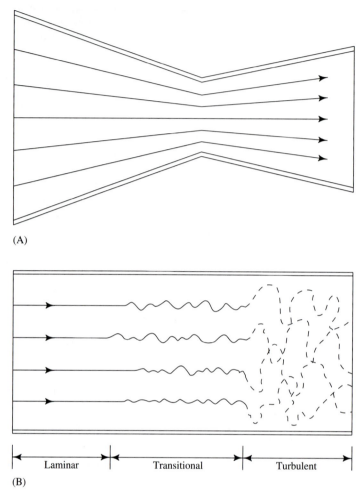

(A)

(B)

|← Laminar →|← Transitional →|← Turbulent →|

Figure 6–6   Streamlines in typical pipe flow: (A) streamlines in laminar flow; (B) streamlines in turbulent flow.

that in both cases fluid close to the pipe wall moves considerably slower than fluid close to the centerline of the pipe.

The difference between laminar and turbulent flow is shown in Figure 6–6 and is best understood by examining the paths followed by individual particles flowing through the pipe. These paths, called *streamlines*, are quite different for the two types of flow. In laminar flow, the streamlines are relatively smooth and always remain parallel. Their spacing can vary as the fluid speeds up or slows down, but the lines will not cross. Turbulent flow, however, is characterized by a "mixing effect," in which the particles of liquid follow random paths across the section while continuing their overall flow along the pipe. Laminar

Table 6–1   Kinematic Viscosities for Hydraulic Liquids at 40°C

Liquid	Type	Use	Kinematic Viscosity (cSt)
Sun 2105	Paraffin-base oil	General-purpose hydraulic fluid	40.4
Sunsafe F	Invert* emulsion	Fire-resistant hydraulic fluid	79.0
Royco 717	Mineral oil	General-purpose hydraulic fluid	26.0
Royco 756	Petroleum base	Aircraft, ordnance, missile hydraulics	11.6[†]
Royco 782	Synthetic hydrocarbon	Fire-resistant, aircraft, ordnance, missiles	14.3
Sunoco Ultra Super C Gold	Petroleum oil	Equivalent to SAE 15W-40 motor oil	15.5[‡]
Sunoco Ultra	Petroleum oil	SAE 30 motor oil	11.5[‡]
Sunoco Type F		Automobile transmission fluid	7.8[‡]
Sunep 68		Gear oil	68.0

*Water in mineral base oil.
[†]At 54°C.
[‡]At 100°C.

flow often becomes turbulent somewhere within a flow conductor, and usually passes through an area of *transitional flow* in the process.

The existence of laminar or turbulent flow in a circular pipe depends upon three factors: pipe diameter, velocity of the liquid, and liquid viscosity. This relationship, first described by Osborne Reynolds in 1883, today defines the flow parameter named in his honor. The value of *Reynolds number, Re,* for any flow can be computed from

$$\text{Re} = \frac{v \times D}{v} \tag{6–5}$$

where v represents liquid velocity, D is the pipe diameter, and v is the *kinematic* viscosity of the liquid. Like Saybolt viscosity, kinematic viscosity is determined by the time required for a specified volume of liquid to flow under gravity through a calibrated viscosimeter at a given temperature (D-445). It also, however, includes the effect of density variations with temperature and can have units of ft²/s, m²/sec, stokes (st), or centistokes (cSt). Unit conversions can be accomplished using the following definitions:

$$1 \, \frac{\text{ft}^2}{\text{s}} = 0.0929 \, \frac{\text{m}^2}{\text{s}} = 929 \, \frac{\text{cm}^2}{\text{s}} = 929 \, \text{St}$$

Typical values of kinematic viscosity for various hydraulic oils are given in Table 6–1, while temperature behaviors for several common liquids and gases are presented in

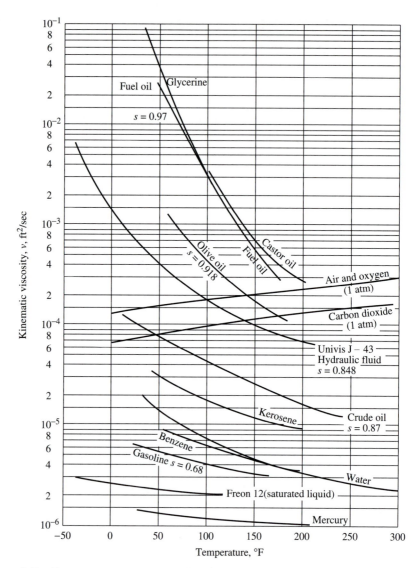

Figure 6–7 Kinematic viscosities of common liquids and gases. *(From Essentials of Engineering Fluid Mechanics by R. M. Olson. Copyright © 1980 by Reuben M. Olson. Reprinted by permission of HarperCollins Publishers, Inc.)*

Figure 6–7. You should verify that in both U.S. customary and SI units Re is a dimensionless number.

Flows that occur at Reynolds numbers of 2000 or less are considered to be laminar; those with Reynolds numbers between 2000 and 4000 are said to be transitional; and those at values above 4000 are fully turbulent. Although Re provides a quantitative method for predicting the type of flow that exists, one of its more important uses is in the computation of frictional losses within fluid power systems. Such calculations are discussed later in this chapter.

EXAMPLE 3

Hydraulic oil flows in a 1/2-in. diameter hose.

(a) If the temperature of this oil is 60°F, find the maximum fluid velocity at which flow is laminar.

(b) Compute the Reynolds number for the flow velocity of part (a) if the liquid temperature is 150°F.

Solution

(a) From Figure 6–7, the kinematic viscosity for oil is approximately 3.2×10^{-4} ft²/s. Rearranging equation (6–5) yields:

$$v = \frac{\text{Re} \times \nu}{D} = \frac{2000 \times 3.2 \times 10^{-4} \text{ ft}^2/\text{s}}{0.0417 \text{ ft}}$$

$$= \textbf{15.3 ft/s}$$

(b) At 150°F, $\nu = 1.0 \times 10^{-4}$ ft²/s. Therefore,

$$\text{Re} = \frac{v \times D}{\nu} = \frac{15.3 \text{ ft/s} \times 0.0417 \text{ ft}}{1.0 \times 10^{-4} \text{ ft}^2/\text{s}}$$

$$= \textbf{6380}$$

Does the existence of laminar or turbulent flow invalidate the principle of continuity and the resulting method of computing flow rates? For the type of applications covered in this book, the effects of velocity variations between an ideal liquid and a real liquid are small and can be neglected, particularly if average velocities are used in the equations presented.

Problem Set I

1. Two 2-in.-diameter hoses are wye-connected to a 4-in.-diameter hose, as shown in Figure 6–8. If each 2-in. hose carries 150 gpm, find the liquid velocities in the 2-in. and 4-in. hoses.

2. Water flows at a velocity of 1.2 m/s in a pipe whose diameter is 48 cm. Further along its length, the pipe narrows to a diameter of 20 cm.
   (a) Find the fluid velocity at the 20-cm section.
   (b) If the kinematic viscosity of water is equal to 1.13×10^{-6} m²/s, find Re at the 48-cm section. Is the flow laminar? Does Re increase or decrease at the 20-cm section?

3. Hydraulic oil flows in a 0.75-in.-diameter hose at the rate of 12.5 gpm.
   (a) Find the liquid velocity in the hose.
   (b) If the oil has a specific gravity of 0.86, find the weight flow rate in lb/s.
   (c) At 100°F, the kinematic viscosity of this oil is 1.82×10^{-4} ft²/s. Find Re.

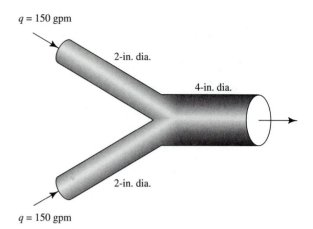

Figure 6–8 Wye-connected hoses for problem 1.

4. Water flows in a 1.0-in.-diameter pipe at a velocity of 2.5 ft/s.
   (a) Compute the flow rate in gpm and lb/s.
   (b) If the pipe is connected in series to a 2-in.-diameter supply line, find the liquid velocity in the 2-in. line.
   (c) For a kinematic viscosity of 1.22×10^{-5} ft²/s, calculate Re in both lines.

5. Combining equations (6–3) and (6–5) for flow through circular pipes yields:

$$q = 0.785 \times D \times \text{Re} \times v$$

   (a) Modify this equation so that q is in gpm if D is in inches and v is in units of ft²/s.
   (b) Write a short computer program (or develop an appropriate spreadsheet) that allows as input the desired value of Re and the value of v for a specific liquid. Use this program to generate tables of maximum flow, in gpm, through pipes of different diameters.

6.2 Energy in Liquids

In Chapter 3, energy was defined as the ability to perform work, and work, in turn, was specified as any activity equivalent to raising a weight through some vertical distance. Using these definitions, it is possible to determine whether an object or a material (such as a fluid) possesses energy and in what form the energy exists.

Fluids can contain energy in a number of forms. If a fluid is combustible, for example, it possesses stored *chemical energy*. Gasoline is routinely burned in internal combustion engines that can then be used to drive an automobile to the top of a hill or lift an airplane into the sky. Fluids that are very hot contain *thermal energy*. Steam can be used to operate a turbine, which drives a generator, which produces electrical energy that can be used to run an electric winch or an elevator. There are other energy forms, such as *nuclear energy*, which a fluid can possess, but none of these, including the ones mentioned above, are common in the typical fluid power systems being studied here.

The three major forms of energy found in a liquid are all categorized as *mechanical energy* and are represented by three physical conditions or states of the liquid, namely its *pressure, velocity,* and *elevation*. Figure 6–9 illustrates how each of the energy states

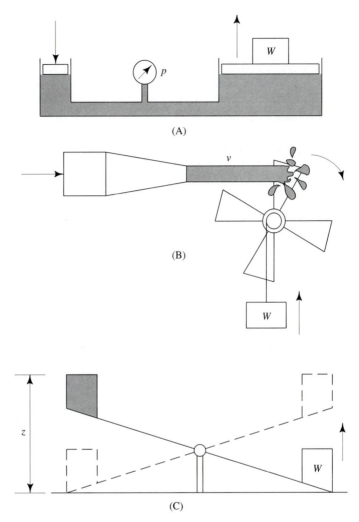

Figure 6–9   Energy forms in a liquid: (A) pressure as energy; (B) velocity as energy; (C) elevation as energy.

could be used to perform some useful work. Using a pressurized liquid (A), for example, the two-piston system analyzed in Chapter 3 is certainly capable of raising a weight through a vertical distance. Likewise, a stream of liquid moving at high velocity (B) could be used to turn a paddle wheel or propeller, thus winding cable on a drum and raising the weight as shown. In a similar manner, liquid at some elevation above its surroundings (C) could raise a counterweight using a simple lever arrangement. Although the devices shown may not be particularly useful, they do demonstrate that the pressure, velocity, and elevation of a liquid are all capable of performing work and therefore qualify as separate energy forms that a liquid may possess.

Both pressure and elevation energy are often referred to as *potential energy,* energy that may be stored in a static condition for use at some later date. Pressurizing a liquid, for example, is very similar to storing mechanical energy in a coiled spring, the energy being able to remain in that state virtually indefinitely. Similarly, water stored in a tank on the roof of a building can, because of its elevation and the effects of gravity, be used in an emergency to supply the fire-sprinkler system within the building. Both pressure and elevation are convenient forms in which to store energy within a liquid.

By contrast, energy in the form of liquid velocity is usually referred to as *kinetic energy,* or energy due to motion. For a variety of reasons, it is difficult to store kinetic energy in its dynamic state. One of the few mechanical devices to do so is the flywheel, and this form of storage is generally used only for short periods of time.

In Chapter 3, the units for both energy and work were defined as the product of force units times distance units. The most common such unit in the U.S. customary system is the foot-pound (ft-lb), and in SI units, it is the newton-meter (N·m). Typical U.S. customary units used to describe pressure, velocity, and elevation of a liquid, however, are lb/ft^2, ft/s, and ft, respectively. Not only are each of these units different, thus making it difficult to compare or combine the energies contained in each form, but none of these units appears to be exactly equal to the ft-lb. It is necessary, then, to develop a consistent set of units for the three forms of mechanical energy found in a liquid.

Elevation Energy

In the study of physics, it is conventional to define the potential elevation energy of an object as:

$$\text{Potential elevation energy} = W \times h$$

where W is the weight of the object and h represents the height of the object above its surroundings. Consider, for example, the 5-gal pail of water shown in Figure 6–10. If this pail contains 40 lb of water and is located on a rooftop 24 ft above the ground, the liquid's total elevation energy is 40 lb $\times$ 24 ft, or 960 ft-lb. This definition of elevation energy is convenient where the behavior of discrete objects having fixed weights is being analyzed. In physics, these objects might include blocks, rockets, footballs, or pails of water. For a fluid power system involving a flowing liquid, however, the liquid at a specific location is being replaced continuously by other liquid, and the total amount of liquid at that location depends upon the flow rate of the system. Because of this, it is much more useful to define the *average* elevation energy of a liquid, or in other words the energy per unit weight of liquid. For the pail of water, 40 lb of liquid contain 960 ft-lb of energy, so the average elevation energy, z, of this liquid is:

$$z = \frac{(960 \text{ ft-lb})}{40 \text{ lb}} = 24 \, \frac{\text{ft-lb}}{\text{lb}}$$

This figure indicates that every pound of water in the pail contains 24 ft-lb of energy and that every pound of *any liquid* at that location would possess 24 ft-lb of elevation energy. It is conventional and algebraically correct to cancel pound units from the numerator and denominator of the answer, so that the average elevation energy, z, is 24 ft. Notice that this

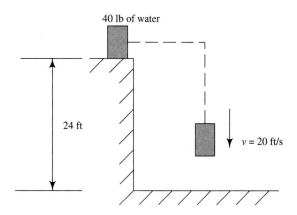

Figure 6–10   Elevation and velocity energy forms. The stationary 40-lb weight has *potential energy* due to its elevated position. This energy may be converted to *kinetic energy* of motion as the weight falls under the influence of gravity.

is simply the height of the liquid above its surroundings and is commonly referred to as the *elevation head.* It is standard practice to use the lowest point in a fluid power system as a reference level, that is, to let $z = 0$ at that point. Both the velocity and pressure forms of energy in a liquid will also be converted to feet when U.S. customary units are involved. In all cases, this unit merely represents an abbreviated form of the average number of *ft-lb of energy per pound of liquid.*

Velocity Energy

The kinetic energy contained in an object due to its motion is defined as:

$$\text{Kinetic energy} = \frac{W \times v^2}{2g}$$

where W is the weight of the object, v is the velocity of the object, and g is the acceleration of gravity. If the pail of water in Figure 6–10 falls off the edge of the building and reaches a velocity of 20 ft/s, then its kinetic energy, K.E., is:

$$\text{K.E.} = \frac{40 \text{ lb} \times (20 \text{ ft/s})^2}{2 \times 32.2 \text{ ft/s}^2} = 248 \text{ ft-lb}$$

The *average* kinetic energy of the liquid is then:

$$\frac{248 \text{ ft-lb}}{40 \text{ lb}} = 6.2 \frac{\text{ft-lb}}{\text{lb}} = 6.2 \text{ ft}$$

Notice that this is the same result that would be obtained by evaluating $(v^2/2g)$, since the water's weight of 40 lb was first used in the numerator of the fraction to find *total* kinetic energy and then in the denominator to compute *average* kinetic energy.

If a 40-lb pail of liquid traveling at 20 ft/s were to collide with another object, the force of impact would be considerable. A stream of liquid moving at this velocity, however, would exert not an impact force, but a steady *pressure* that might also be considerable. For this reason, the average kinetic energy of a liquid represented by $(v^2/2g)$ is called the *dynamic pressure head,* and its units are consistent with those for the average elevation head discussed above.

Pressure Energy

In Chapter 3, Pascal's principle for a confined liquid was applied to multipiston systems, in which the liquid velocity was essentially zero and the difference in elevation between the highest and lowest points of the system was negligible. From an energy point of view, then, Pascal found that all particles of liquid in the system possessed the same average energy as represented by static pressure. This phenomenon was based on the fact that particles of fluid, unlike those in a solid, are able to move freely and adjust to external forces so that all particles are equally "energized" and in a state of pressure equilibrium. The principle also applies to other fluid systems as well. Consider the tank of liquid shown in Figure 6–11. This tank is open to the atmosphere and contains liquid of specific weight γ_f to a depth h. Since a particle of liquid on the surface at A has zero velocity and zero gauge pressure, its energy is entirely in the form of elevation energy, z. At this point $z = h$ so the average energy of a particle here is also equal to h. By contrast, particle B at the bottom of the tank has both zero velocity and zero elevation but is at some pressure, p, due to the standing column of liquid. From fluid statics, this pressure may be computed as $p = \gamma_f \times h$, which may be rearranged to yield $h = p/\gamma_f$. Assuming that each particle in the tank possesses the same *energy, h,* then (p/γ_f) must represent the average energy of any liquid at some pressure p. If the tank in Figure 6–11, for example, contains water to a depth of 50 ft, then $z = 50$ ft at the top surface and p at the bottom of the tank is 21.7 psig or 3120 psfg. Evaluating (p/γ_f) results in:

$$\frac{3120 \text{ lb/ft}^2}{62.4 \text{ lb/ft}^3} = 50 \text{ ft}$$

Every pound of liquid in the tank, therefore, contains 50 ft-lb of energy. This form of energy resulting from confinement of the liquid and represented by (p/γ_f) is called the *static pressure head* and has units consistent with the other two energy forms already presented.

The sum of static pressure head, dynamic pressure head, and elevation head represents the total energy in a liquid and is often referred to as the *total head* of the liquid.

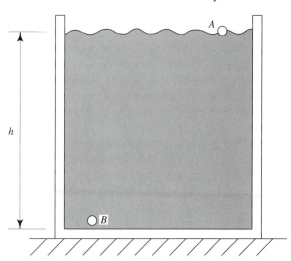

Figure 6–11   All particles in tank have equal energy.

EXAMPLE 4

Water flows in a pipe located 28 ft above the ground. If the pressure in this pipe is 40 psig and the liquid flows at a velocity of 15 ft/s:

(a) Find the total energy in the liquid.

(b) Repeat part (a) if the liquid is oil, $S = 0.7$.

Solution   (a) The static pressure in the pipe is 5760 psfg and for water γ_f is 62.4 lb/ft^3.
Evaluating the three energy forms:

$$\left(\frac{p}{\gamma_f}\right) = \frac{5760 \text{ lb/ft}^2}{62.4 \text{ lb/ft}^3} = 92.3 \text{ ft}$$

$$\left(\frac{v^2}{2g}\right) = \frac{(15 \text{ ft/s})^2}{2 \times 32.2 \text{ ft/s}^2} = 3.49 \text{ ft}$$

$$z = 28.0 \text{ ft}$$

Total energy, E, in the liquid is then:

$$E = 92.3 \text{ ft} + 3.49 \text{ ft} + 28.0 \text{ ft} = \textbf{124 ft}$$

Each pound of liquid flowing through this pipe contains a total of 124 ft-lb of energy, most of which is in the form of static pressure and the least of which is in the velocity, or dynamic pressure, form.

(b) For oil, $\gamma_f = 0.7 \times 62.4 \text{ lb/ft}^3 = 43.7 \text{ lb/ft}^3$. Only static pressure head is affected by specific weight, so:

$$\frac{5760 \text{ lb/ft}^2}{43.7 \text{ lb/ft}^3} = 132 \text{ ft}$$

and:

$$E = 132 \text{ ft} + 3.49 \text{ ft} + 28.0 \text{ ft} = \textbf{163 ft}$$

EXAMPLE 5

What velocity must a liquid have in order for its dynamic pressure head to equal its static pressure head? Find the water velocity that corresponds to a static pressure of 65 psig.

Solution
$$\left(\frac{p}{\gamma_f}\right) = \left(\frac{v^2}{2g}\right) \text{ only if } v = \sqrt{\frac{2gp}{\gamma_f}}$$

If $p = 65$ psig $= 9360$ psfg, and $\gamma_f = 62.4$ lb/ft^3, then the equivalent velocity is:

$$v = \sqrt{\frac{2 \times 32.2 \text{ ft/s}^2 \times 9360 \text{ lb/ft}^2}{62.4 \text{ lb/ft}^3}} = \textbf{98.3 ft/s} = \textbf{67.0 mph}$$

Problem Set 2

1. According to the rules of baseball, an official ball must weigh not less than 5 oz and not more than 5.25 oz.
   (a) Most major-league pitchers have no difficulty throwing a baseball at speeds of 90 mph. Find the kinetic energy of a baseball moving at this speed.
   (b) If the ball of part (a) is thrown vertically upward, how high must it rise in order to convert all its kinetic energy into potential energy?
   (c) Does the actual weight of the ball affect your answers in parts (a) and (b)?

2. (a) If the water leaving a garden hose has a velocity of 15 ft/s, to what vertical distance will this stream of water reach?
   (b) What if the velocity in part (a) is 20 ft/s?
   (c) What if the liquid in part (b) is oil, $S = 0.65$?

3. Water flows through an 8-in.-diameter pipe at a velocity of 12 ft/s.
   (a) Find the dynamic pressure head of the liquid.
   (b) If the pipe narrows to a 3-in. diameter, find the new dynamic pressure head.

4. In SI units the specific weight of water is 9807 N/m^3.
   (a) Compute the static pressure head equivalent to 125 kPa.
   (b) What velocity must a liquid have in order that its dynamic pressure head is equivalent to the static pressure head of part (a)?

5. Oil ($S = 0.8$) flows in a 1.25-in.-diameter supply line at a rate of 18.4 gpm and a static pressure of 500 psig. If the line is located 40 ft above the floor of a manufacturing facility, find the total head of the liquid.

6. The static pressure at the 8-in.-diameter section of problem 3 is 90 psig.
   (a) Compute the static pressure head at the 8-in. section.
   (b) Assuming that the pipe is horizontal and that the total head remains constant, find the static pressure at the 3-in.-diameter section.

6.3   Bernoulli's Equation

In 1738, Daniel Bernoulli published an analysis dealing with the conservation of energy in a liquid. The following year, his father Johann published similar results, which Johann claimed predated the work of his son. The long and bitter controversy that ensued between the two was never resolved. Although historians generally credit Daniel for the fundamental work done in this area, the now famous result is known simply as *Bernoulli's equation*.

As with energy conservation in other mechanical systems, the basic concept is as follows: If a system, or material, initially possesses a given amount of energy, and no energy is added to or removed from the system, then the total amount of energy in the system re-

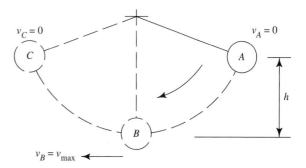

Figure 6–12   Although the total energy of a simple pendulum remains constant, this energy continuously shifts between potential and kinetic forms as the elevation and velocity of the pendulum change.

mains constant. Although energy may shift from one form to another within the system or material, the total energy does not change.

A good example of a mechanical system that demonstrates the conservation of energy is a simple pendulum clock such as the one shown in Figure 6–12. As the pendulum swings to a peak position at A, its velocity decreases to zero and all its energy is in the form of potential energy, as determined by distance h. When the pendulum swings down to position B, this potential energy is completely converted to kinetic energy and the pendulum is here moving at its greatest velocity. As the pendulum continues to swing from position A to B to C and back again, it is constantly shifting energy from potential to kinetic forms. At any intermediate points between A and B, or B and C, the pendulum's energy is a combination of potential and kinetic, but, ideally, the total remains constant, thus satisfying the conservation of energy.

In reality, because of friction and the effects of gravity, a little energy is lost with each swing of the pendulum. Eventually the pendulum would come to rest if the "lost" energy were not replaced. This replacement is accomplished through the potential energy stored in a battery or wound spring contained within the clock. Over a period of time, the stored energy is slowly fed into the system to maintain a constant energy level in the pendulum; once this stored energy is depleted, the battery must be replaced or the mainspring rewound.

When the conservation of energy principle is applied to a liquid, it results in Bernoulli's equation:

$$\frac{p_1}{\gamma_f} + \frac{v_1^{\,2}}{2g} + z_1 = \frac{p_2}{\gamma_f} + \frac{v_2^{\,2}}{2g} + z_2 \qquad (6\text{--}6)$$

For a flowing liquid where no energy is added, removed, or lost, the total energy, as represented by the sum of elevation, static pressure, and dynamic pressure heads, remains constant for any particle of liquid in the system. When used in conjunction with the theory of continuity, Bernoulli's equation provides a powerful tool for predicting the behavior of a flowing liquid.

Consider, for example, the horizontal pipe shown in Figure 6–13(A). Liquid flowing in this pipe maintains a constant total head as it moves from section 1 to section 2. Applying equation (6–6) between points on the pipe's centerline ($z_1 = z_2$) at these two sections yields:

$$\frac{p_1}{\gamma_f} + \frac{v_1^{\,2}}{2g} = \frac{p_2}{\gamma_f} + \frac{v_2^{\,2}}{2g}$$

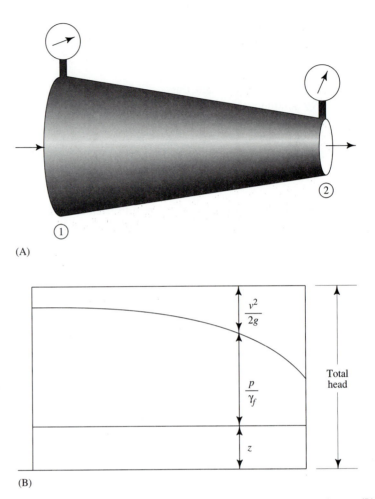

(A)

(B)

Figure 6–13 Depictions for Example 6: (A) flow through a horizontal pipe; (B) energy distribution in the liquid along pipe centerline. Because the pipe is horizontal, elevation energy, z, remains constant. Because flow area is decreasing from left to right, continuity demands an increase in fluid velocity and thus a corresponding increase in kinetic energy, $(v^2/2g)$. With a uniform total head, this increase in kinetic energy comes at the expense of static pressure, (p/γ_f), so gauge readings decrease from point 1 to point 2.

But continuity predicts that v_2 will be higher than v_1 since the cross-sectional area of the pipe is smaller at that section. If the dynamic pressure head must increase and the total head remains constant, some of the static pressure head must be converted into velocity. As a result, the static pressure read by the gauge at section 2 is less than that indicated on the gauge at section 1.

This conversion of energy from one form to another is shown qualitatively in Figure 6–13(B). Total head of the liquid is represented by the fixed distance between the top and bottom lines on the diagram. Because the pipe is horizontal, elevation head of the liquid remains constant and is indicated by the uniform interval measured from the bottom of the graph. From continuity, both velocity and dynamic pressure head increase along the pipe as shown by a smoothly increasing distance measured down from the top of the diagram. The interval between elevation and dynamic pressure heads represents the behavior of static pressure in the system. This diagram graphically illustrates Bernoulli's equation. A vertical line drawn anywhere along the diagram shows the relative distribution between the three forms of energy in the liquid at that section of pipe. Numerical values for these quantities may be obtained using equation (6–6), as illustrated by the following example.

EXAMPLE 6

Water flows in the pipe of Figure 6–13(A). The velocity, static pressure, and diameter at section 1 are 4 m/s, 250 kPa, and 14 cm respectively. If the diameter at section 2 is 8 cm, find the pressure gauge reading at that section. For water, $\gamma_f = 9800$ N/m^3.

Solution   From continuity,

$$v_2 = 4 \text{ m/s} \times \left(\frac{14 \text{ cm}}{8 \text{ cm}}\right)^2 = 12.3 \text{ m/s}$$

which represents a dynamic pressure head of:

$$\frac{v_2^2}{2g} = \frac{(12.3 \text{ m/s})^2}{(2 \times 9.8 \text{ m/s}^2)} = 7.72 \text{ m}$$

At section 1, the static and dynamic pressure heads are:

$$\frac{v_1^2}{2g} = \frac{(4 \text{ m/s})^2}{(2 \times 9.8 \text{ m/s}^2)} = 0.816 \text{ m}$$

$$\left(\frac{p_1}{\gamma_f}\right) = \frac{(250,000 \text{ N/m}^2)}{(9800 \text{ N/m}^3)} = 25.5 \text{ m}$$

Since the pipe is horizontal, $z_1 = z_2$, and the elevation heads cancel when Bernoulli's equation is written between points 1 and 2. Putting the results in tabular form yields:

	Section 1	Section 2
Static pressure head	25.5 m	?
Dynamic pressure head	0.816 m	7.72 m
Total head	26.3 m	26.3 m

If the total pressure head remains constant due to the conservation of energy, then static pressure head at point 2 must equal (26.3 m − 7.72 m) = 18.6 m, which translates into:

$$p_2 = \gamma_f \times h = \left(9800 \ \frac{\text{N}}{\text{m}^3}\right) \times 18.6 \text{ m} = \textbf{182 kPa}$$

Example 6 illustrates the interaction between continuity and the conservation of energy. If the area of a conductor decreases:

1. Continuity predicts that fluid velocity will increase.

2. Bernoulli's equation states that this increase in velocity can occur only at the expense of the two remaining energy forms, namely, elevation and static pressure.

In the sections that follow, these principles are used to analyze several flow conditions and measuring devices often found in open or continuous flow systems.

Problem Set 3

1. Sketch the energy distribution for the flow system shown in Figure 6–14.

2. Sketch the flow system whose energy distribution varies as shown in Figure 6–15.

3. Static pressure at the 4-in.-diameter section of Figure 6–3 is 75 psig. Find the static pressures at the 3-in. and 6-in. sections.

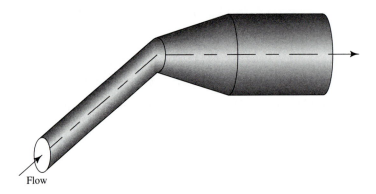

Flow

Figure 6–14 Flow system for Problem 1.

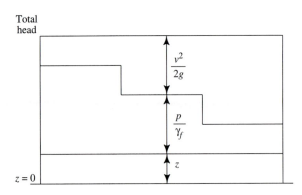

Total head

$\dfrac{v^2}{2g}$

$\dfrac{p}{\gamma_f}$

z

$z = 0$

Figure 6–15 Energy distributions for problem 2.

4. The flow of oil ($S = 0.75$) through the pipe shown in Figure 6–13(A) is summarized next. Find the missing data.

	Section 1	Section 2
D	?	12 in.
υ	28 ft/s	?
p	84 psig	65 psig

5. Seawater ($S = 1.03$) flows through the pipe section shown in Figure 6–13. The diameter at section 1 is 34 in. and at section 2 it is 22 in. If the liquid velocity at the 34-in. section is 17.5 ft/s, find the difference in pressure gauge readings at the two sections.

6.4   Torricelli's Theorem

In 1644, an Italian philosopher named Evangelista Torricelli investigated the velocity of liquid discharging from an opening in the side of a tank, as shown in Figure 6–16. His results disclosed that this escape velocity was proportional to the square root of distance h, the vertical distance from the opening to the surface of the liquid. Not until the development of Bernoulli's equation almost a century later, however, was the exact relation between velocity and column height formulated.

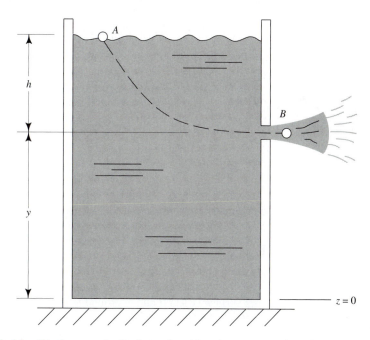

Figure 6–16   Discharge velocity from the side of an open tank is determined by the conversion of potential energy to kinetic energy for a liquid particle as it moves from point A to point B.

Consider a particle of liquid on the surface at point A. The energy states for this liquid are as follows:

$$p_A = 0 \quad \text{(Tank is open to the atmosphere.)}$$

$$v_A = 0 \quad \text{(Particle is assumed to be at rest on surface.)}$$

$$z_A = (y + h) \quad \text{(Bottom of tank is reference level.)}$$

If the particle moves from point A along the dotted streamline and is eventually discharged through the opening, its energy states outside the tank at B are:

$$p_B = 0 \quad \text{(Outside tank, liquid is unconfined and subject only to atmospheric pressure.)}$$

$$v_B = ? \quad \text{(The unknown velocity of discharge.)}$$

$$z_B = y$$

Applying Bernoulli's equation between points A and B yields:

$$0 + 0 + (y + h) = 0 + \frac{(v_B)^2}{2g} + (y)$$

Solving this equation for v_B produces the following result, universally known as *Torricelli's theorem:*

$$v_B = \sqrt{2gh} = (2gh)^{0.5} \tag{6-7}$$

Several points deserve mention here. Notice from equation (6–7) that discharge velocity is independent of the liquid's specific weight as well as the size of the opening in the tank. Also, although the particular case originally studied by Torricelli finds somewhat limited application in today's fluid power systems, his results may be used to predict the discharge velocity of any pressurized liquid by first computing the height of a standing column of that liquid which produces the same pressure.

EXAMPLE 7

A vented tank containing water to a depth of 28 ft is attached to a 95-ft vertical standpipe, as shown in Figure 6–17.

(a) Find the maximum liquid discharge velocity possible at ground level.

(b) At what height above the ground would the velocity be half that of part (a)?

Solution   (a) The liquid surface is 123 ft above ground level. Using this value for h in equation (6–7) yields:

$$v = (2 \times 32.2 \text{ ft/s}^2 \times 123 \text{ ft})^{0.5} = \textbf{89.0 ft/s}$$

(b) Solving equation (6–7) for h produces $h = v^2/2g$. Since the required velocity is v = 44.5 ft/s,

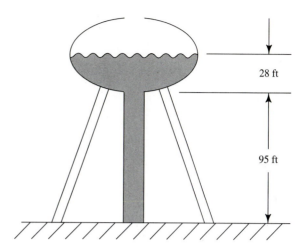

Figure 6–17 Water tank for
Example 7.

28 ft

95 ft

$$h = \frac{(44.5 \text{ ft/s})^2}{(2 \times 32.2 \text{ ft/s}^2)} = \textbf{30.7 ft}$$

Note that this is *not* half of the total column height.

EXAMPLE 8

A hydraulic line contains oil ($S = 0.8$) pressurized to 2500 psig. If this line develops a
small leak, find the velocity of the escaping liquid.

Solution   Although this line pressure is almost certainly being maintained by a pump, it *could* be
created by an equivalent standing column of oil. Referring to equation (5-5), the height,
h, necessary to produce 2500 psig at its base may be computed as:

$$h = \frac{p}{0.433 \times S} = \frac{2500 \text{ psig}}{0.433 \times 0.8} = 7220 \text{ ft}$$

Substituting this value of h into equation (6–7) yields

$$v = (2 \times 32.2 \text{ ft/s}^2 \times 7217 \text{ ft})^{0.5} = \textbf{682 ft/s}$$

This velocity corresponds to almost 465 mph and illustrates the potential safety hazard
that can exist in the vicinity of common hydraulic circuits. Therefore, it is strongly recommended that safety glasses with side shields always be worn by personnel working
near lines containing pressurized liquids or gases.

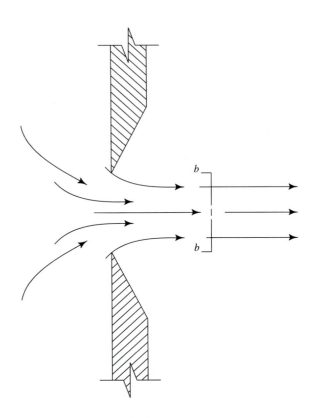

Figure 6–18   Flow through standard orifice.

Torricelli's theorem may also be incorporated into flow equation (6–3) to yield the following result, which allows the *loss rate* to be computed for a leak or other opening:

$$q = A \times v = A \times (2gh)^{0.5} \qquad (6\text{–}8)$$

This equation predicts the *ideal* flow rate. In reality, however, several frictional effects occur that reduce the actual flow.

Figure 6–18 shows a sharp-edged hole with beveled edges known as a *standard orifice*. Fluid discharging through such an opening has its ideal velocity reduced by a factor known as the coefficient of velocity, C_v. This loss is caused by friction on the inside of the tank in the vicinity of the opening and can be influenced both by the condition of the tank walls and the way the opening is formed. If a valve is mounted on the tank, for example, or a hydrant-type opening is used, the inside edges of the opening can differ considerably from that of a standard orifice whose value of C_v is approximately 0.98.

In addition, once outside the tank, the stream of fluid decreases in area as shown at section *b-b* of Figure 6–18. This reduction, known as a *vena contracta*, causes further energy losses in the fluid, resulting in an additional drop in discharge velocity. The effect is accounted for through a coefficient of contraction, C_c, which for the standard orifice is generally taken to be 0.62.

The effects of friction and contraction are commonly lumped together into a single discharge coefficient, C_d, which is the product of C_v time C_c. To predict actual flow rates, C_d is simply factored into the standard flow equation as follows:

$$q = C_d \times A \times (2gh)^{0.5} \qquad\qquad (6\text{–}9)$$

The standard orifice has a C_d value of about 0.61. By comparison, the types of openings common to fire hydrants have discharge coefficients ranging from 0.70 to 0.90, while a smooth, tapering nozzle has a value close to 0.98.

EXAMPLE 9

If the opening through which liquid discharges in Example 8 is a standard orifice of $^1/8$-in. diameter, find:

(a) The actual discharge velocity.

(b) The actual loss rate of liquid from the leak.

Solution

(a) For a standard orifice, $C_d = 0.61$. Applying this coefficient to equation (6–7) yields the actual velocity

$$v = C_d \times (2gh)^{0.5} = 0.61 \times 682 \text{ ft/s} = \textbf{416 ft/s}$$

(b) The $^1/8$-in.-diameter opening has a cross-sectional area of 0.00008522 ft². Using equation (6–8) with the *actual* liquid velocity computed in part (a):

$$q = (0.00008522 \text{ ft}^2) \times 416 \text{ ft/s} = 0.0355 \text{ ft}^3/\text{s}$$

$$= \textbf{15.9 gpm}$$

Problem Set 4

1. A municipal water tank is used to provide drinking water in a rural midwestern community. The tank has a diameter of 80 ft, is vented to the atmosphere, and contains water to a depth of 135 ft. If this tank springs a leak 52 ft above its base:
    (a) What is the velocity of escaping liquid?
    (b) What is the loss rate in gpm if the leak consists of a cylindrical hole 1.5 in. in diameter?
    (c) How far would the water level drop in this tank if the flow rate of part (b) could be maintained continuously for one hour with no water being added to the tank during that period?
    (d) If the opening in part (b) is a standard orifice, compute the actual flow rate.

2. A vertical standpipe contains oil ($S = 0.65$) to a depth of 45 m. Calculate values of v for values of h ranging from 0 through 45 m at intervals of 5 m. Using these data, plot a graph of v versus h.

3. Referring to Example 8 in the text, if the pressurized line contains water at 2500 psig, what is the discharge velocity? (*Note:* The velocity predicted by Torricelli's theorem is the same for all liquids at the *same depth h*. This problem and Example 8, however, involve two different liquids at the *same pressure,* each requiring a different column height to produce that pressure.)

6.5   Flow Measurement

Liquid velocities and flow rates in open systems are commonly measured using either a pitot gauge or a venturi meter. Although these devices find limited application in closed fluid power systems, their operation—based upon the principles of continuity and energy conservation—clearly illustrate the concepts presented earlier in this chapter. Other types of flow meters will be discussed in Chapter 11.

Figure 6–19(A) shows an open tank whose discharge flows through a section of pipe located at the base of the tank. This pipe contains pressure gauges at *A, B,* and *C,* and a valve at *D* to control flow through the pipe. An enlarged view of this section is given in Figure 6–19(B), which shows a "no-flow" condition with the valve completely shut.

Notice that the inlet position for each pressure gauge is slightly different. Assuming that flow will occur from the tank to atmosphere, that is from left to right in the figure, gauge *A* is facing upstream, gauge *B* is inserted perpendicular to the direction of flow, and gauge *C* faces downstream. Each particle of liquid in the tank has essentially the same total energy and each is assumed to be at rest (velocity equal to zero). This energy, then, can be all in the form of elevation (top surface of liquid), all in static pressure (base of tank), or some combination of elevation and static pressure (anywhere else in tank).

With no flow in the discharge pipe, gauges *A, B,* and *C* all read the same pressure, p_o, regardless of their inlet position. Dividing p_o by the specific weight of the liquid results in h_o, the height of the standing column of liquid. This height is referred to as the *total head* and is used as a measure of the *total energy* contained in each particle of liquid.

If the valve is now opened and flow established in the pipe section (see Figure 6–19(C)), what happens to the readings of gauges *A, B,* and *C*? Since no energy has been added to the liquid, its total energy must remain the same as it was for the no-flow condition. In order to get the liquid moving, however, some of that total energy must be converted from static pressure to velocity. Gauges *B* and *C* do not "see" or sense this velocity, but measure only the diminished static pressure. The readings at *B* and *C*, then, are identical but less than p_o once flow begins.

Gauge *A*, however, faces upstream and feels the full effect of the liquid velocity in addition to the reduced static pressure in the pipe. No liquid actually flows into the inlet tube of gauge *A* since the tube dead-ends at the gauge movement itself, but the liquid particle at the open end of the tube is constantly buffeted by the mainstream flow within the pipe. Since the particle at point 1 has no velocity, it is called a *stagnation point* and represents a point at which the velocity energy of the flowing liquid is entirely converted to a dynamic pressure. The sum of the static and dynamic pressures at point 1 equals the total energy in the liquid, as did p_o, so that the gauge reading at *A* remains unchanged from its no-flow value.

This phenomenon is similar to that of a mercury thermometer used outdoors on a sunny day. In the shade of a tree or on the north side of a building, the thermometer is shielded from the direct rays of the sun. Under these conditions, the thermometer registers the ambient or "static" temperature regardless of the instrument's orientation. If this thermometer is now moved out of the shade and faced directly towards the sun, a higher temperature is indicated due to the "dynamic" effect of the incoming radiation.

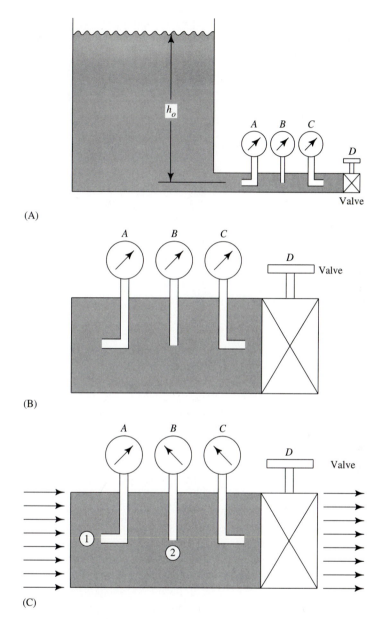

Figure 6–19 Tank and pressure gauges: (A) tank with metered discharge tube; (B) enlargement showing alignment of gauges—"no-flow" condition; (C) behavior of gauges with flow through system.

For the flow conditions shown in Figure 6–19(C), liquid velocity may be computed using indicated pressures at A and B (or A and C). Applying Bernoulli's equation between points 1 and 2 yields:

$$\frac{p_1}{\gamma_f} + 0 + 0 = \frac{p^2}{\gamma_f} + \frac{v_2^2}{2g} + 0$$

Solving this equation for v_2 produces the following result:

$$v_2 = \sqrt{\frac{2g(p_1 - p_2)}{\gamma_f}} \tag{6–10}$$

A device that measures fluid velocity based on the difference between total pressure and static pressure is called a *pitot gauge* or *pitot tube*. Named after Henri Pitot who developed such an instrument in 1735, the pitot gauge is widely used to measure the speed of ships and airplanes and to determine available flow rates in water supply systems. Figure 6–20 shows examples of these gauges, while Figure 6–21 shows such a device being used to measure discharge rate of a hydrant. Pitot gauges are also used to measure flow capabilities of the pumps found on fire trucks. In each case the gauge is placed squarely in the flow and reads dynamic pressure directly.

EXAMPLE 10

The tank shown in Figure 6–19 contains water and has a discharge pipe whose inside diameter is 4.0 in. If the gauge at A reads 24.8 psig and the gauge at B reads 23.8 psig, find:

(a) The velocity of fluid in the pipe.

(b) The flow rate through the pipe, in gallons per minute.

Solution   (a) Converting the pressures to psfg yields $p_1 = 3570$ psfg and $p_2 = 3430$ psfg. For water, $\gamma_f = 62.4$ lb/ft^3. Substituting into equation (6–10) yields:

$$v_2 = \left[\frac{(2 \times 32.2 \text{ ft/s}^2) \times (3570 - 3430) \text{ lb/ft}^2}{62.4 \text{ lb/ft}^3} \right]^{0.5}$$

or

$$v_2 = \left[149 \frac{\text{ft}^2}{\text{s}^2} \right]^{0.5} = \textbf{12.2 ft/s}$$

(b) The flow area, A, is computed as:

$$A = 0.7854 \times (4.0 \text{ in.})^2 = 12.6 \text{ in.}^2 = 0.0873 \text{ ft}^2$$

Computing the flow rate as $q = A \times v$ results in:

$$q = (0.0873 \text{ ft}^2) \times (12.2 \text{ ft/s}) = 1.07 \text{ ft}^3/\text{s} = \textbf{480 gpm}$$

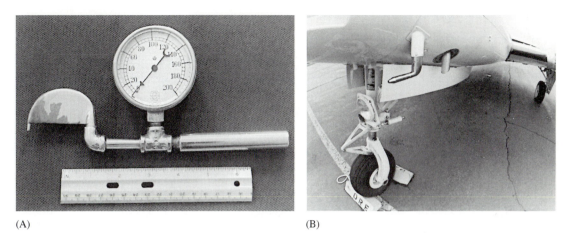

(A)   (B)

Figure 6–20 Pitot gauges. (A) Typical pitot gauge used to measure water flow. A small-bore sensing tube follows the perimeter of the semicircular knife-edged blade. (B) Shiny **L**-shaped pitot gauge mounted to fuselage measures air speed of this plane.

Figure 6–21 A pitot gauge is used to determine flow rate from a hydrant. The nozzle attached to this hydrant helps collimate the flow and provides an accurate-sized opening for the flow calculation.

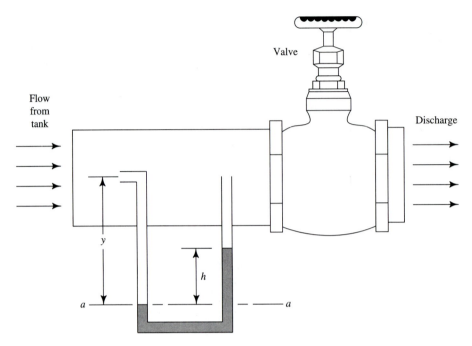

Figure 6–22   A simple pitot gauge (see Example 11).

Notice that the flow rate computed in Example 10 is quite substantial even though the actual pressure difference is small. Such a differential pressure would be difficult to measure using mechanical gauges, so in applications where greater sensitivity is required a manometer of the type shown in Figure 6–22 is often used. Using fluid statics to compute pressures at level *a-a* in the left and right legs of the manometer yields:

$$p_1 + [(\gamma_f \times y)] = p_2 + [\gamma_f \times (y - h)] + [\gamma_m \times h]$$

where γ_m represents the specific weight of liquid used in the manometer, and γ_f the specific weight of the flowing liquid. Solving this equation for the differential pressure produces:

$$(p_1 - p_2) = h \times (\gamma_m - \gamma_f)$$

Substituting into equation (6–10) and simplifying results in:

$$v_2 = \sqrt{2gh \left(\frac{\gamma_m}{\gamma_f} - 1\right)} \qquad (6\text{–}11)$$

In this equation, the ratio of specific weights may be replaced by a ratio of corresponding specific gravities (S_m/S_f) if desired.

EXAMPLE 11

If a mercury manometer were used in Example 10 to replace gauges A and B (Figure 6–22), what height, h, would exist between the upper and lower levels of the manometer fluid?

Solution   The specific gravity of mercury , S_m, is 13.6, and for water, S_f, is 1.0. Solving equation (6–11) for h yields:

$$h = \frac{v_2^{\,2}}{2g \times \left(\dfrac{S_m}{S_f} - 1\right)} = \frac{(12.2 \text{ ft/s})^2}{2 \times 32.2 \text{ ft/s}^2 \times \left(\dfrac{13.6}{1.0} - 1\right)}$$

or:

$$h = 0.183 \text{ ft} = \textbf{2.20 in.}$$

Example 11 illustrates that even with the use of a manometer the actual linear distances measured are small. Some magnification is possible if a liquid having a lower specific gravity than mercury is used, but the choice of such liquids is quite limited. Inclined or well-type manometers (see Figure 5–17) also provide some degree of magnification, but the trend in recent years has been toward the use of sensitive electronic devices, called differential pressure cells, whose output may be fed directly into digital displays, data recorders, or appropriate electronic control circuitry. Cells whose output *voltages* are proportional to the differential pressure are called *transducers;* those whose output *currents* are proportional to differential pressure are known as *transmitters.* A typical pressure transmitter is shown in Figure 6–23.

Another fluid device that utilizes differential pressure is the *venturi tube* or *venturi meter,* developed through work done in the late eighteenth century by the Italian physicist G. B. Venturi. Although any flow conductor whose cross-sectional area varies may act as a venturi, engineers and technologists generally use the term in reference to devices whose walls taper linearly as shown in Figure 6–24. Consider points 1 and 2 in the venturi of that figure. From continuity of flow through circular pipes:

$$v_2 = v_1 \times \left(\frac{D_1}{D_2}\right)^2$$

Assuming that the pipe is horizontal and its centerline is at zero elevation, energy conservation between points 1 and 2 yields:

$$\frac{p_1}{\gamma_f} + \frac{v_1^{\,2}}{2g} + 0 = \frac{p_2}{\gamma_f} + \frac{v_2^{\,2}}{2g} + 0$$

Squaring the first equation, substituting into the second equation, and solving for v_1 gives the following result:

$$v_1 = \sqrt{\frac{2g(p_1 - p_2)}{\gamma_f\left[\left(\dfrac{D_1}{D_2}\right)^4 - 1\right]}} \tag{6–12}$$

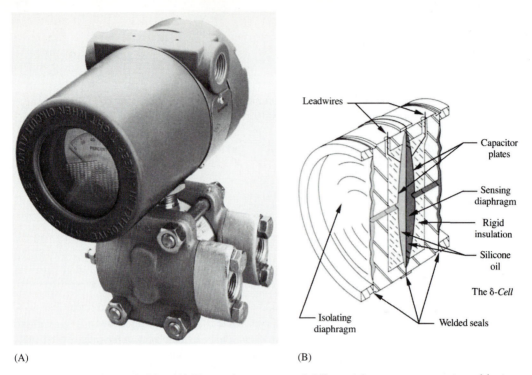

(A)   (B)

Figure 6–23   (A) Photo shows a typical differential pressure transmitter. Maximum scales may range from 0.5 to 750 in. of water with an accuracy of ± 0.2%. (B) As shown in the cutaway drawing, the sensing diaphragm acts as one plate in a capacitor. The difference in pressure causes this diaphragm to deflect, resulting in a capacitance change that is read electronically. *(Rosemount Inc., Eden Prairie, Minn.)*

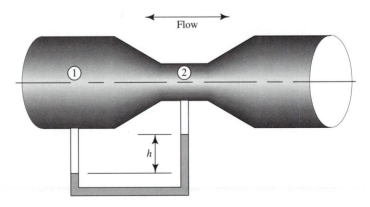

Figure 6–24   A simple venturi operates on the principle demonstrated in Example 6 (see also Figure 6–13). Because continuity requires that the flow velocity at point 2 be greater than that at point 1 while total energy in the liquid remains constant, static pressure decreases from point 1 to point 2.

Figure 6–25   The venturi shown was originally fabricated for use at the 1893 World's Columbian Exposition held in Chicago, and was later used as a secondary calibration standard until 1972. This flow meter decreases from a 36-in. diameter to a throat diameter of 16 in. and has a maximum flow rate of approximately 45,000 gpm.

Therefore, the measured differential pressure ($p_1 - p_2$) can be used to compute flow velocity, and thus flow rate, through the pipe. Such devices find extensive use in industries that must monitor large flows of various liquids; these include chemical production, power generation, petroleum refining, waste water treatment, and food processing industries. Venturi meters may also be used to measure the flow of gases and steam. Figure 6–25 shows a typical flow meter used to monitor the flow of water.

If a manometer is used to measure the differential pressure, then:

$$(p_1 - p_2) = h \times (\gamma_m - \gamma_f)$$

Substituting this result into equation (6–12) and simplifying:

$$v_1 = \sqrt{\frac{2gh(\gamma_m/\gamma_f - 1)}{[(D_1/D_2)^4 - 1]}} \tag{6–13}$$

Either manometers or differential pressure cells are used in the calibration of venturi meters, as shown in Figure 6–26. Although most meters are calibrated using water flows, appropriate correction factors may be applied to extend their use to other liquids and gases.

The narrowest part of a venturi is called the *throat,* and reduced pressures in this section offer the potential for cavitation. On the other hand, low throat pressures provide an excellent means of introducing material to be mixed with the main flow or of atomizing a liquid if the flowing fluid is a gas. Examples of this include vaporization of gasoline in a carburetor, dispersion of paint in a spray gun, and atomization of water in commercial humidifiers. Many such applications utilize the venturi effect within a nozzle, including most sprays of household cleaners, disinfectants, and toiletries dispensed from cans of compressed gas. For this reason, venturi tubes find a variety of uses in areas other than flow measurement.

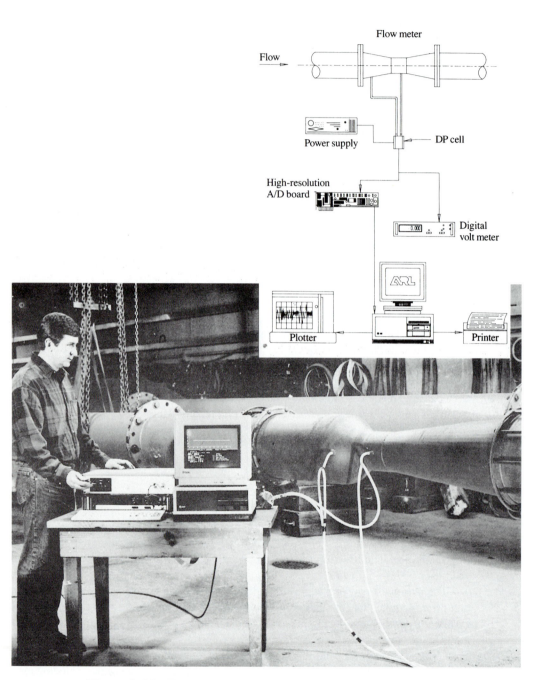

Figure 6–26 Typical equipment configuration for computer acquisition of data during calibration of a flow meter. "DP cell" is a differential pressure cell such as the one shown in Figure 6–23. *(Alden Research Laboratory, Holden, Mass.)*

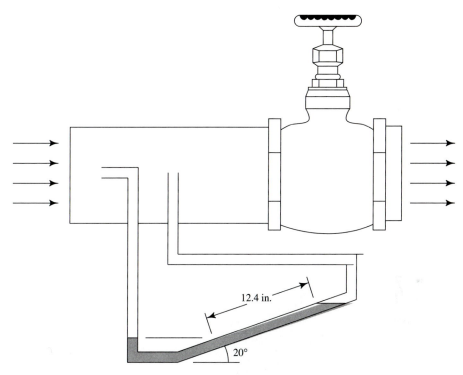

12.4 in.

20°

Figure 6–27   Inclined manometer for problem 2.

Problem Set 5

1. Water flow in a pipe is measured using a pitot gauge that registers a differential pressure of 17.3 kPa. Find the fluid velocity in meters per second.

2. Figure 6–27 shows an installation that uses an *inclined manometer.* This device magnifies the linear distance used to indicate differential pressure sensed by the pitot tube.
   (a) For the conditions shown, what is the water velocity in the pipe if the manometer liquid is mercury?
   (b) If the manometer's angle of inclination is raised to 45°, what linear distance appears on the inclined leg for the same flow velocity as part (a)?

3. Oil with a specific gravity of 0.65 flows in a pipe at a velocity of 23.7 ft/s. If water is used as the manometer fluid, what differential height will be produced in the device by this flow?

4. Find the height, h, of a mercury manometer used as shown in Figure 6–24, for the venturi configuration and maximum water flow rate of the meter pictured in Figure 6–25.

6.6   Fluid Power

Most practical fluid power systems involve at least one change in the total energy of a liquid. Consider Figure 6–28(A), for example, in which liquid enters the system at point 1 with zero total energy and leaves the system at point 2 with a combination of pressure, velocity,

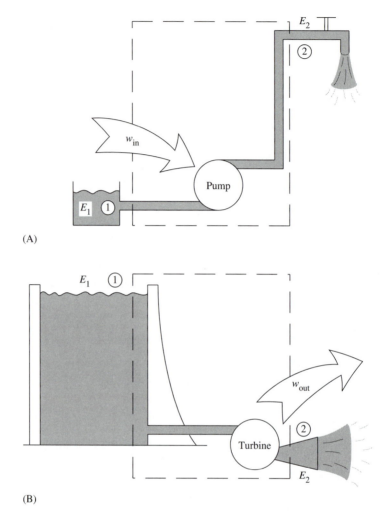

Figure 6–28   Energy transfer in open systems. (A) Work done *on* system increases energy of liquid. (B) Work done *by* system decreases energy of liquid.

and elevation energies. Here, work done on the system by a pump is converted into an increase in the various energy forms of the liquid.

Conversely, liquid with some initial energy can pass through a system and give up this energy by performing useful work. In Figure 6–28(B), liquid at point 1 in the reservoir has elevation energy that is transformed into pressure and velocity as the liquid flows through the system. This energy may be extracted by the turbine and used to perform work such as that involved in the generation of electricity.

Although the two systems shown in Figure 6–28 both represent open systems in which the liquid is not recirculated, the same principle applies to the closed systems analyzed in

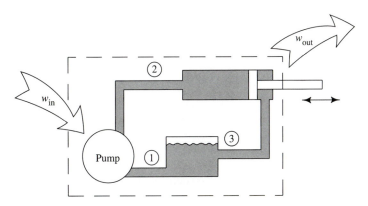

Figure 6–29 Energy transfer in a closed system. A pump takes hydraulic liquid that has no initial energy (point 1) and energizes this liquid using input work, w_{in}, supplied by an external source (motor). The energized fluid (point 2) then passes through a linear or rotary actuator, which extracts energy from the liquid and performs useful work, w_{out}, outside the fluid power system. Finally, the spent liquid (point 3), which is once again without significant energy, returns to the reservoir. Notice that the entire fluid power system acts simply as an intermediary, a convenient means of transferring work or energy *from* one external point (source) *to* another (output).

Chapter 4, an example of which is shown in Figure 6–29. Liquid in the reservoir at point 1 has essentially zero energy, but work done on the system by the pump adds energy to the liquid. This energized liquid at 2 performs useful work through the hydraulic cylinder and eventually arrives at point 3 with little or no energy. For an ideal closed system of this type, the *work done on the system, w_{in},* equals the *work done by the system, w_{out}.* By computing the system's maximum mechanical output during its work cycle, it is possible to determine the necessary energy (and power) input for the system. This concept was applied in Chapter 4.

For an open system, however, liquid delivered with specific energy levels may find a variety of uses that are not as easy to analyze as the repetitive motion of a piston. When water is the working fluid, these uses could include watering crops, fighting a fire, or operating a dishwasher. For such a system, then, the work done on, or done by, a liquid is determined from the change in total energy of the liquid. This is accomplished by modifying Bernoulli's equation to the general form given below, where E_1 and E_2 represent total head at points 1 and 2 respectively, and h_L is the energy loss due to friction. Note that the equation acts like an energy "checking account" for the liquid by allowing for deposits and withdrawals of energy, in the form of work, between points 1 and 2:

$$E_1 \quad + \quad w_{in} \quad - \quad w_{out} \quad - \quad h_L \quad = \quad E_2 \qquad (6\text{–}14)$$

| Initial energy "balance" | Energy "deposit" | Energy "withdrawal" | "Service charge" | Final energy "balance" |

Most open systems involve a single energy exchange, and if the liquid is considered ideal, no frictional losses occur. For these conditions, equation (6–14) simplifies to either:

$$w_{in} = E_2 - E_1$$

or:

$$w_{out} = E_1 - E_2$$

EXAMPLE 12

The pump in Figure 6–28(A) delivers water at the rate of 200 gpm through a 2.5-in.-diameter hose to a point 48 ft above the ground. Residual static pressure in the hose at that point is to be 30 psig. Water entering the pump at ground level comes from an open reservoir at zero pressure and velocity. Find the energy that must be added by the pump to deliver liquid at the specified energy levels.

Solution   Since $p_1 = 0$, $v_1 = 0$, and $z_1 = 0$, then the total initial energy in the liquid, E_1, is zero. For the given flow rate and conductor size, the final velocity, v_2, may be computed as 13.1 ft/s. Static pressure, p_2, is 30 psig or 4320 psfg, and elevation, z_2, is 48 ft. Total final energy, E_2, is then equal to:

$$E_2 = \frac{(4320 \text{ lb/ft}^2)}{(62.4 \text{ lb/ft}^3)} + \frac{(13.1 \text{ ft/s})^2}{(2 \times 32.2 \text{ ft/s}^2)} + 48 \text{ ft}$$

or:

$$E_2 = 69.2 \text{ ft} + 2.66 \text{ ft} + 48 \text{ ft} = 120 \text{ ft}$$

Then,

$$w_{in} = E_2 - E_1 = 120 \text{ ft} - 0 \text{ ft} = \mathbf{120 \text{ ft}}$$

This represents the work, in ft-lb, that the pump must perform on every pound of liquid to be delivered at point 2 with the desired energy content.

From Example 12, it should be apparent that energy added to a liquid in the form of work does not, by itself, represent the *power* required for specified performance of the system. Power is dependent not only upon the energy added to, or subtracted from, a liquid, but also upon the *rate* at which this energy transfer is made. The time-related quantity that determines how fast energy must move into, or out of, a system is the liquid flow rate. In Example 12, every pound of liquid moving through the system receives 120 ft-lb of energy from the pump. If the amount of liquid flowing through the system increases, so does the rate at which these 120 ft-lb "packets" of energy must be supplied by the pump. This results in a corresponding increase in power necessary to perform the desired function.

Generally then, the required power input, or available power output, of a fluid power system may be defined as:

$$P = w \times Q \qquad\qquad (6\text{–}15)$$

where P represents power, w is the work done to bring about the desired energy change in the liquid, and Q is the weight flow rate defined by equation (6–4).

EXAMPLE 13

Find the power required by the pump in Example 12.

Solution The flow rate may be computed as:

$$q = \frac{(200 \text{ gpm})}{449 \text{ gpm/cfs}} = 0.445 \text{ cfs}$$

Then,

$$Q = q \times \gamma_f = (0.445 \text{ ft}^3/\text{s}) \times (62.4 \text{ lb/ft}^3)$$

$$= 27.8 \text{ lb/s}$$

and

$$P = w_{in} \times Q = (120 \text{ ft}) \times (27.8 \text{ lb/s})$$

$$= \textbf{3340 ft-lb/s}$$

From Chapter 3, 1 hp $= 550$ ft-lb/s, so P becomes

$$P = \frac{(3340)}{550} \text{ hp} = \textbf{6.07 hp}$$

EXAMPLE 14

Find the maximum power available in the flowing water of Example 6.

Solution Computing the flow rate at section 1 yields:

$$q = 0.7854 \times (0.14 \text{ m})^2 \times (4 \text{ m/s}) = 0.0615 \text{ m}^3/\text{s}$$

Then:

$$Q = q \times \gamma_f = 0.0615 \text{ m}^3/\text{s} \times 9800 \text{ N/m}^3$$

$$= 603 \text{ N/s}$$

Total work, w, that could be extracted from the liquid is its head of 26.3 m, so:

$$P = 603 \text{ N/s} \times 26.3 \text{ m} = 15{,}900 \text{ N·m/s} = \textbf{15.9 kw}$$

Equation (6–15) may also be used to determine power required for the operation of a closed system. Since elevation and velocity energies are generally small in such a system, the equation may be modified to use just the flow rate and pressure ratings of the pump, as illustrated by the following example.

EXAMPLE 15

Modify equation (6–15) to yield fluid horsepower as a function of flow rate q, in gpm, and rated pump pressure p, in psig.

Solution   From equation (6–15):

$$P \text{ (ft-lb/s)} = Q \text{ (lb/s)} \times w \text{ (ft)} \qquad \text{(a)}$$

If liquid velocity at the pump output is small, as is usually the case with closed systems, it may be neglected, and all the energy assumed to be in the form of static pressure. The liquid in the reservoir has zero initial energy, so the pump puts energy into the liquid in the form of work, w, which may be computed as:

$$w \text{ (ft)} = \frac{[p \text{ (lb/in.}^2) \times (144 \text{ in.}^2/\text{ft}^2)]}{\gamma_f \text{ lb/ft}^3} \qquad \text{(b)}$$

Considering flow rate:

$$Q \text{ (lb/s)} = q \text{ (ft}^3/\text{s)} \times \gamma_f \text{ (lb/ft}^3)$$

$$= \left[\frac{q \text{ (gpm)}}{449 \text{ gpm/cfs}} \right] \times \gamma_f \text{ (lb/ft}^3) \qquad \text{(c)}$$

$$= 0.00223 \times q \text{ (gpm)} \times \gamma_f \text{ (lb/ft}^3)$$

Substituting (b) and (c) into equation (a) results in:

$$P \text{ (ft-lb/s)} = 0.00223 \times q \text{ (gpm)} \times 144 \times p \text{ (psig)}$$

$$= 0.3207 \times q \text{ (gpm)} \times p \text{ (psig)}$$

Since 1 hp = 550 ft-lb/s,

$$P \text{ (hp)} = \frac{P(\text{ft-lb/s})}{550} = \frac{(0.3207)}{550} \times q \text{ (gpm)} \times p \text{ (psig)}$$

or:

$$P \text{ (hp)} = 0.000583 \times q \text{ (gpm)} \times p \text{ (psig)} \qquad \text{(6–16)}$$

Equation (6–16) may be used to evaluate the fluid power output of *any* pump if static pressure p is replaced by stagnation pressure p_o. However, it is most usefully applied to closed systems where pump output pressure is high and velocity effects are negligible, a condition assumed when a pump's rated pressure and its specified flow rate are used. This equation should seem familiar to you, since it was presented in Chapter 4 as equation (4-5) and used to evaluate pump efficiencies.

EXAMPLE 16

Using equation (6–16), verify the horsepower requirements that were obtained in Chapter 4 from an analysis of cylinder behavior during the system's work cycle. These results were presented in Table 4–5.

Solution   Cylinder A required a pump capable of moving 17.8 gpm against a pressure of 3024 psig. From equation (6–16):

$$P \text{ (hp)} = 0.000583 \times 17.8 \text{ gpm} \times 3024 \text{ psig} = \textbf{31.4 hp}$$

This agrees with the 31.5 hp presented in Table 4–5. Similarly for cylinders B and C:

$$P \text{ (hp)} = 0.000583 \times 26.2 \text{ gpm} \times 1935 \text{ psig} = \textbf{29.6 hp}$$

$$P \text{ (hp)} = 0.000583 \times 38.7 \text{ gpm} \times 1344 \text{ psig} = \textbf{30.3 hp}$$

These values compare with table values of 29.4 hp and 30.3 hp, respectively. Any variations in the third significant figure are caused by numerical round-off in the two methods of computation. For this example, the differences are all considerably less than 1%.

Problem Set 6

1. (a) How much work is required to raise 1 lb of water 85 ft in the air? To raise 1 gal of water the same distance?
   (b) How much work is required to raise 100 lb of water 85 ft? If the work is done in 8 s, how much power is required? If done in 12 s?
   (c) If a pump can raise 10 gpm of water to a height of 85 ft, how much power is required?

2. (a) How much work is required to pressurize 1 lb of oil ($S = 0.8$) to 40 psi? To pressurize 1 gal?
   (b) How much power is required to pressurize 100 gpm of oil to 40 psi?

3. A pump is required to deliver 0.5 ft^3/s of water to the twelfth floor (144 ft) of a high-rise student dormitory at a residual faucet pressure of 20 psi through a 2-in.-diameter pipe. Find the power output of this pump.

4. Find the hp available to a hydraulic motor in the flowing oil of problem 5 in problem set 2 of this chapter.

5. A pump at the bottom of a 200-ft mine shaft must pump 3 ft^3/s of water to the surface through 3-in.-diameter pipe. What horsepower is required? Find the horsepower if 4-in. pipe is used.

6. The Canadian side of Niagara Falls (known as Horseshoe Falls) is 170 ft high. Between April and October every year, daytime flow rate over these falls is an estimated 675,000 gpm of water. If all

the power contained in this flow could be converted into mechanical or electrical form, how many horsepower would be produced? How many kilowatts?

7. If the hydraulic gear pump shown and described in Figure 7–23 operates against a continuous pressure of 1500 psi, find:
   (a) The ideal input horsepower required by this pump.
   (b) The ideal input torque for these conditions.

6.7 Sizing the Open System

Performance criteria for open systems are generally specified by the amount of liquid (*flow rate*) to be delivered at a particular location (*elevation*) with given levels of energy (*velocity* and *pressure*). If the initial energy of liquid entering the system is known, the total change in energy, called the *net head,* that must be added to the liquid by the pump/motor may be computed. Together, the flow rate and net head determine the required pump characteristics, and equation (6–15) predicts minimum input power for the system.

Although this method works equally well for closed systems, it is not often used, since pump characteristics must be matched to the hydraulic actuator selected. The actuator, in turn, must be capable of producing some definite linear or rotary output that is generally not specified in terms of the liquid's energy level. Input power is instead determined in one of two ways:

1. By calculating the mechanical power output of the system during the most demanding portion of its work cycle.

2. If velocity and elevation effects may be neglected, by using equation (6–16) or its SI equivalent with the flow rate and output pressure of the pump selected to obtain desired system performance.

As a final comparison, open systems frequently involve the movement of large volumes of liquid at relatively low pressures. The pumps used to accomplish such tasks are generally different from the positive displacement pumps used in the closed fluid power systems of Chapter 4. One common type of pump found in open systems is the centrifugal pump, which not only has high-flow-rate–low-pressure capabilities, but is characterized by a lower efficiency than typical gear or vane type pumps. Average efficiency for a centrifugal pump is 60% to 75%, and unless data are available for a particular pump a value of $\eta = 0.6$ is often assumed. Various types of pumps are discussed in Chapter 7, where we shall see that centrifugal pumps are also used frequently in closed systems to circulate a coolant or working fluid such as water. Examples of this include residential forced hot-water heating systems, swimming pool filtration systems, and automobile cooling systems.

EXAMPLE 17

The centrifugal pump on a fire truck delivers 300 gpm of water to the top of a four-story building 48 ft high. The water is flowing through a 1.25-in.-diameter nozzle at a residual static pressure of 40 psig. Assuming that liquid flowing into the pump has zero initial energy, find the horsepower required to drive this system.

Solution   The flow rate converts to $q = $ (300 gpm)/449 = 0.668 cfs, and flow area is $A = 0.7854$ $\times (1.25/12)^2\text{ft}^2 = 0.00852$ ft². Then flow velocity becomes $v = q/A = 78.4$ ft/s. Computing the energy forms in the liquid:

$$\left(\frac{p}{\gamma_f}\right) = \frac{(5760 \text{ lb/ft}^2)}{(62.4 \text{ lb/ft}^3)} = 92.3 \text{ ft}$$

$$\left(\frac{v^2}{2g}\right) = \frac{(78.4 \text{ ft/s})^2}{(2 \times 32.2 \text{ ft/s}^2)} = 95.4 \text{ ft}$$

Total energy, or net head, added to the liquid is then:

$$w = 92.3 \text{ ft} + 95.4 \text{ ft} + 48 \text{ ft} = 236 \text{ ft}$$

Since the weight flow rate, Q, is:

$$Q = (0.668 \text{ ft}^3/\text{s}) \times (62.4 \text{ lb/ft}^3) = 41.7 \text{ lb/s}$$

Then:

$$P = Q \times w = (41.7 \text{ lb/s}) \times 236 \text{ ft} = 9840 \text{ ft-lb/s}$$

The ideal input horsepower is $P_i = $ (9840/550) hp = 17.9 hp. Assuming an efficiency of 60% for the centrifugal pump, the actual input power required becomes:

$$P_a = \left(\frac{17.9 \text{ hp}}{0.6}\right) = \textbf{29.8 hp}$$

Similarly, the performance of an existing open system can be analyzed. The procedure, in fact, is merely a quantitative evaluation that closely follows the one shown graphically in Figure 6–13(B). Flow rate determines velocity head and the location of pipe or hose dictates elevation head. When these are subtracted from the total head produced by the pump, only residual or static pressure head remains.

EXAMPLE 18

A centrifugal pump located on the bank of a river can move 200 gpm against a total head of 65 psig. Water from this pump moves uphill a vertical distance of 35 ft and feeds a single 2-in.-diameter irrigation pipe.

(a) Find the horsepower required to drive the pump to its maximum capacity.

(b) What is the residual static pressure in the irrigation pipe?

Solution   (a) For water, the pump's total head capability is:

$$\left(\frac{p_o}{\gamma_f}\right) = \frac{(9360 \text{ lb/ft}^2)}{(62.4 \text{ lb/ft}^3)} = 150 \text{ ft}$$

The pump flow rate of 200 gpm or 0.445 cfs corresponds to $Q = 27.8$ lb/s. Ideal horsepower becomes:

$$P_i = \left(\frac{150 \times 27.8}{550}\right) \text{hp} = 7.58 \text{ hp}$$

and

$$P_a = \frac{P_i}{\eta} = \left(\frac{7.58}{0.6}\right) \text{hp} = \textbf{12.6 hp}$$

Notice that the same result would have been obtained using equation (6–16):

$$P_i = 0.000583 \times q \times p = 0.000583 \times 200 \text{ gpm} \times 65 \text{ psi}$$
$$= 7.59 \text{ hp}$$

(b) Flow area of the 2-in. pipe is 0.0218 ft², so liquid velocity becomes:

$$v = \frac{(0.445 \text{ ft}^3/\text{s})}{0.0218 \text{ ft}^2} = 20.4 \text{ ft/s}$$

This yields a velocity head of:

$$\frac{v^2}{2g} = \frac{(20.4 \text{ ft/s})^2}{(2 \times 32.2 \text{ ft/s}^2)} = 6.46 \text{ ft}$$

Since the total head is 150 ft and the elevation head is 35 ft, residual static pressure head becomes:

$$\frac{p}{\gamma_f} = 150 \text{ ft} - 35 \text{ ft} - 6.46 \text{ ft} = 109 \text{ ft}$$

or:

$$p = 0.433 \times 109 \text{ ft} = \textbf{47.2 psig}$$

Problem Set 7

1. The preceding method of analysis must sometimes be used on closed systems where mechanical output is not specified. For example, hydraulic chain saws such as the type shown in Figure 6–30 are commonly used by power line crews because the devices are light, quiet, reliable, and generally safer than gas-powered saws. To operate at full capacity, the saw must receive 8 gpm of oil ($S = 0.75$) at a static pressure of 2000 psi.
   (a) Using equation (6–16), compute the ideal input power required.
   (b) The saw is to be used from the bucket of a cherry picker 40 ft above the ground, with working fluid carried to that location through a 1.5-in.-diameter hose. Compute the pressure, velocity, and elevation heads for the liquid, and use equation (6–15) to compute the ideal input horsepower required. How does this answer compare with that of part (a)?

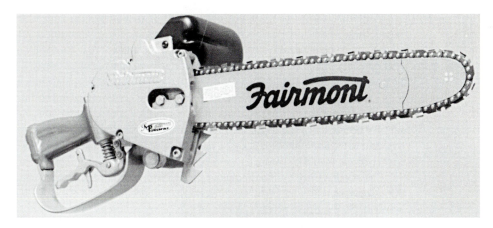

Figure 6–30   Hydraulic chain saw for problem 1. *(Greenlee Fairmont Textron, Greenlee Textron, Inc./Subsidiary of Textron, Inc., Fairmont, Minn.)*

2. A centrifugal pump moves 120 gpm at 200 psi.
    (a) Estimate the actual input power required.
    (b) If the pump moves water through 1.0-in.-diameter pipe located 20 ft above the pump, find the ideal residual pressure in this pipe.

6.8   Losses Due to Friction

In equation (6–14), frictional energy losses in a flowing liquid were represented by the term h_L. Frictional losses occur because both the liquid and the flow conditions are imperfect. Although our analyses in the last few sections ignored such losses, they can become significant when a real liquid flows through an actual fluid power system.

Total energy losses due to friction are generally caused by one or more of the following:

1. In any laminar or streamline flow, or in turbulent flow through smooth pipes, friction losses are due entirely to viscosity of the liquid as layers of fluid slide over one another.

2. For turbulent flow through rough pipes, frictional losses are a combination of viscosity effects within the liquid and the effect of surface roughness as the liquid interacts with the pipe wall.

3. In addition to the above effects for flow through straight pipe of uniform area, any obstructions to flow will cause additional losses. These obstructions might include bends or elbows, valves, or sudden expansions and contractions in flow area, each of which causes the flow to "stumble" over itself as it tries to adjust to the abrupt change. This frequently causes the formation of eddies or flow pockets, inefficiencies that remove useful energy from the liquid.

Although losses due to friction cause a decrease in the total energy, or total head, of the liquid, these losses always appear as a drop in static pressure. Continuity requires that fluid velocity must be maintained throughout the system, and the pipe location dictates elevation energy at any given location, so the only energy form capable of sustaining a loss is the static pressure. It does, in fact, behave much like the spring in our pendulum clock of section 6.3. There the stored potential energy of the spring was gradually fed into the system to overcome the effects of friction and maintain the level of kinetic/elevation energy necessary for operation of the clock. In a fluid power system, static pressure is depleted in much the same way, and although it is considered "lost" as far as useful work is concerned, this energy remains in the system in the form of heat.

The total energy loss, or head loss, due to friction, then, is a combination of frictional effects in straight pipe flow and frictional effects due to flow obstructions. In other words:

$$h_L = h_f + h_b \tag{6-17}$$

where h_L is the total frictional loss, h_f represents losses in head due to flow through straight pipe sections, and h_b is the head loss due to obstructions such as pipe bends and valves.

Values of h_f for liquid flowing through pipe of length L and diameter D at a velocity v may be computed from the Darcy-Weisbach equation:

$$h_f = f \frac{L}{D} \frac{v^2}{2g} \tag{6-18}$$

Quantity f is known as the *friction factor* and may be computed for three different types of flows.

For laminar flow:

$$f = \frac{64}{\text{Re}} \tag{6-19}$$

For turbulent flow in smooth pipes:

$$f = \frac{0.316}{(\text{Re})^{0.25}} \tag{6-20}$$

For fully developed turbulent flow in rough pipes:

$$f = \frac{1}{\left[2 \log \left(\frac{D}{2K}\right) + 1.74\right]^2} \tag{6-21}$$

In equations (6-19) and (6-20), Re represents the Reynolds number of section 6.1, and in equation (6-21) K represents a *surface roughness factor*, experimental values of which are given in Table 6-2. Values of f may also be determined graphically from Figure 6-31 if Re and *relative roughness factor* (K/D) are known.

Table 6–2 Typical Surface Roughness Factors	Pipe Material	K (in.)	K (m)
	Brass	Smooth	Smooth
	Cast iron	0.0102	0.000259
	Concrete	0.0120	0.000305
	Copper	Smooth	Smooth
	Galvanized iron	0.0060	0.000152
	Steel (riveted)	0.0360	0.000914
	Steel (commercial)	0.0018	0.000046

EXAMPLE 19

Hydraulic oil ($S = 0.75$) at 150°F flows through 1-in.-diameter cast iron pipe at 65 gpm. For this flow, find the pressure drop, in psi, per 100 ft of pipe.

Solution

(a) First compute fluid velocity, v. Since the flow area, A, is equal to 0.785 in.$^2 =$ 0.00545 ft^2 and flow rate $q = \dfrac{65}{449}$ cfs $= 0.145$ cfs, then velocity, v, is:

$$v = \frac{q}{A} = \frac{(0.145 \text{ ft}^3/\text{s})}{(0.00545 \text{ ft}^2)} = 26.6 \text{ ft/s}$$

(b) Next compute the Reynolds number, Re, as Re $= \dfrac{v \times D}{v}$, where $D = \frac{1}{12}$ ft $=$ 0.0833 ft, and for oil at 150°F the kinematic vicosity $v = 1.0 \times 10^{-4}$ ft^2/s.

$$\text{Re} = \left(\frac{26.6 \text{ ft/s} \times 0.0833 \text{ ft}}{1.0 \times 10^{-4} \text{ ft}^2/\text{s}} \right)$$

$$= 22{,}200$$

(c) Determine the surface roughness factor from Table 6–2. For cast iron, $K = 0.0102$ in.

(d) Compute the relative roughness factor (K/D).

$$\left(\frac{K}{D} \right) = \frac{0.0102 \text{ in.}}{1 \text{ in.}} = 0.0102$$

(e) Find f using Figure 6–31 and the results of parts (b) and (d). For Re $= 22{,}200$ and (K/D) $= 0.0102$, the approximate friction factor is $f = 0.041$. Note that use of equation (6–21) yields a value of $f = 0.0381$. It should be obvious to the reader

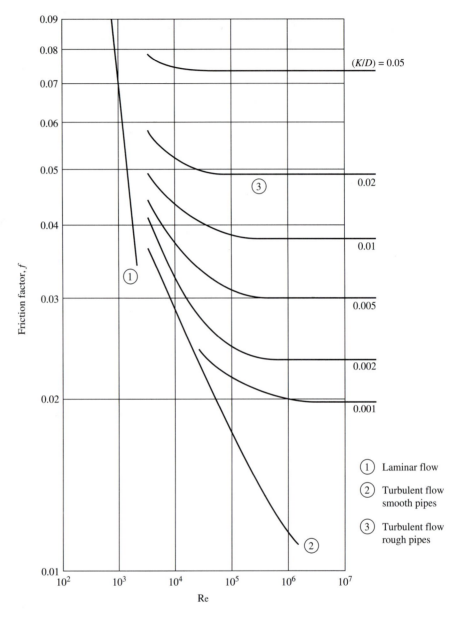

Figure 6–31 Friction factors. *(From data presented in "Friction Factors for Pipe Flow," by L. F. Moody, Trans. ASME, Vol. 66, 1944. Reproduced with permission of the publishers, The American Society of Mechanical Engineers.)*

that neither value can ever be completely accurate; each contains some degree of uncertainty and should be regarded as a reasonable approximation.

(f) Using equation (6–18), compute the head loss, h_f, for 100 ft of pipe.

Since $L = 100$ ft $= 120$ in., then $\left(\dfrac{L}{D}\right) = \left(\dfrac{1200 \text{ in.}}{1 \text{ in.}}\right) = 1200$. Therefore,

$$h_f = f \times \left(\frac{L}{D}\right) \times \frac{v^2}{2g}$$

$$= 0.041 \times 1200 \times \frac{(26.6 \text{ ft/s})^2}{(2 \times 32.2 \text{ ft/s}^2)}$$

$$= 541 \text{ ft}$$

For oil, this is equivalent to a pressure drop of:

$$p = 0.75 \times 0.433 \text{ psi/ft} \times 541 \text{ ft} = \textbf{176 psi}$$

Friction losses due to obstructions are accounted for in a similar manner through the use of experimentally determined loss coefficients, C_b, typical values of which are given in Table 6–3. The magnitude of head loss is related primarily to liquid velocity through the fitting, so the formula for h_b becomes:

$$h_b = C_b \left(\frac{v^2}{2g}\right) \tag{6–22}$$

Table 6–3 Typical Loss Coefficients for Obstructions to Flow	**Obstruction Type**	C_b
	Gate valve, wide open	0.19
	Globe valve, wide open	10.0
	Swing check valve, wide open	2.3
	Ball check valve, wide open	70.0
	45° bend	0.50
	90° bend	0.90
	180° bend	2.0
	Abrupt expansion	$\left[1 - \left(\dfrac{A_1}{A_2}\right)\right]^2$
	Abrupt contraction	$0.5\left[1 - \left(\dfrac{A_2}{A_1}\right)\right]$
	Gradual pipe entrance	0.05

For abrupt expansions or contractions in which liquid velocity changes, equation (6–22) takes the form:

$$h_b = C_b \frac{(v_1^2 - v_2^2)}{2g} \tag{6–23}$$

The head loss through a fitting is often expressed as the *equivalent length* of pipe, L_e, which would produce the same loss. Combining equations (6–18) and (6–22) yields:

$$L_e = D\left(\frac{C_b}{f}\right) \tag{6–24}$$

EXAMPLE 20

The flow of Example 19 makes a 90° bend.

    (a) What is the drop in static pressure across this bend?

    (b) Find the equivalent length of 1-in.-diameter cast iron pipe that produces the same pressure drop.

Solution   (a) From Example 19, $v = 26.6$ ft/s, and from Table 6–3, the loss coefficient is $C_b = 0.9$. Equation (6–22) then yields:

$$h_b = 0.9 \times \frac{(26.6 \text{ ft/s})^2}{(2 \times 32.2 \text{ ft/s}^2)}$$

$$= 9.89 \text{ ft}$$

or:

$$p = 0.75 \times 0.433 \text{ psi/ft} \times 9.89 \text{ ft} = \textbf{3.21 psi}$$

    (b) For the flow of Example 19, friction factor $f = 0.041$. From equation (6–24):

$$L_e = D \times \left(\frac{C_b}{f}\right) = 0.0833 \text{ ft} \times \left(\frac{0.9}{0.041}\right)$$

$$= \textbf{1.83 ft}$$

Problem Set 8

    1. If the flow rate of Example 19 is doubled, find the pressure drop in 100 ft of pipe. Has the pressure drop doubled?

    2. If the 1-in.-diameter pipe of Example 19 is replaced by 2-in.-diameter pipe and the flow rate remains the same, find the pressure drop in 100 ft of pipe.

3. What length of 1-in.-diameter copper tubing would produce the same pressure drop (176 psi) as the 100-ft cast iron pipe of Example 19?

4. The 1-in.-diameter pipe of Example 19 abruptly enlarges to a diameter of 2.50 in. Find the frictional pressure loss caused by this sudden expansion.

Questions

1. The equation of continuity expresses which of these physical conditions?
   (a) Conservation of energy.
   (b) Conservation of velocity.
   (c) Conservation of momentum.
   (d) Conservation of mass.

2. What is an ideal fluid?

3. Identify the three types of liquid flow classified by the behavior of their streamlines.

4. Why does a boundary layer form during flow of a real liquid?

5. What does Re represent? How is this quantity used?

6. What is the approximate value of Re for which a liquid flow ceases to be laminar? Is this an upper or lower limit for Re in laminar flow?

7. Which of these changes in flow conditions or liquid properties will cause a *decrease* in Re?
   (a) Increase in flow velocity.
   (b) Decrease in flow velocity.
   (c) Increase in liquid temperature.
   (d) Decrease in liquid temperature.
   (e) Increase in pipe diameter.
   (f) Decrease in pipe diameter.

8. Bernoulli's equation expresses which of these physical conditions?
   (a) Conservation of energy.
   (b) Conservation of velocity.
   (c) Conservation of momentum.
   (d) Conservation of mass.

9. For the flow of a real liquid, small changes in thermal energy often occur but are generally ignored. Can you think of any ways in which such temperature changes could be used to perform useful work?

10. The terms coefficient of velocity, coefficient of contraction, and discharge coefficient are used to describe flow of a liquid through an orifice. Explain these three terms.

11. What is a stagnation point?

12. Explain the operation of a pitot gauge. Where are such devices commonly used?

13. Which two physical principles govern the behavior of a venturi tube?

14. Why does the flow from one faucet change when another faucet is opened in the same building?

15. The Bernoulli effect is used to draw gasoline into the throat of an automobile carburetor, as shown in Figure 6–32. Can you think of any other applications using moving liquids or gases where this mixing or atomizing effect would be useful?

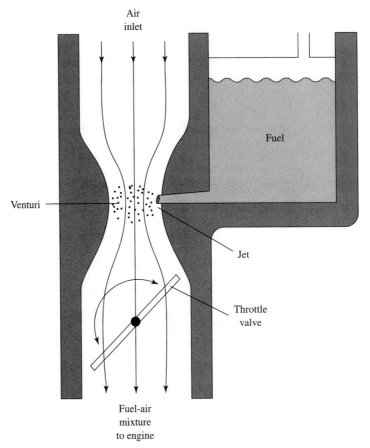

Air
inlet

Fuel

Venturi

Jet

Throttle
valve

Fuel-air
mixture
to engine

Figure 6–32   Low pressure occurs at the throat of an automobile carburetor due to the Bernoulli effect on a moving stream of air. As a result, gasoline is drawn into the flow, where it becomes atomized and forms a combustible mixture (question 15).

16. During electrical blackouts, most urban apartment dwellers still have a supply of running water. Under these conditions, would a miniature electric generator powered by water from a faucet be practical to operate a hotplate, light, or radio? (See suggested activity 2.)

17. After extended use, fluid conductors often accumulate deposits on their inside walls. If the effective diameter of a circular pipe is decreased 5% by such deposits, estimate:
    (a) The change in velocity required to maintain the original (clean pipe) flow.
    (b) The combined effect on head loss described by the factors (L/D) and $(v^2/2g)$ in equation (6–18).
    (c) What effect does this have on the pump used to move liquid around the system? (What if the conductors are arteries or veins?)

18. As the flow rate through a pipe increases, so does Re for the flow. Figure 6–31 shows that as Re increases values of friction factor decrease and that turbulent flow through smooth pipe has a con-

siderably smaller friction factor than laminar flow through the same pipe. Can you think of any possible explanation for this phenomenon?

19. Explain the difference between the surface roughness factor and the relative roughness factor.

Review Problems

1. Equation (6-5) is valid only for circular pipes. For flow through noncircular conductors, the *hydraulic diameter*, D_h, is used in place of pipe diameter D. By definition, the hydraulic diameter is computed as:

$$D_h = \frac{(4 \times \text{cross-sectional area})}{(\text{perimeter of cross-sectional area})}$$

where A is the cross-sectional area of the conductor. Note that for a circular conductor,

$$D_h = \frac{4 \times 0.7854 \times D^2}{3.142 \times D} = D$$

   (a) For the flow conditions of problem 4 in problem set 1, find the dimension, s, of a square conductor that has the same cross-sectional area as the 2-in. pipe.
   (b) Compute D_h for this square conductor.
   (c) Find Re for the specified gpm flow rate through this square conductor.

2. Kerosene ($S = 0.82$) flows in a horizontal pipe at a static pressure of 130 kPa. If the pipe, whose cross section is constant, slopes downward, find the static pressure at a location 25 m below the horizontal section.

3. A 36-in.-diameter water main buried 4 ft below street level contains liquid at 60 psig. The main supplies an apartment 42 ft above street level with 5 gpm of liquid through pipe which is $^1/_2$ in. in diameter. What is the static pressure in the line for the given flow rate? Repeat for $^3/_8$-in.-diameter pipe.

4. For a venturi, the ratio of throat diameter, D_2, to main pipe diameter, D_1, is called the *beta ratio*. Numerical values of D_2/D_1 close to 0.5 are common. For a venturi whose beta ratio is 0.5, flowing oil whose specific gravity is 0.8, what flow velocity corresponds to a differential pressure of 186 kPa?

5. (a) Incorporate equation (6–12) into flow equation (6–3) and modify the results so that flow rate (in gpm) is a function of differential pressure (in psi).

   (b) Write a short computer program that evaluates the equation derived in part (a) and allows as input the main pipe diameter (in inches), beta ratio for the venturi, and specific weight of the flowing liquid (in lb/ft³).

   (c) Generate a table of flow rates (gpm) versus pressure differential (psi) at 2-psi intervals from 20 psig to 80 psig.

   (d) Investigate the effects of beta ratio and specific weight of the liquid by varying these inputs, one at a time, and repeating part (c).

6. Each turbine at the Hoover Dam operates under a total head of 480 ft and produces 115,000 hp. Assuming a 95% efficiency, what must the flow rate be through each machine?

7. A pump moves liquid at the rate of 0.612 m³/sec against a pressure of 425 kPa. Compute the ideal input power, in kilowatts, required to accomplish this task. (Hint: See review problem 1 in Chapter 4.)

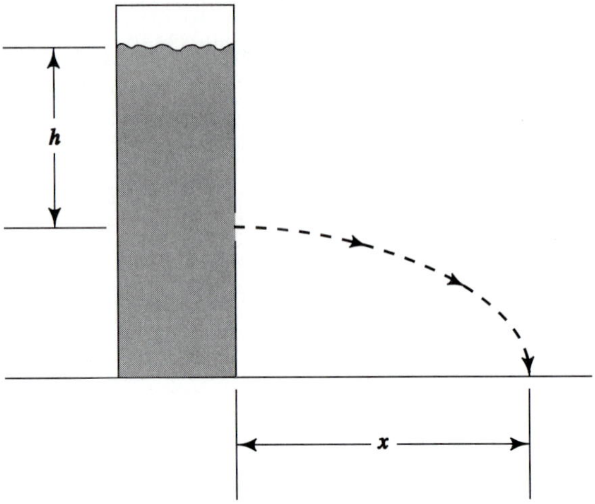

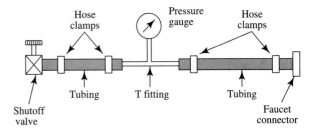

Figure 6–33 Water-filled
container for suggested activity 1.

Figure 6–34 Apparatus for
suggested activity 2.

Suggested Activities

1. Drill a very small hole in the side of a tall, smooth-walled container. Holding your finger tightly over the hole, fill the container with water. Remove your finger from the hole, and measure distance x, the distance from the base of the container at which the escaping liquid falls (Figure 6–33). Repeat using different initial liquid levels, h, and plot a graph of h versus x.

2. Modify the apparatus of Figure 5–31 to include a tee fitting and a small shut-off valve (Figure 6–34), and attach this apparatus to any convenient faucet.
   (a) With the shutoff valve closed, open the tap fully and measure static pressure at the faucet for *no-flow* conditions.
   (b) Open the shutoff valve fully, and record the static pressure for this *maximum flow* condition.
   (c) With the flow still at its maximum, measure the time required to fill a container of known volume. Using the time and volume, as well as the inside diameter of the tee fitting used on your apparatus, compute the fluid velocity in the fitting.
   (d) Calculate the static and dynamic pressure heads for this maximum flow condition. Does their sum equal the static or total head for the no-flow condition? Why not?
   (e) Investigate the effect of velocity on head loss, h_L, by repeating steps (b) through (d) for *partial flows* less than the maximum condition.

7

Hydraulic System Components: Symbols and Hardware

In Chapters 3 through 6, theoretical aspects related to the design of hydraulic circuits were presented. No introduction to fluid power systems would be complete, however, without a discussion of the actual components and hardware commonly found in such systems, as well as the practical considerations involved in their use.

Technologists working with fluid power systems are also expected to be familiar with the symbols used to represent specific pieces of hardware. Combined in schematic diagrams, these symbols show the relationship between components and simplify the analysis of system behavior; circuit diagrams are particularly useful in troubleshooting systems that do not operate as anticipated.

In this chapter, then, we will examine those hydraulic components typically found in fluid power systems. Because such a staggering number of products are available that incorporate slight variations in construction, operation, and appearance, only a general overview is possible within the framework of this book. Schematic representations established by the American National Standards Institute (ANSI) will be correlated to the actual devices they represent. In Chapter 8, these symbols will be combined into diagrams for complete systems. First, however, let us begin with a few comments on the process of component selection.

7.1 Selecting Components

When faced with the responsibility of selecting components for a specific application, most technologists are overwhelmed by the incredible number and variety of products available. Since many of these components are quite expensive, it is important to evaluate equipment options carefully before actually purchasing any hardware; trial-and-error solutions can become very costly. By following a few simple steps, however, most of the uncertainty can be eliminated from the selection process.

1. Locate several companies that manufacture the type of component being selected. Do not restrict your search to local distributors since their product line can be limited to a single manufacturer. One excellent reference for locating components and their manufacturers is the *Thomas Register of American Manufacturers*. So indispensable is this annual, multivolume publication that it is found in most public and university libraries, as well as in the technical libraries or purchasing departments of many industrial firms. Two other extremely helpful publications are the *Fluid Power Handbook & Directory* and the Annual Designers' Guide to Fluid Power Products (an issue of *Hydraulics & Pneumatics Magazine*), both available

from Penton Publishing (see Figure 1–9). Information about specific companies may also be obtained from the Fluid Power Distributors Association (Figure 1–9) or any recent volume of the *U.S. Industrial Directory.* Another starting point can be the yellow pages section of any large regional telephone directory.

2. Obtain specific information about a manufacturer's product line or about a particular component of interest. Most companies provide such data on their products via catalogs or application handbooks, generally at no charge to potential customers. With the proliferation of computers, many of these same companies now supply free software for the design, simulation, and cost analysis of hydraulic circuits that utilize their specific products. Additional information is also available from the manufacturer through one or more of the following sources:

   • *Local distributors* who actually sell the components are often quite knowledgeable about the behavior of these products in specific applications. They can also provide catalogs, performance data, and prices in a timely fashion.

   • *Regional offices* are staffed by sales and field engineers who can answer most technical questions and who are quite happy to sit down with the customer to discuss his or her particular needs. They can usually recommend appropriate equipment for any given application.

   • *Engineering departments,* generally located at the manufacturer's home offices, are made up of people who design and test the company's products; they can therefore answer the most difficult of questions pertaining to the behavior of a component under any given operating conditions.

Although the suitability of hardware for a specific application can be self-evident, the resources listed above are just a telephone call away. For this reason, technologists should not hesitate to seek detailed information as needed.

7.2   Reservoirs

Although the primary function of a reservoir is to store liquid, it has several important secondary functions that are often overlooked. These include:

   • cooling of the liquid
   • elimination of air and settling of contaminants
   • access to liquid for maintenance operations

Each function should be considered in the design of a reservoir.

The schematic symbol for a reservoir is shown in Figure 7–1. Like many graphic symbols, its appearance closely resembles the actual configuration of the device, in this case a vented tank or receptacle containing return/feed lines.

Use of the reservoir symbol is similar to that for ground connections in schematic diagrams of electronic circuits such as the one shown in Figure 7–2(A). To simplify such diagrams and eliminate long lines that cross and recross, ground symbols (denoted with shading) are used wherever convenient, even though they represent a common connection.

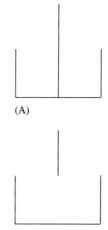

(A)

Figure 7–1 Schematic symbols for reservoirs. (A) Line enters tank below liquid level (generally used for lines carrying fluid out of reservoir). (B) Line enters tank above liquid level (used only for return lines bringing liquid into reservoir).

(B)

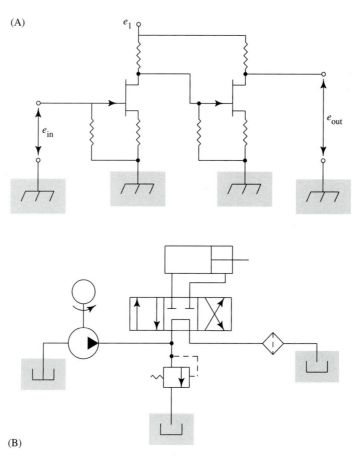

(A)

e_1

e_{in}

e_{out}

(B)

Figure 7–2 Comparative uses of ground symbol and reservoir symbol in electronic and fluid power schematics. (A) Use of ground symbols in schematic diagram of electronic circuit. (B) Use of reservoir symbol in fluid power schematic diagram.

211

Figure 7–3   Typical styles of commercial reservoirs. *(Buyers Products Company, Mentor, Ohio)*

Similarly, though most fluid power systems contain a single reservoir, the symbol for this device is used wherever necessary to simplify or clarify a fluid power circuit. An example is shown in Figure 7–2(B) (shading).

Although commercial units are available (Figure 7–3), many fluid power systems contain reservoirs that have been individually fabricated for their specific installations. Such tanks, as shown in Figure 7–4, should incorporate the following features to adequately store and maintain the hydraulic fluid.

1. Adequate Capacity   To maintain the desired viscosity of the liquid and reduce sludge formation caused by oxidation, temperature of the liquid must not become excessively high. One way to minimize the degrading effects of heat on fluid power systems is to increase the amount of liquid in the system. A rule of thumb for determining adequate reservoir size is as follows:

$$\text{Required volume (gallons)} = [2.5 \times \text{pump flow rate in gpm}]$$
$$+ [\text{volume (gallons) of cylinders/lines}]$$

Unfortunately, in many applications reservoir size is determined solely by economic considerations. For example, a typical commercial log splitter utilizes an 11-gpm pump and only a 3.5-gal reservoir. During continuous operation of this system, the same liquid passes through the pump approximately three times each minute. From the rule of thumb given above, reservoir size here should be at least 27.5 gal, which would allow liquid to flow through the system only once every 2.5 min. Although increasing the reservoir size from 3.5 gal to 27.5 gal would prolong the life of such a system, increased tank size and 96 additional quarts of hydraulic fluid might increase the price of the unit by as much as 25%. Industrial fluid power systems, however, cannot afford to sacrifice long-term reliability for short-term savings. As far as reservoir size is concerned, biggest is best.

2. Provisions for Heat Dissipation   Several measures can be taken to increase conduction of heat away from the liquid. These include locating the reservoir away from objects that can block the flow of air around the unit, placing the reservoir on legs (typically pieces

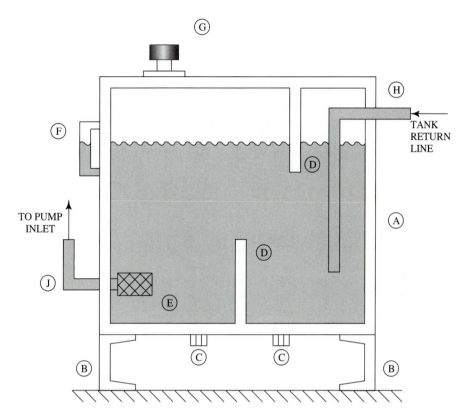

Figure 7–4 Construction of a typical hydraulic reservoir, showing its components: A, tank; B, supporting legs; C, drain plugs; D, baffles; E, strainer; F, liquid level indicator with thermometer; G, breather-filler cap with tank access; H, tank inlet; and J, tank outlet.

of channel or angle iron) to promote air flow, increasing reservoir surface area by attaching external cooling fins, and enhancing air flow by use of an auxiliary fan. For extreme cases, the reservoir can be used in conjunction with a heat exchanger whose cooling coils are placed directly in the stored liquid.

3. Vented Filler Cap During operation of a fluid power system, the volume of liquid stored in the reservoir can vary with time. If the amount of gas (air) above the liquid is held constant by sealing the tank, then any changes in liquid level may result in gas pressures that are slightly above or below atmospheric pressure. In an extreme case, if enough liquid is drawn from the tank by a pump and little or none is returned (perhaps because of leaks elsewhere in the system), sufficient vacuum can develop within the tank to cause its collapse. (See Figures 3–11 and 5–10(B).) To prevent this from happening and to eliminate the effects of variations in gas pressures above the liquid, most reservoirs are vented or open to the atmosphere. This allows the tank and its contents to breathe but also provides a means by which airborne dirt, moisture, and other contaminants may enter the system. Venting is therefore best achieved through a filler cap that contains an air filter. Such filler caps are readily available commercially and are generally part of a complete assembly, including strainer, filler neck, and mounting plate which is easily installed on the top or side of the reservoir (Figure 7–5). If bolted rather than welded to the tank, such an assembly can also provide access to the tank interior for routine maintenance operations.

4. Appropriate Locations of Tank Openings   The reservoir *outlet,* which carries fluid from the tank to the pump inlet, is normally located well below liquid level to minimize the possibility that a drop in level will cause the pump to run itself dry. The outlet should also incorporate a strainer on the inside of the tank to filter out large particles that, if allowed to pass into the system, could severely damage the pump or other major components. This feature is an absolute necessity on tanks that have been fabricated, since that process generally results in significant amounts of weld spatter attached to the inside of the tank. These little beads of solidified metal occasionally break away from the walls and are swept up by liquid flow through the tank; if allowed to reach the pump, their effect is usually disastrous.

Fluid returning to the reservoir passes through the tank *inlet,* which may be located either above or below liquid level. Some fluid power experts believe that this return line should enter below the surface of the liquid to prevent frothing at the surface; others believe it should enter above the surface so as not to stir up any contaminants that have settled to the bottom of the tank. In either configuration, *baffles* are generally used to diminish agita-

Figure 7–5   Several variations of tank filler-breather assemblies. *(Flow Ezy Filters, Inc., Ann Arbor, Mich.)*

tion of the liquid within the reservoir. Horizontal baffles are used to dampen surface motion, while vertical baffles are used to retain settled contaminants in one portion of the tank and to control the flow of liquid through the reservoir (see Figure 7–4). Baffles typically consist of thin metal plates that are sometimes perforated.

One or more *drain plugs* should be located on the reservoir bottom to allow the tank to be emptied. An *access hatch* must be provided on top of the tank so that routine maintenance operations such as cleaning the tank and replacing strainer elements can be performed. As indicated earlier, this access hatch is frequently part of the filler cap installation.

5. Proper Location Relative to Pump   Most pump manufacturers specify the maximum vertical distance above the liquid surface in the reservoir at which a particular pump can be located. This distance is related to the pump's net positive suction head (NPSH), and is determined by the pump's lifting capability as well as the flow characteristics of the system in which it is used (see section 5.4). Obviously, then, NPSH controls the relative positions of pump and reservoir, and may, in fact, require that the pump be located below the elevation of the liquid surface. Typical pumps found in fluid power systems have positive values of lift and are often mounted adjacent to, or on top of, the reservoir. In such installations, however, care should be taken that the pump itself does not become a source of heat contamination for the liquid in the system.

7.3   Strainers and Filters

Strainers and filters are both represented by the schematic symbol shown in Figure 7–6. Although the purpose of each element is to remove impurities from the liquid, their construction and use within a fluid power system are significantly different.

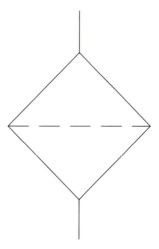

Figure 7–6   Both strainers and filters are represented by this symbol.

Table 7–1 Relative Particle Sizes of Some Common Objects *(Parker Filtration, Metamora, Ohio)*		**Size**	
Substance		**(μm)**	**(in.)**
Grain of table salt		100	0.0039
Human hair		70	0.0027
Lower limit of visibility		40	0.00158
White blood cells		25	0.001
Talcum powder		10	0.00039
Red blood cells		8	0.0003
Bacteria (average)		2	0.000078

Figure 7–7   Typical sump-type strainer for mounting inside reservoir, as shown in Figure 7–4. *(Flow Ezy Filters, Inc., Ann Arbor, Mich.)*

Strainers are used to remove "large" contaminants, those generally visible to the naked eye. Quantitatively, these consist of particles larger than 50 micrometers. (One *micrometer* (μm), sometimes referred to by the term *micron,* is equal to one-millionth of a meter, or 0.0000394 in.). To help put the sizes of these particles in perspective, Table 7–1 lists some common objects whose average dimensions fall in the micrometer range.

Strainers are generally constructed of wire mesh formed into a cylindrical sleeve or basket that can be cleaned periodically by scrubbing, back-flushing, or exposing the element to ultrasonic waves. Flow through the strainer essentially follows a straight line and because the captured particles are relatively large the pressure drop across a strainer is small. For these reasons, pump cavitation due to a clogged inlet is unlikely, so strainers are most commonly located somewhere along the pump suction line. Sump type strainers (Figure 7–7) are mounted inside the reservoir (refer to Figure 7–4), while in-line strainers are used in the suction line between tank and pump (Figure 7–8). Either type may incorporate magnets (Figure 7–9) to attract and hold small metal particles such as filings or weld spatter.

Table 7–2 lists the typical clearances found in common fluid power components. Since particles smaller than 50 μm will pass through most strainers, the potential for wear and scoring of metal surfaces is high. Obviously a second line of defense is needed to protect these precise, expensive components.

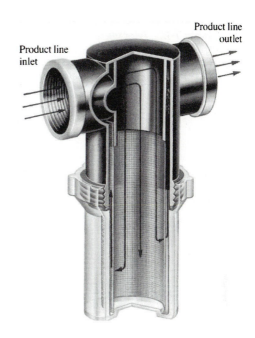

Product line inlet

Product line outlet

Figure 7–8   An in-line strainer, usually mounted in pump suction line. *(Ron-Vik, Inc., Minneapolis, Minn.)*

Figure 7–9   This strainer incorporates ceramic magnets to trap small metal particles. *(Schroeder Industries, Inc., McKees Rocks, Penn.)*

Filters are used to remove particles below 50 μm in size. In addition to capturing solid particles, many filters are capable of removing undesirable liquids from a system's working fluid, such as eliminating water from hydraulic oil. Depending upon the liquid's path through the filtration element, filters are classified as either *surface filters* or *depth-type filters*. In a surface filter, the liquid path is essentially a straight line, with particles trapped at the outside edge or surface of the filter. The most common type of surface filter, known as the *edge-type filter* (Figure 7–10), contains an element

Component	Clearance (μm)
Slide bearings (vane pump)	0.5
Tip of vane (control valve)	0.5
Roller element bearings	0.1–1
Hydrostatic bearings	1–25
Gears	0.1–1
Gear pump (tooth tip to case)	0.5–5
Vane pump (vane tip)	0.5–1
Piston pump (piston to bore)	5–40
Servo valves (spool sleeve)	1–4

Table 7–2 Average Clearances for Some Fluid Power Components *(Parker Filtration, Metamora, Ohio)*

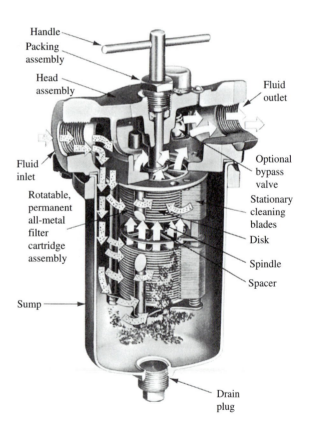

Figure 7–10 Operation of an edge-type filter. *(Cuno Inc., Meriden, Conn.)*

Handle
Packing assembly
Head assembly
Fluid outlet
Fluid inlet
Optional bypass valve
Rotatable, permanent all-metal filter cartridge assembly
Stationary cleaning blades
Disk
Spindle
Spacer
Sump
Drain plug

comprised of stacked metal plates in which fixed plates alternate with movable plates. As fluid flows radially between these thin disks, particles are held at the outside edges of the stack. The movable blades are attached to a handle that extends outside the filter housing. When this handle is rotated, the movable blades pass over a set of stationary cleaning blades that scrape away the captured particles (Figure 7–11). These contaminants then drop to the bottom of the filter where they may be conveniently removed through a drain plug at the bottom of the housing. Since edge-type filters generally remove particles larger than 35 μm, they are only slightly better than strainers and are often used in their place.

Depth-type filters generally consist of a relatively thick layer of some porous material such as paper, cellulose, polyester, fiberglass, or sintered metal that has either been pleated (Figure 7–12) or formed into a solid hollow tube (Figure 7–13). Liquid follows an intricate and tortuous path through such elements, with particles trapped to a considerable depth in the filter's many tiny passageways. The size of the particles removed from the fluid is determined by porosity of the element; filtration down to 0.1 μm is possible.

Magnets may also be used on depth-type filters and are often inserted along or wrapped around the outside perimeter of the filter element. In addition, some depth-type filters are manufactured of materials that possess, or will accept, a slight positive electrostatic charge. Since most contaminant particles are negatively charged, they adhere to the passageway walls in a process known as *electrokinetic adsorption,* which is particularly effective for particles of submicrometer size. One manufacturer of such equipment offers systems that will remove particles of 0.01 μm and smaller for 2000 to 10,000 hours of maintenance-free operation.

Although depth-type filters often consist of a removable element located within a permanent housing, many of these filters are of the "spin-on" variety (Figure 7–14) in which the element is sealed within its housing and the entire unit discarded after use. Removable elements may often be cleaned either chemically or ultrasonically.

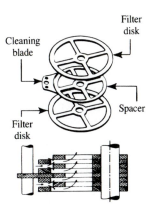

Figure 7–11   Arrangement of stacked plates in an edge-type filter. *(Cuno Inc., Meriden, Conn.)*

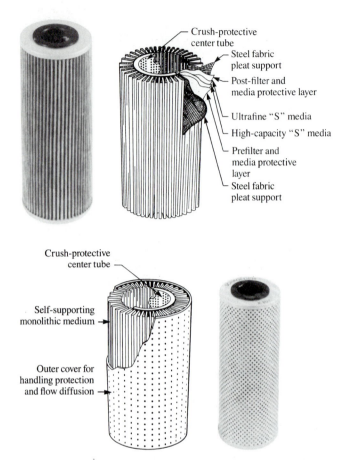

Crush-protective
center tube

Steel fabric
pleat support

Post-filter and
media protective layer

Ultrafine "S" media

High-capacity "S" media

Prefilter and
media protective
layer

Steel fabric
pleat support

Crush-protective
center tube

Self-supporting
monolithic medium

Outer cover for
handling protection
and flow diffusion

Figure 7–12   Construction and appearance of pleated depth-type filters. *(Schroeder Industries, McKees Rocks, Penn.)*

One of the major problems relating to the use of filters is determining the condition of the filtration element. In an automobile, for example, mileage is used to estimate when the lubrication system's oil filter should be replaced. Typical fluid power systems, however, have no such guideline since few include a timer to indicate the number of hours that the system has been in operation. Instead, most manufacturers build some type of signaling device into the filter head (the hardware to which housing and filter element are attached) that indicates the condition of the filtration element. Mechanical indicators may consist of a lever, gauge, or button; electrical indicators that can sound alarms or even shut down the entire system are also available. Virtually all signals work on the principle that pressure on the upstream side of the filter increases, as the element becomes clogged; this increased pressure triggers the indicator. Most filter heads also contain a bypass valve that opens when a preset pressure has been reached. This safety feature allows liquid to flow through the system even if the filter has become completely clogged.

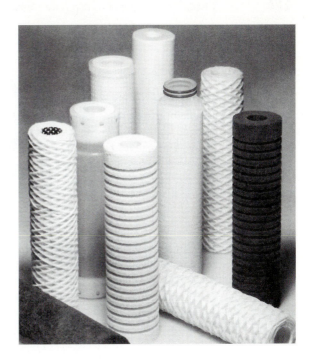

Figure 7–13 Disposable elements for depth-type filters. *(Cuno Inc., Meriden, Conn.)*

Figure 7–14 Spin-on filters with elements sealed in housings. *(Flow Ezy Filters, Inc., Ann Arbor, Mich.)*

Figure 7–15   Although this flow diffuser resembles a strainer, it provides no particle removal; instead it helps reduce the velocity of liquid entering the reservoir. *(Flow Ezy Filters, Inc., Ann Arbor, Mich.)*

One other piece of hardware that resembles a filter is the *flow diffuser.* This is generally a perforated metal tube whose purpose is to slow the velocity and diffuse the flow of liquid returning to the reservoir. Diffusers are mounted inside the tank or reservoir and serve to reduce the foaming and agitation that might otherwise occur. They are not intended to provide any degree of filtration, but are sometimes used to eliminate the need for baffles within the tank. A typical diffuser is shown in Figure 7–15.

7.4   Pumps

Pumps are the components that actually move liquid through a fluid power system. As we saw in Chapter 4, their performance is generally specified quantitatively by the *flow rate* that they can deliver against a specified total *pressure* or *head.* Pumps can also be classified or described qualitatively according to their operational characteristics of *motion, delivery* (flow rate), and *displacement.*

Rotary pumps use the motion of rotating mechanical elements to move liquid, while *reciprocating* pumps achieve the same result using one or more sliding pistons. Originally these designations were used to indicate important differences in flow; rotary pumps are known to deliver a smooth continuous flow, while reciprocating pumps can be characterized by an intermittent or pulsating flow. In recent years, however, technology has reduced these differences in pump characteristics to such an extent that the distinction may now be more of academic interest than practical application.

Pumps can also be classified as *fixed capacity* (flow rate) or *variable capacity* (flow rate) machines, either of which may be capable of *reversible* or *nonreversible* operation. Four combinations of these characteristics are possible, as shown by the schematic representations for each (Figure 7–16). Reversibility is indicated by *two* solid triangles pointing toward the *outside* of the pump envelope, while variable flow is represented by a diagonal arrow across the schematic symbol. As we shall see shortly, the flow delivered by most pumps varies somewhat with operating pressure and speed, but fixed capacity machines offer no other means of flow control. Delivery of liquid from variable capacity pumps, however, can be controlled by using valves located outside the pump or by altering the internal geometry of the pump itself.

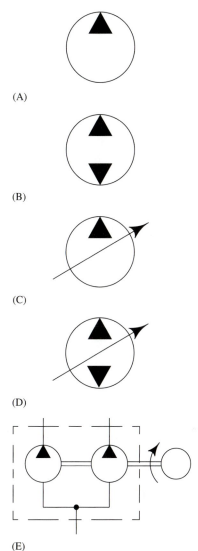

Figure 7–16   Schematic symbols for pumps with various combinations of operating characteristics: (A) fixed capacity, nonreversible; (B) fixed capacity, reversible; (C) variable capacity, nonreversible; (D) variable capacity, reversible; and (E) two-stage pump.

Reversible rotary pumps will function safely in either direction—clockwise or counterclockwise—but nonreversible pumps should be operated only in the direction specified. Rotation is taken to be that viewed from the free end of the pump shaft (Figure 7–17(A)), and is an important consideration when assembling a motor/pump combination. If the pump and motor/engine used to drive it both turn in the same direction, then pulleys and belts or some type of gearbox must connect the two (Figure 7–17(B)); if pump and motor turn in opposite directions, then a simple direct-drive coupling may be

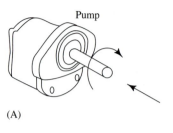

(A)

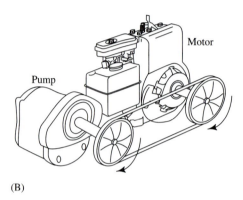

(B)

Figure 7–17 Pump and motor rotations affect coupling method: (A) direction of rotation taken from end view of shaft; (B) pump and motor with same direction of rotation, requiring pulley drive; (C) pump and motor with opposite directions of rotation, allowing direct drive via coupling.

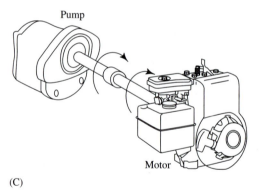

(C)

used (Figure 7–17(C)). Many types of pumps are available in both reversible or nonreversible models; the latter type generally allow the customer to specify a preferred direction of rotation.

Perhaps the most important indicator of a pump's suitability for a particular application is its designation as a nonpositive displacement or positive displacement machine. *Nonpositive displacement* pumps contain large clearances between moving parts. This allows them to be mass produced economically and sold at relatively inexpensive prices. Such pumps are also quite durable, since an obstruction or flow control valve in the pump outlet creates

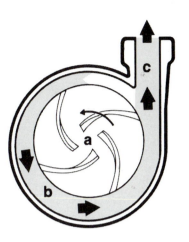

Figure 7–18   Operation of a centrifugal pump. Liquid enters the pump housing at point a, is swept by the rotating impeller along flow path b, and discharges at outlet c. Note the large radial clearance between the impeller and the pump housing. *(ITT JABSCO, Costa Mesa, Calif.)*

Figure 7–19   External appearance of centrifugal pump mounted on electric motor. *(Price Pump Company, Sonoma, Calif.)*

both back pressure and internal slippage that are easily tolerated by the large clearances. This allows for operation over a wide range of conditions without damage to the pump.

Perhaps the most common nonpositive displacement pump is the *centrifugal* type shown in Figure 7–18. Liquid enters the housing at point a and is pushed by the rotating impeller along the housing wall to the discharge port at c. Physical appearance of a typical centrifugal pump is shown in Figure 7–19. The lack of close tolerances in these pumps limits the maximum pressure against which a given pump can work, as evidenced by the performance curves of Figure 7–20. Hence, most nonpositive displacement pumps are best suited to moving large volumes of liquid at low pressures. They find wide application in liquid transfer and heat-exchange systems such as automobile cooling systems (Figure 7–21), swimming pool filtration systems, and irrigation and water supply systems, as well as home appliances such as dishwashers and washing machines.

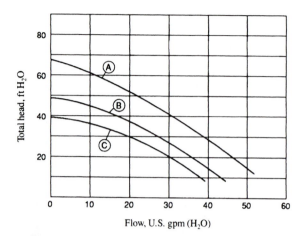

Figure 7–20   Performance curves for centrifugal pumps. *(ITT JABSCO, Costa Mesa, Calif.)*

Figure 7–21   Typical 3-in.-diameter impeller from automobile water pump.

In contrast, *positive displacement* pumps contain very close tolerances and are able to work against high pressures, but produce relatively low flow rates. Fixed capacity pumps deliver a specified volume of liquid with each revolution of the pump shaft and are often rated in in.^{3}/rev or ml/rev. With little or no internal slippage, all positive displacement pumps are subject to damage if output flow is obstructed in any way.

The two most common types of positive displacement pumps found in closed fluid power systems are the *gear pump* and the *vane pump*. Operation of an external gear pump is shown in Figure 7–22. As the gears rotate, liquid is drawn into the housing, compressed between gear teeth and housing, and delivered at pressure to the discharge port. A disassembled gear pump is shown in Figure 7–23, while typical operating curves are presented in Figure 7–24.

Figure 7–22 Operation of an external gear pump. (A) Fluid is drawn into pump. (B) Teeth carry fluid through pump. (C) Fluid is discharged. *(B S M Pump Corporation, North Kingstown, R.I.)*

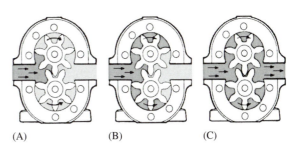

(A)   (B)   (C)

Figure 7–23 The disassembled gear pump shown here operates at 1750 rpm and is capable of moving 1.14 gpm of liquid against a continuous pressure of 1500 psi or intermittent pressures as high as 2500 psi.

Vane pumps contain a number of radial vanes located in a rotor within the pump housing (Figure 7–25(A)). The vanes are free to slide within this rotor and are held in contact with the inner wall of the pumping chamber by a combination of centrifugal force and fluid pressure. A cam ring can be used as shown to provide variable delivery. Pumping takes place only if this ring is offset relative to the center of the vaned rotor shaft. Liquid collected between the vanes where radial clearance is the greatest is pressurized as this radial clearance

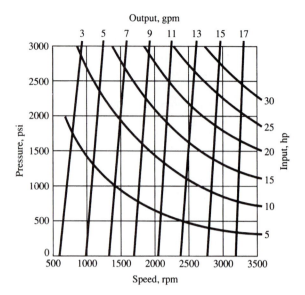

Figure 7–24 Typical performance curves for a gear pump. *(Eaton Corporation, Hydraulics Division, Eden Prairie, Minn.)*

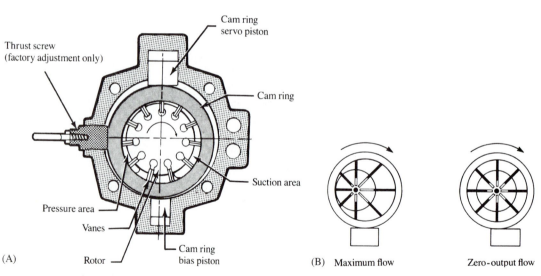

(A)

(B)   Maximum flow

Zero-output flow

Figure 7–25 Construction and operation of a variable delivery vane pump: (A) cross section showing radial vanes in rotor; (B) flow rate determined by position of cam ring relative to rotor. *(Parker Hannifin Corporation, Fluidpower Pump Division, Otsego, Mich.)*

decreases in the direction of rotation. The vanes slide radially within their individual sleeves to accommodate these changes in clearance. If the cam ring is centered on the rotating shaft, no pumping takes place (Figure 7–25(B)). By varying cam ring offset, both flow rate and output pressure may be increased or decreased to maintain preset levels. This is accomplished automatically during pump operation by an electrohydraulic controller mounted on the housing.

Another pump that operates on the same principle as the vane pump is the *radial piston pump* shown in Figure 7–26. The stroke ring offset may be controlled by a variety of mechanical or electrohydraulic compensators that sense either flow rate or pressure during pump operation. The actual pumping action takes place in a series of small radial pistons that are acted upon by slipper pads as shown in Figure 7–27. These positive displacement pumps are generally reversible and are noted for their durability as well as their distinctive appearance (Figure 7–28).

Another type of piston pump is the *axial piston pump* shown in Figure 7–29. The cylinder is attached to an input shaft and houses a series of axial pistons similar to the arrangement of bullets in a revolver. As this assembly rotates, the pistons maintain contact with a fixed plate, called a *swash plate,* which is set at an angle to the input shaft. This causes the pistons to move in and out of their individual cylinders, creating a pumping action. Springs

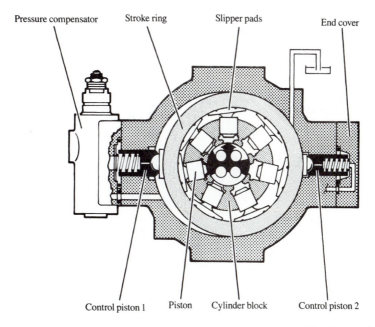

Figure 7–26   Construction of a radial piston pump. *(Robert Bosch Fluid Power Corp., Racine, Wis.)*

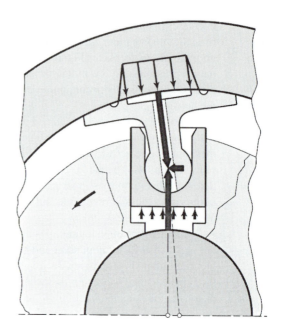

Figure 7–27 Detail of a slipper pad and piston. Pumping takes place due to reciprocating piston movement caused by changes in radial clearance. *(Robert Bosch Fluid Power Corp., Racine, Wis.)*

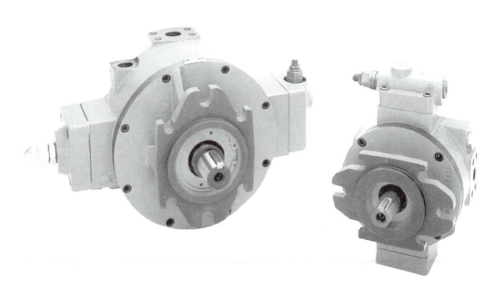

Figure 7–28 Typical radial piston pumps. *(Robert Bosch Fluid Power Corp., Racine, Wis.)*

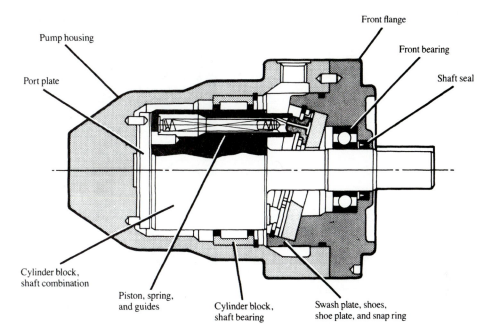

Figure 7–29   Construction of an axial piston pump where flow rate is controlled by swash plate angle. Because each piston assembly rides on the surface of this plate, no pumping occurs when the plate and shaft axis are perpendicular. However, if the plate is tipped (as shown in figure), each piston reciprocates while following the variations in axial position of points along its path on the swash plate. *(Parker Hannifin Corporation, Fluidpower Pump Division, Otsego, Mich.)*

and hydraulic pressure ensure contact between pistons and swash plate, while flow rate is controlled by the angle of the plate itself. If the swash plate is perpendicular to the input shaft, no axial piston movement, and therefore no pumping, takes place. Various compensators are available to control swash plate angle.

Commercial units are available in which pumps, reservoirs, and associated controls are combined to form complete hydraulic power supplies (Figure 7–30). Although any type of pump can be used, vane and gear pumps are most common for this application.

A different type of axial piston pump is shown in Figure 7–31. The unit shown, sometimes called an *intensifier,* consists of three pistons mounted on a common shaft; it actually has two separate fluid systems. Pressurized hydraulic fluid from an external source is applied to one side of the large motive piston located at the center of the shaft. The total force developed on this piston is then applied to a smaller piston at one end of the shaft, developing a higher pressure in the working liquid being pumped. At the same time, suction created by the piston at the opposite end of the shaft draws in a new charge

Figure 7–30 Pumps, reservoirs, and associated controls are often combined to form complete hydraulic power supplies. *(Norman Equipment Company, Bridgeview, Ill.)*

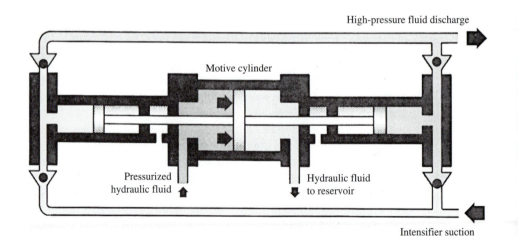

Figure 7–31 Axial piston pump acting as a pressure intensifier. *(Hydro-Pac, Inc., Fairview, Penn.)*

Figure 7–32   A complete
high-pressure intensifier unit.
(Hydro-Pac, Inc., Fairview, Penn.)

of working liquid. Through appropriate valving, the motive piston is next driven in the opposite direction, boosting the pressure in this new charge of liquid.

A commercial intensifier unit is shown in Figure 7–32. Used primarily for specialized applications such as the fluid cutting jet shown in Figure 1–3, these devices are capable of pressures as high as 200,000 psi (1380 MPa), though flow rates at such pressures are generally less than 1 gpm.

7.5   **Actuators**

Any device that converts the power of a pressurized liquid into linear or rotary motion is called an *actuator*. As we saw in Chapter 4, one of the most useful actuators is the hydraulic cylinder. This fluid power component is known as a *linear actuator* because it provides straight line motion and forces that may be used in a variety of manufacturing operations, including lifting, clamping, pressing, punching, shearing, and bending.

Hydraulic cylinders can be of two general types: *single-acting* or *double-acting*. Schematic symbols for each are shown in Figure 7–33. For a single-acting cylinder, the piston is pressure

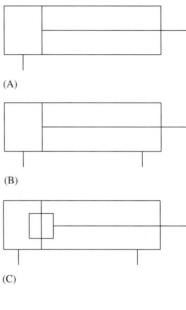

(A)

(B)

(C)

Figure 7–33 Schematic symbols for hydraulic cylinders: (A) single-acting; (B) double-acting; (C) double-acting—fixed cushions both directions; and (D) double-acting—adjustable cushion one direction.

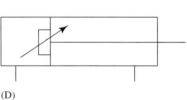

(D)

driven in one direction only and is returned to its original position either by springs or gravity. The blade on a snowplow, for example, is lifted by the action of a pressurized liquid in a single-acting cylinder; when this pressure is released, only the plow's own weight causes the piston to retract and allows the blade to lower. Double-acting cylinders are pressure driven in both directions and are the most common type found on industrial fluid power systems.

Cylinders also come in several variations as shown in Figure 7–34. *Ram-type* cylinders are single-acting units that are generally mounted vertically and are useful where long strokes are required; *tandem* cylinders allow increased output where mounting space is limited; *duplex* cylinders, which may also be constructed with pistons back-to-back, are often used in automated systems to provide three-position operation.

Construction of a typical hydraulic cylinder is shown in Figure 7–35. Pistons are often of cast iron, steel, or aluminum, while the output rod is usually high-strength steel, hardened and plated or polished to resist corrosion and scoring. Notice also the cushions attached to the rod on either side of the piston; these precision parts slide into mating holes in the cylinder end caps to decelerate the piston as it reaches the end of its stroke during both extension and retraction.

Cylinders are a primary source of contamination in hydraulic systems. As the piston rod extends and retracts, it carries a variety of contaminants, including dust and moisture,

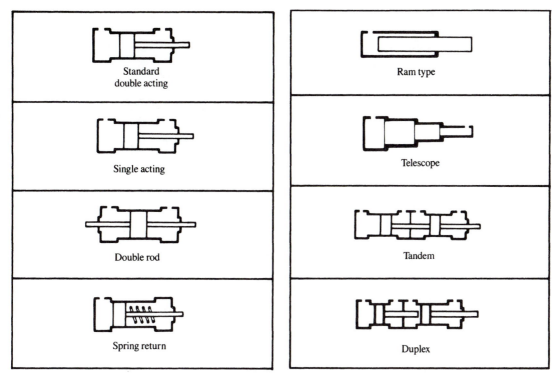

Figure 7–34   Variations in cylinder construction. *(Parker Fluidpower, Cylinder Division, Des Plaines, Ill.)*

into the cylinder. Although most cylinders contain both rod wipers and seals to clean the rod as it retracts, some contamination invariably occurs. Over an extended period of time, this leads to degradation of the seals, scoring of the cylinder, and ultimate failure of the cylinder, hastened by the usual wear that occurs during normal operation of this component. Replacement seal kits are generally available for most cylinders, and scored cylinder walls can be improved by honing if disassembly of the unit is possible.

The cylinder in Figure 7–35 utilizes *tie-rod* construction; square end caps are attached by long threaded rods that extend the length of the cylinder (Figure 7–36). This configuration is common in industry because of the many mounting options available through different styles of end caps and because these removable caps provide easy access to the interior of the cylinder. However, threaded rods and square end caps are not well suited to mobile applications in harsh environments such as those endured by construction and agricultural equipment. Cylinders used under such conditions are often of the *welded end* type or employ retaining rings threaded to the outside of the cylinder at the rod end (Figure 7–37).

Whatever the type of construction, linear actuators are available in specialty forms ranging from subminiature sizes, for use in servo-control systems, to heavy manufacturing assemblies such as the one pictured in Figure 7–38. However, even the largest and most

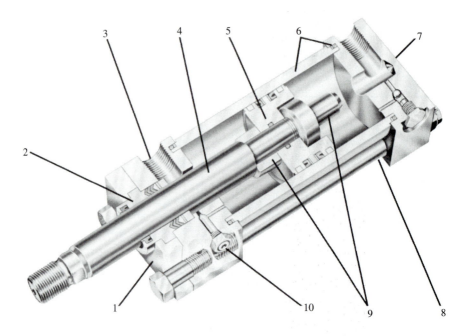

Figure 7–35 Assembly of double-acting hydraulic cylinder—tie-rod construction: 1, retainer plate; 2, rod bearing and seals; 3, inlet/outlet ports; 4, piston rod; 5, piston; 6, cylinder barrel and seals; 7, end caps; 8, tie rods and nuts; 9, cushions; and 10, cushion adjustment. *(Milwaukee Cylinder, Cudahy, Wis.)*

Figure 7–36 Double-acting cylinder—tie-rod construction. *(Hydro-Line Manufacturing Company, Rockford, Ill.)*

Figure 7–37 Double-acting cylinder for mobile applications—threaded retaining ring at rod end. *(Martner Products Mfg. Co., Montgomery, Ill.)*

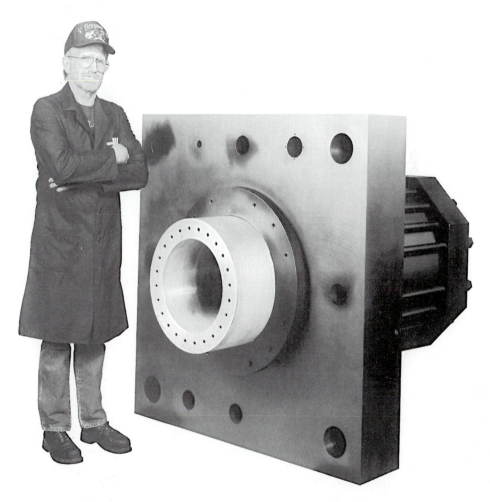

Figure 7–38 Hydraulic cylinders for special applications range from miniature sizes up to units 3 ft in diameter and 36 ft long. The cylinder shown here, for instance, has a 24-in. bore and 20-in. stroke and is used on a machine that produces structural foam. With its 52-in. × 52-in. × 10-in.-thick mounting plate/end cap, the entire assembly weighs 17,000 lb and is capable of generating forces in excess of 1.3 million pounds. *(Hydro-Line, Inc., Rockford, Ill.)*

Figure 7–39   Although linear actuators are capable of generating extremely large axial forces, their thin-walled cylinders are often susceptible to side loads. The damage shown here occurred when a bolt broke during operation of a machine, causing the cylinder and its frame assembly to separate.

powerful of these actuators can suffer extensive (and expensive!) damage from side-loads caused by improper installation or usage (see Figure 7–39).

Continuous rotary motion is obtained from *hydraulic motors*. These devices are similar in construction to the pumps discussed in the previous section, but while pumps convert the mechanical power of a rotating shaft into the fluid power of a pressurized liquid, motors reverse this process by converting fluid power into mechanical form.

Motors may be of various types, including piston, vane, and gear, but small hydraulic motors commonly utilize a gerotor (or generated rotor) of the type shown in Figure 7–40. Similar in operation to an internal gear pump or motor, the star-shaped gerotor rotates within its housing under the influence of a pressurized liquid and transmits its rotary motion to a splined output shaft. Motors of this type offer several advantages: they are quite compact and have good mechanical and volumetric efficiencies, and most allow rotation of the output shaft to be reversed instantly merely by reversing the flow of liquid.

As you might expect, the schematic symbol for a hydraulic motor (Figure 7–41(A)) is very similar to that of a pump. However, since most hydraulic motors essentially act as

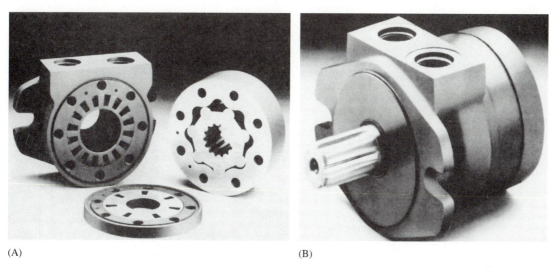

(A)   (B)

Figure 7–40   Gerotor-type hydraulic motor: (A) disassembled and (B) assembled.
(Parker Fluidpower, Nichols Motor Operation, Gray, Maine)

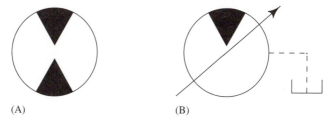

(A)   (B)

Figure 7–41   Schematic symbols for (A) reversible, fixed-delivery hydraulic motor and (B) nonreversible, variable-delivery motor with internal (case) drain.

small enclosed turbines (extracting power from a moving fluid), the primary flow of liquid is *into* the device rather than out of it. For this reason, the solid triangles point toward the center of the symbol instead of away from the center as for a pump. Diagonal arrows may also be used to represent variable displacement units.

In addition, hydraulic motor symbols will frequently have an attached reservoir symbol drawn in dotted or dashed lines (Figure 7–41(B)). This indicates that the device contains an internal or *case drain* for the spent liquid; often this consists of special porting and check valves that prevent the high pressure on the inlet side of the motor from being directly applied to the low-pressure lines on the outlet or drain side of the unit. This potentially damaging situation can easily occur, for example, if direction of the motor is reversed frequently during its normal mode of operation.

Figure 7–42   Schematic symbol for oscillating hydraulic motor.

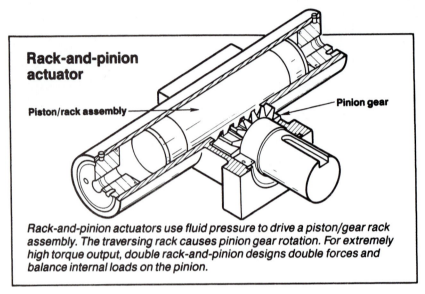

Rack-and-pinion actuator

Piston/rack assembly ——————→

Pinion gear ←————

Rack-and-pinion actuators use fluid pressure to drive a piston/gear rack assembly. The traversing rack causes pinion gear rotation. For extremely high torque output, double rack-and-pinion designs double forces and balance internal loads on the pinion.

Figure 7–43   A rack-and-pinion rotary actuator. *("Machine Design" magazine, Penton Publishing, Inc., Cleveland, Ohio)*

Other devices are also available that allow for oscillating linear or rotary motion as necessary in specific manufacturing operations. Limit switches sense the position of a cylinder's piston or output rod and feed this information back to valves that control the flow of liquid into the cylinder. In this manner, oscillating motion or variations in stroke can be obtained from the cylinder. Similar controls can be used to produce sequential or oscillating motion in hydraulic motors (see Figure 7–42), while rack-and-pinion actuators (Figure 7–43) offer the ability to convert the linear motion of a pressure-driven rack into the rotary motion of a pinion.

In addition, specialty products such as the hydraulic swivel (Figure 7–44) are available to accommodate unusual operating conditions. This particular device, also called a rotary manifold, consists of a grooved spool within a housing and is used to link moving components to stationary supply lines. The spool remains fixed and is connected to one or more supply lines; hydraulic fluid flows in the grooves and out through ports in the housing as it rotates in sequence with devices that are moving relative to the spool.

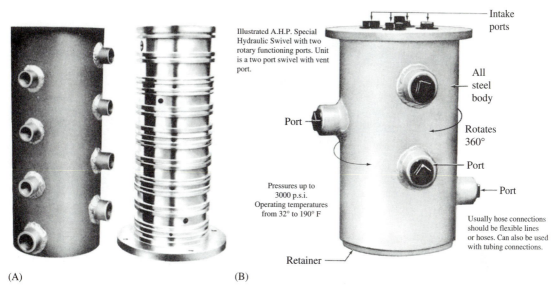

Housing can rotate continously 360° around stationary mounting flange

Illustrated A.H.P. Special
Hydraulic Swivel with two
rotary functioning ports. Unit
is a two port swivel with vent
port.

Intake
ports

All
steel
body

Port

Rotates
360°

Port

Pressures up to
3000 p.s.i.
Operating temperatures
from 32° to 190° F

Port

Usually hose connections
should be flexible lines
or hoses. Can also be used
with tubing connections.

Retainer

(A)   (B)

Figure 7–44   Typical hydraulic swivel: (A) housing and spool for an eight-function swivel; (B) an assembled hydraulic swivel. *(Air & Hydraulic Power, Inc., Wyckoff, N.J.)*

7.6   Valves

Valves are used in fluid power circuits to control the movement of fluid between components. These devices can generally be classified according to their function: *directional control, flow control,* or *pressure control.*

The most common type of directional control valve is the spool valve shown in Figure 7–45. A metal spool inside the valve is machined in a series of lands and grooves that cover various combinations of inlet and outlet ports as the spool slides within the valve body. Positioning of the spool can be accomplished by a variety of manual, electric, or hydraulic/pneumatic actuators; Figure 7–46 shows a single-spool valve activated by hand lever.

Directional control valves are represented schematically by rectangular "envelopes," as shown in Figure 7–47: the number of valve positions is equal to the number of compartments within the envelope. Internal flow paths as determined by spool configuration are shown within compartments for each valve position. The two-way, two-position valve shown in Figure 7–48, for instance, acts essentially as an on-off switch. Notice that this valve has two ports (an inlet and an outlet), and that if it is used with a positive displacement pump some provision must be made for pressure relief during the no-flow condition to avoid damage to the pump. This relief can be provided externally, although many spool valves have internal provisions for releasing excess pressure.

Single-acting cylinders are most commonly controlled by a three-way valve as shown in Figure 7–49. This device has three ports (one inlet from pump, two outlets) and also

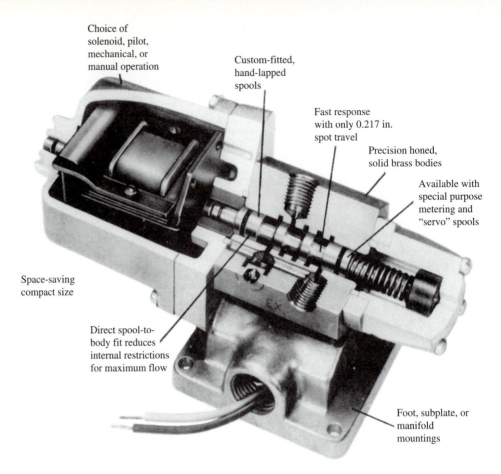

Choice of solenoid, pilot, mechanical, or manual operation

Custom-fitted, hand-lapped spools

Fast response with only 0.217 in. spot travel

Precision honed, solid brass bodies

Available with special purpose metering and "servo" spools

Space-saving compact size

Direct spool-to-body fit reduces internal restrictions for maximum flow

Foot, subplate, or manifold mountings

Figure 7–45 Internal construction of a solenoid-operated spool valve. *(Lexair, Inc., Lexington, Ky.)*

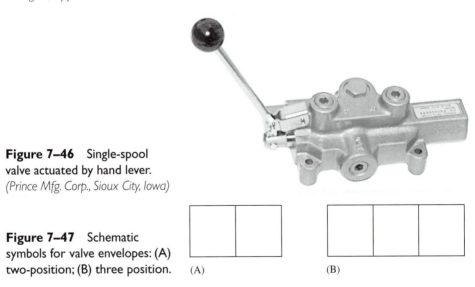

Figure 7–46 Single-spool valve actuated by hand lever. *(Prince Mfg. Corp., Sioux City, Iowa)*

Figure 7–47 Schematic symbols for valve envelopes: (A) two-position; (B) three position.

(A)   (B)

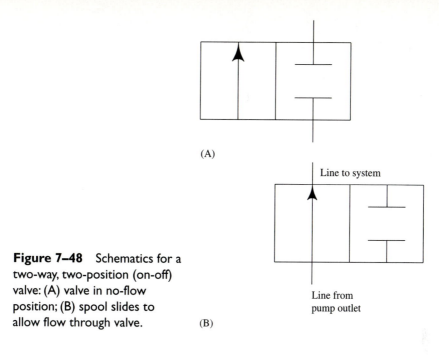

(A)

Line to system

Line from
pump outlet

(B)

Figure 7–48 Schematics for a two-way, two-position (on-off) valve: (A) valve in no-flow position; (B) spool slides to allow flow through valve.

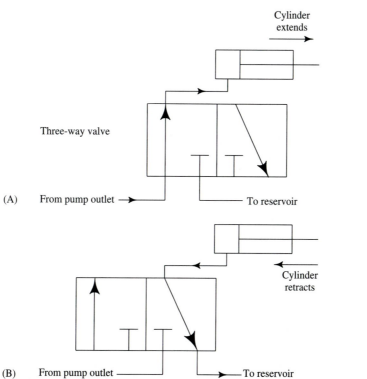

Cylinder
extends

Three-way valve

(A) From pump outlet →   To reservoir

Cylinder
retracts

(B) From pump outlet   To reservoir

Figure 7–49 Schematic representation and use of a three-way, two-position valve: (A) cylinder extends; (B) cylinder retracts.

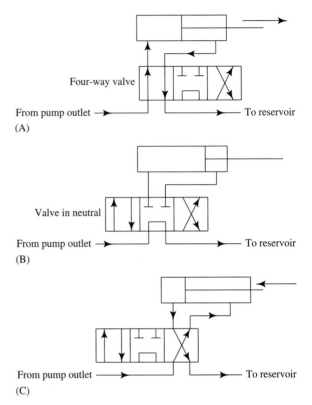

Four-way valve

From pump outlet →   ▶ To reservoir

(A)

Valve in neutral

From pump outlet →   ▶ To reservoir

(B)

Figure 7–50 Schematic representation and use of four-way, three-position valve: (A) double-acting cylinder extends; (B) cylinder locked in position; (C) cylinder retracts.

From pump outlet ———▶   ▶ To reservoir

(C)

requires some form of pressure relief during retraction of the cylinder. Figure 7–50 illustrates the use of a four-way valve to control the motion of a double-acting cylinder. This type of valve has four ports (one inlet from pump, three outlets), and because the unit represented in the figure contains a neutral position it is a three-position valve. Although the neutral, or center, position is available in a variety of configurations, the design shown locks the piston in position and allows free flow from pump to tank. (Before proceeding with this section, you should be able to trace the flow of liquid through three-way and four-way valves in all positions, and correlate this flow with motion of the piston.)

Figure 7–51 presents schematic representations for some of the many types of valve actuators available, while Figure 7–52 illustrates one particular spool valve with various actuators attached.

Metering and control of liquid flow rate can be accomplished through the use of an adjustable valve that represents a variable restriction in the line of flow. These valves typically contain a spherical or tapered element that is manually raised or lowered to enlarge or reduce the flow area. When the element is lowered onto a matching seat, the resulting seal restricts flow completely. Valves of this type include gate, ball, needle, and globe valves, all of which can be indicated by the schematic symbols shown in Figure 7–53.

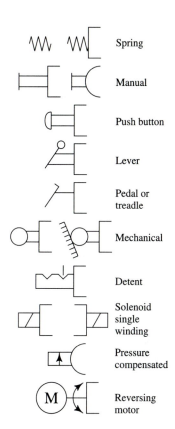

Figure 7–51 Schematic symbols for various types of valve actuators.

Accurate flow regulation can be accomplished using either *needle valves* or *metering valves* (Figure 7–54) of *angle* or *globe* (straight) design, while *gate valves* and *plug valves* generally function as on-off devices. The rotary plug valve (Figure 7–55) contains a cylindrical element or plug that has a through hole drilled across its diameter. When this hole is aligned with the pipe axis, flow occurs through the valve; rotating the plug 90° completely obstructs the flow area. *Pinch valves* (Figure 7–56) contain flexible rubber or elastomer sleeves that are pinched shut by rods mounted on the ends of plungers; these valves are often automated to dispense fixed amounts of fluid, particularly when corrosive liquids or abrasive slurries are involved.

Pressure control valves can serve a variety of functions in fluid power systems. The most obvious use of these devices is to prevent the build-up of excessive pressure, thereby ensuring the *safety* of equipment and personnel. As we shall see in Chapter 8, however, pressure control valves can also be used to achieve certain types of system behavior relating to the application and removal of loads.

The most common pressure control valve is the *relief valve,* whose symbol and assembly are shown in Figure 7–57. These are generally *poppet valves,* consisting of a ball, piston, or uniquely shaped plug element (called a poppet) held on its seat by a fixed or

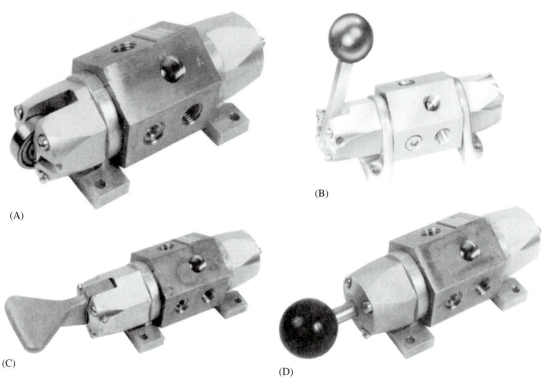

Figure 7–52 Spool valve with different actuators attached. (A) Cam follower. (B) Hand lever. (C) Pedal. (D) Knob. *(Lexair, Inc., Lexington, Ky.)*

Figure 7–53 Schematic symbols for (A) variable restriction such as flow-control, throttling, or metering valve and (B) shutoff valve.

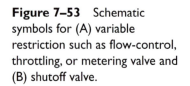

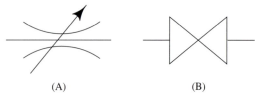

adjustable spring. The relief valve is normally closed, but as system pressure increases, the poppet is lifted off its seat, allowing liquid to flow through the valve to reservoir. *Direct-acting* relief valves (Figure 7–58(A)) that simply lift off a seat are characterized by fast response times and are widely used for protection against intermittent shock overloads. *Pilot-operated* relief valves (Figure 7–58(B)) use a relatively small flow through a sensing or *pilot* line to control the movement of the valve element. (Pilot lines are most commonly represented by dashed lines on schematic symbols for valves.) These pilot-operated devices are able to perform at higher pressures for a given valve size, help to maintain uniform pressures over a variety of operating conditions, and are well suited to continuous relief opera-

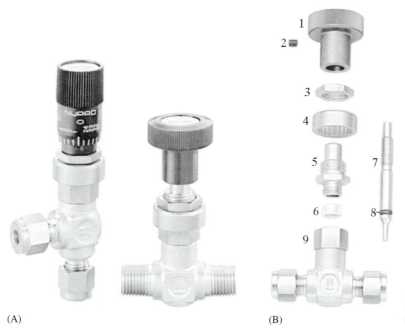

(A)   (B)

Figure 7–54   Appearance and construction of typical metering valve. (A) Metering valves of *angle* and *globe* designs often utilize vernier dials for accurate flow control. (B) Components of globe design include 1, handle; 2, handle screw; 3, panel mount nut; 4, bonnet sleeve; 5, bonnet; 6, guide ring; 7, stem; 8, stem O-ring; and 9, valve body. *(SWAGELOK® Co., Solon, Ohio)*

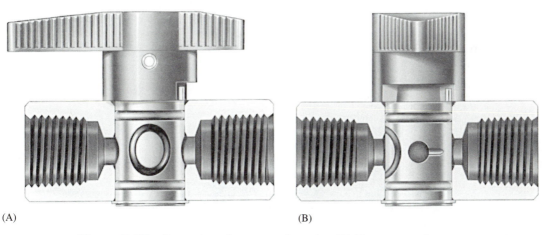

(A)   (B)

Figure 7–55   Operation of a rotary plug valve. (A) Flow occurs (horizontally in figure) when a hole in the plug is aligned with the valve's inlet/outlet. (B) Turning the plug 90° places this hole perpendicular to the inlet/outlet, so flow is blocked. *(SWAGELOK® Co., Solon, Ohio)*

Figure 7–56 Construction of a manually operated pinch valve. The valve's flow passage is lined with a durable rubber or elastomer sleeve to protect it from corrosive liquids and abrasive slurries. On this unit, the handwheel at top is used to raise or lower a vertical metal rod, which pinches the flexible sleeve until the desired flow rate is reached. *(Robbins & Myers, Inc., RKL Controls®, Lumberton, N.J.)*

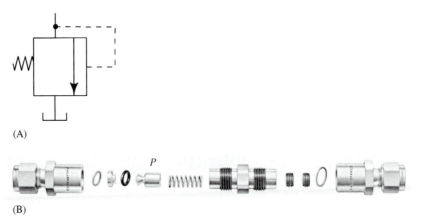

(A)

(B)

Figure 7–57 Pressure relief valve; (A) symbol; (B) typical assembly, element *P* being the poppet. *(SWAGELOK® Co., Solon, Ohio)*

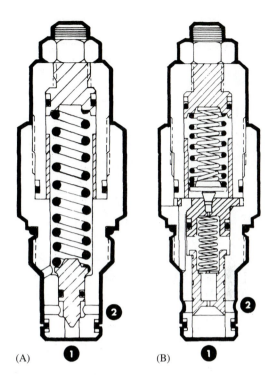

Figure 7–58 Cartridge-type relief valves: (A) direct-acting; (B) pilot-operated. *(Danfoss Fluid Power, a division of Danfoss, Inc., Racine, Wis.)*   (A)   (B)

tion. In the pilot-operated valve of Figure 7–58(B), for example, the piston at the bottom of the valve contains a small longitudinal hole or orifice that allows equal pressures to exist on the top and bottom faces of the piston. As system pressure increases, however, the cone-shaped element in the center of the valve lifts off its seat, causing pressure to drop on the top face of the piston. This pressure *difference* across the piston causes it to slide upward, uncovering port 2 and allowing liquid to flow through the valve. Various relief valve configurations are shown in Figure 7–59.

Several other valves with similar construction and operation characteristics are the sequencing valve, the pressure-reducing valve, and the unloading valve, whose applications we will examine in Chapter 8. The *sequencing valve* (Figure 7–60(A)) acts like the relief valve but sends liquid to a secondary hydraulic circuit rather than back to the reservoir. It provides sequential operation by preventing activity in the secondary circuit until some initial activity has first been completed in the primary circuit. Sequencing valves find wide use in "clamp-and-work" circuits in which the workpiece must first be clamped (primary circuit) before a tool is applied to it (secondary circuit).

The *pressure-reducing valve* (Figure 7–60(B)), which is normally open, senses pressure in the downstream or outlet line. As pressure in this line increases, the valve begins to close, thus slowing the flow of pressurized liquid. This valve is used to maintain relatively

	VALVE TYPE	DUTY CYCLE	CHARACTERISTICS	FUNCTION
	SPRING LOADED BALL DIRECT ACTING	Intermittent	Fast Acting Low Leakage Rates Capacities to 12 gpm (45 l/min) Adj. Pressure Ranges from 500 to 8000 psi (35 to 550 bar)	Thermal or Safety Relief Pilot sections for pilot operated valves
	SPRING LOADED POPPET DIRECT ACTING	Continuous	Dirt Tolerant Reliable Near Zero Leakage Capacities to 60 gpm (227 l/min) Pressure Ranges to 65 psi (4.5 bar)	Low Pressure Relief To insure pilot pressure requirements to other components
	GUIDED POPPET DIRECT ACTING	Continuous	Fast Acting — Reliable Good Pressure vs. Flow Curve Capacities to 25 gpm (95 l/min) Adj. Pressure Ranges to 3000 psi (210 bar)	General applications
	GUIDED PISTON DIRECT ACTING	Continuous	Accurate Fast Responding Quiet Stable Performance Capacities to 40 gpm (151 l/min) Adj. Pressure Ranges from 5–2000 psi (0.3 to 140 bar)	Low Pressure relief
	DIFFEREN-TIAL PISTON DIRECT ACTING	Frequent Intermittent	Good Pressure vs. Flow Curve Long-Life Dirt Tolerant Capacities to 100 gpm (380 l/min) Adj. Pressure to 5000 psi (350 bar)	Shock, surge or overload relief in single or dual relief applications
	PILOT OPERATED SLIDING SPOOL	Continuous	Accurate Under Widely Varying Flow Conditions Remote or Vent Feature Available for Versatility Capacities to 100 gpm (380 l/min) Adj. Pressures from 25–6000 psi (1.7 to 415 bar)	Pressure Control under widely varying flow conditions
	PILOT OPERATED POPPET	Continuous	Versatile, with Remote or Vent Feature Capacities to 100 gpm (380 l/min) Adj. Pressure Ranges from 50–4000 psi (3.5 to 275 bar)	Close pressure control under widely varying flow conditions

Figure 7–59 Relief valve selection guide. *(Danfoss Fluid Power, a division of Danfoss, Inc., Racine, Wis.)*

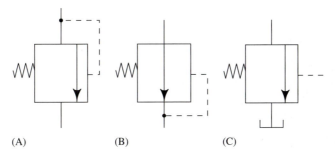

(A)　　　　　　　　　　(B)　　　　　　　　　　(C)

Figure 7–60　Schematic symbols for several types of pressure-control valves with pilot lines represented by dashed lines. (A) For the *sequencing valve,* pressure must first build up to a desired level in the primary circuit (solid line at top of valve) before the valve opens to pressurize secondary circuit (solid line at bottom of valve). (B) The *pressure-reducing valve* compares pressure in the output line (bottom) to a preset value and opens the valve to its no-flow position when output pressure exceeds this reference pressure. (C) Through its pilot line, the *unloading valve* senses pressure at a desired point in the system and opens at a preset value to dump fluid from (unload) a selected portion of the circuit.

constant output pressures for varying input pressures, although output pressures must always be less than those at the valve input.

　　The *unloading valve* (Figure 7–60(C)) decreases pump load during periods of low demand; it is commonly used to bypass the high-volume–low-pressure section of a two-stage pump (see section 4.4) back to reservoir once system pressure has increased to the point at which the low-volume–high-pressure stage becomes dominant. In addition, these valves can also be used together with components known as *accumulators* to reduce loads on a pump during various portions of the work cycle. Accumulators are nothing more than tanks that contain a piston (Figure 7–61) or a bladder/diaphragm (Figure 7–62) that can be loaded by a spring, weight, or compressed gas. Schematic symbols for accumulators are shown in Figure 7–63.

　　The interaction between pump, accumulator, and unloading valve is a straightforward one. As pressure in a system rises, some liquid flows into the accumulator, where its pressure compresses the spring or gas charge or raises the weight. At maximum system pressure, the unloading valve opens and reduces pressure demand on the pump; the system is then fed with pressurized liquid driven from the accumulator by gas, spring, or weight.

　　Use of accumulators can significantly increase pump life, and the "shock absorber" effect of these devices also tends to reduce pulsations in a fluid power system. In addition, accumulators can be used to supplement pump delivery during selected portions of a work cycle. This increase in available short-term flow rates often allows use of a smaller pump and motor to obtain desired system performance.

　　Because most pressure relief valves allow flow in only one direction, they are often referred to as *check valves* or *nonreturn valves,* and do, in fact, perform that flow control

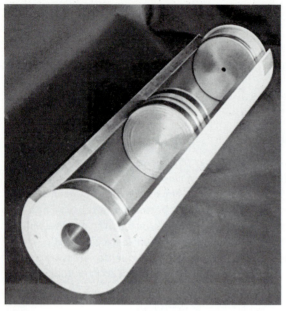

(A)

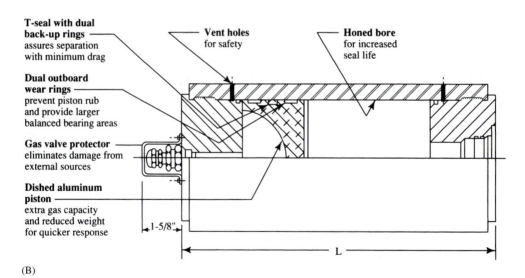

T-seal with dual back-up rings assures separation with minimum drag

Dual outboard wear rings prevent piston rub and provide larger balanced bearing areas

Gas valve protector eliminates damage from external sources

Dished aluminum piston extra gas capacity and reduced weight for quicker response

Vent holes for safety

Honed bore for increased seal life

1-5/8"

L

(B)

Figure 7–61 Piston-type accumulator: (A) cutaway view; (B) internal construction. *(Victor Fluid Power, Granite Falls, Minn.)*

Figure 7–62 Bladder-type
accumulator. *(Wilkes & McLean,
Ltd. (NACOL), Elk Grove Village, Ill.)*

Figure 7–63 Accumulator
symbols: (A) spring-loaded;
(B) compressed gas;
(C) weighted.

(A)

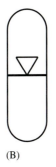

(B)

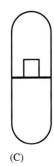

(C)

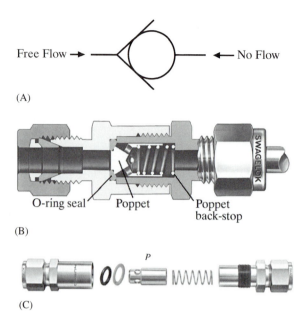

Free Flow →   ← No Flow

(A)

O-ring seal   Poppet   Poppet back-stop

(B)

P

(C)

Figure 7–64   Check valve: (A) schematic symbol; (B) cross-sectional view; (C) disassembled valve. Poppet in (C) has different shape than that shown in (B). *(SWAGELOK ® Co., Solon, Ohio)*

function. In addition, poppets are frequently used as the elements in check valves, and spring-loaded check valves of this type are sometimes used as low pressure relief valves. However, true check valves generally open at system pressures of less than 10 psi and allow virtually unrestricted flow in one direction. Most pressure relief valves found in commercial fluid power systems operate at pressures between 50 psi and 4000 psi and offer substantial resistance to flow. The graphic symbol for a check valve is shown in Figure 7–64.

Although many of the valves described in this section are available in "parts-in-housing" form that allows disassembly for maintenance or repair, use of these components in "cartridge" form, particularly relief and check valves, has grown steadily during the last few years. Cartridge valves are sealed units that generally thread into fittings located at appropriate points in a fluid power system. These disposable devices offer the advantage of easy replacement without disturbing system plumbing.

7.7   Lines, Fittings, and Gauges

All the various components in a fluid power system are connected by hydraulic lines or conductors; on a schematic diagram these lines and their connections are represented as shown in Figure 7–65.

The most common types of hydraulic lines in use today are:

- Steel pipe
- Steel tubing
- Rubber or thermoplastic hose

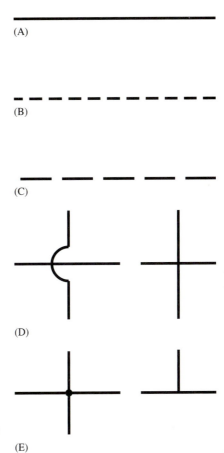

Figure 7–65 Schematic symbols for various hydraulic lines: (A) solid line—primary conducting line; (B) dotted line—exhaust or drain line; (C) dashed line—pilot or control line; (D) lines that cross but do not connect; and (E) lines that connect.

Since each type of line has certain advantages, it is not unusual to find several types used within the same system.

Two conductors that are *not* recommended for use in fluid power systems are galvanized pipe and copper tubing. Galvanized coatings, which often react unfavorably with hydraulic fluids, tend to flake off the internal walls of their conductors, thus offering the potential for system contamination or damage to components. Although copper tubing is highly resistant to corrosion, it tends to promote the formation of sludge and other products of oxidation in petroleum-based liquids such as hydraulic oils.

Steel pipe is most frequently used on permanent installations or those that move high volumes of liquid. A variety of standard pipe fittings (Figure 7–66) is available at any plumbing supply house and can be used to fabricate a system that fits any desired configuration. Since the number of fittings necessary to install a given system is often quite large, leaks at the fittings are common. Although joint compound can be applied to the tapered pipe threads to reduce leakage, the use of sealing tape within the joints is not recommended. Exposed to the flowing liquid, this tape eventually tatters or shreds and enters the system.

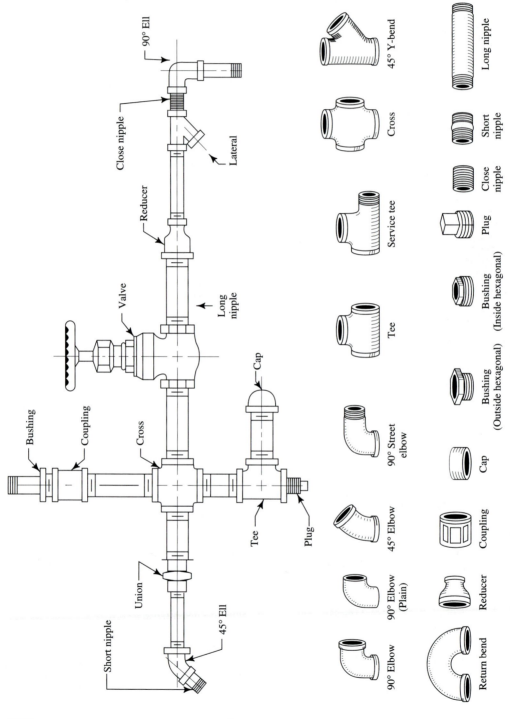

Figure 7-66 Pipe fittings and their applications.

Table 7–3 Suggested Allowable Pressure for Carbon Steel Tubing
(SWAGELOK® Co., Solon, Ohio)

Soft annealed carbon steel hydraulic tubing ASTM A179 or equivalent. Based on ultimate tensile strength 47,000 psi (323,800 kPa). For metal temperatures –20° to 100°F (–29° to 37°C). Allowable working pressure loads calculated from S values (15,700 psi-108,200 kPa) as specified by ANSI B31.3 code.

TUBE O.D. (IN.)	TUBE WALL THICKNESS (INCHES)												
	.028	.035	.049	.065	.083	.095	.109	.120	.134	.148	.165	.180	.220
1/8	8000	10,200						WORKING PRESSURE (PSIG)					
3/16	5100	6600	9600										
1/4	3700	4800	7000	9600									
5/16		3700	5500	7500						NOTE: For tubing for gas service use only tube wall thickness on outside of screened area.			
3/8		3100	4500	6200									
1/2		2300	3200	4500	5900								
5/8		1800	2600	3500	4600	5300							
3/4			2100	2900	3700	4300	5100						
7/8			1800	2400	3200	3700	4300						
1			1500	2100	2700	3200	3700	4100					
1 1/4				1600	2100	2500	2900	3200	3600	4000	4600	5000	
1 1/2					1800	2000	2400	2600	2900	3300	3700	4100	5100
2						1500	1700	1900	2100	2400	2700	3000	3700

Table 7–4 Suggested Allowable Pressure for Stainless Steel Tubing
(SWAGELOK® Co., Solon, Ohio)

Annealed 304 or 316 stainless steel tubing ASTM A269 or equivalent. Based on ultimate tensile strength 75,000 psi (516,700 kPa). For metal temperature from –20° to 100°F (–29° to 37°C). Allowable working pressure loads calculated from S values (20,000 psi 137,800 kPa) as specified by ANSI B31.3 code.

For Seamless Tubing

NOTE: For welded and drawn tubing, a derating factor must be applied for weld integrity: for *double welded* tubing multiply pressure rating by .85 — for *single welded* tubing multiply pressure rating by .80.

TUBE O.D. (IN.)	TUBE WALL THICKNESS (INCHES)															
	.010	.012	.014	.016	.020	.028	.035	.049	.065	.083	.095	.109	.120	.134	.156	.188
1/16	5600	6800	8100	9400	12,000			WORKING PRESSURE (PSIG)								
1/8						8500	10,900									
3/16						5400	7000	10,200								
1/4						4000	5100	7500	10,200							
5/16							4000	5800	8000		NOTE: For tubing for gas service use only tube wall thickness on outside of screened area.					
3/8							3300	4800	6500							
1/2						2400	3500	4700	6200							
5/8							2900	4000	5200	6000						
3/4							2400	3300	4200	4900	5800					
7/8							2000	2800	3600	4200	4800					
1								2400	3100	3600	4200	4700				
1 1/4									2400	2800	3300	3600	4100	4900		
1 1/2										2300	2700	3000	3400	4000	4900	
2											2000	2200	2500	2900	3600	

Steel tubing has several advantages over pipe. Threading reduces the effective wall thickness of pipe, so tubing is considerably lighter than pipe of equivalent flow area and burst pressure. Table 7–3 and Table 7–4 list the suggested allowable pressures for carbon steel and stainless steel tubing respectively. (Because surface abrasion and scratches can reduce sealing ability for tubing, particularly when the working fluid is a gas, only wall thicknesses in the unshaded areas are recommended for gas service.) In smaller diameters,

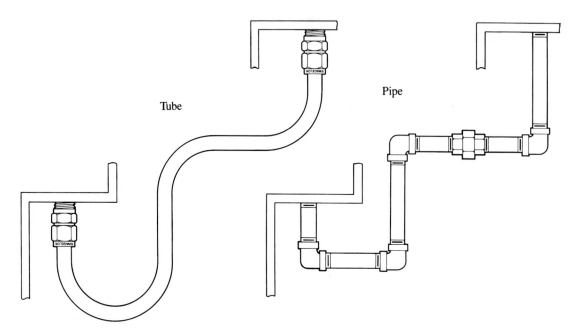

Figure 7–67   Fewer fittings can reduce potential leaks and flow-induced pressure drops. *(SWAGELOK® Co., Solon, Ohio)*

tubing is easily formed, so that the number of fittings required to plumb a given system may be reduced considerably (Figure 7–67).

A wide variety of fittings is available for connecting tubing-to-tubing as well as tubing-to-pipe and tubing-to-hose (Figure 7–68). One type of fitting called a *compression fitting* is shown in Figure 7–69. Since many of these systems of fittings are incompatible, mixed use of ferrules, nuts, and fitting bodies from different manufacturers is not recommended.

Flexible hose offers the ultimate in flexibility and ease of installation. It finds widespread use in systems that are subject to excessive vibration, contain components that undergo significant linear/rotary displacements during use, or include fluid lines that must frequently be connected and disconnected. The symbol for a flexible line is shown in Figure 7–70.

Hoses are typically manufactured of rubber or thermoplastic and contain one or more layers of reinforcing material (Figure 7–71). Suction hose, such as that used on the inlet of a pump, generally contains additional reinforcement to prevent hose collapse that could starve the pump of liquid.

Fittings used on hoses are designated according to the angle of their mating surfaces. Most popular are the NPTF (National Pipe Thread Fitting) at 30°, the JIC 37 (Joint Industry

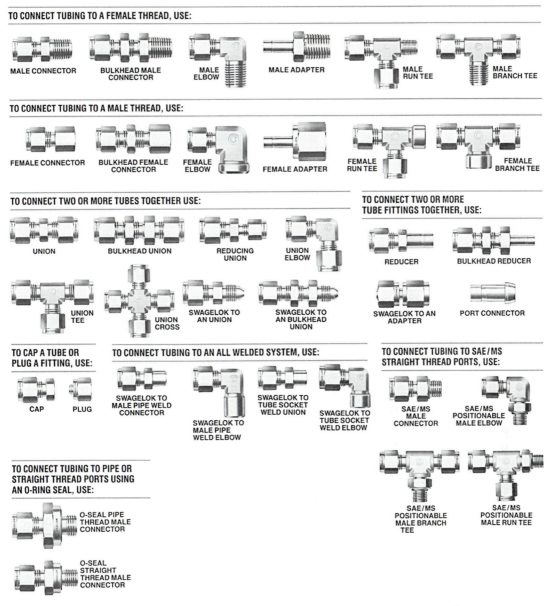

Figure 7–68 Some of the many tube fittings available. *(SWAGELOK® Co., Solon, Ohio)*

SWAGELOK
front
ferrule

SWAGELOK
body

SWAGELOK
back
ferrule

SWAGELOK
nut

(A)

Figure 7–69 Compression fitting. (A) Tubing is inserted through the nut into a loosely assembled fitting whose components are shown above. This fitting is then tightened to produce the leak-free joint pictured in (B). *(SWAGELOK® Co., Solon, Ohio)*

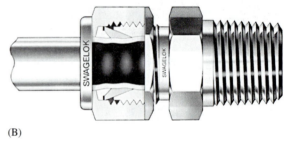

(B)

Figure 7–70 Schematic symbol for a flexible line.

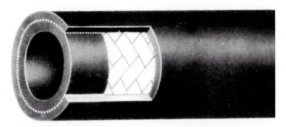

MSHA ACCEPTED (USMSHA NO. 1C-2G-31C/2)

Recommended for: Applications which exceed SAE 100R1 working Pressures. But do not require a two wire braid.
Tube: Black, oil resistant, nitrile (NBR).
Reinforcement: One braid of higher tensile steel wire.
Cover: Black, oil and abrasion resistant neoprene (CR).
Temperature range: – 40 F to +200 F.

MSHA ACCEPTED (USMSHA NO. 1C-2G-31C/2)

Recommended for: High pressure hydraulic oil lines, exceeding requirements of SAE 100R2. Smaller O.D. provides greater flexibility in service.
Tube: Black, oil resistant, nitrile (NBR).
Reinforcement: Two braids of higher tensile steel wire.
Cover: Black, oil and abrasion resistant neoprene (CR).
Temperature range: – 40 F to +200 F.

MSHA ACCEPTED (REF. USMSHA NO. IC-2G-31C)

Recommended for: Petroleum and water based hydraulic fluids, in low pressure and vacuum service. Especially recommended for return lines in hydraulic control systems. Meets or exceeds requirements of SAE 1004R4.
Tube: Black, oil resistant, Nitrile (NBR).
Reinforcement: Two high tensile textile braids, pre-treated for tube and cover adhesion. Helix wires are located between the braids to resist collapse under vacuum or bending.
Cover: Black, oil and weather resistant Neoprene (CR), wrapped, all sizes.
Temperature range: -40°F to +200°F.

Recommended For: Hydraulic oil lines. Medium pressure. Meets or exceeds requirements of SAE 100R7.
Tube: Polyester elastomer.
Reinforcement: Polyester braid.
Cover: Black, wear resistant urethane.
Temperature Range: – 40°F. to +200°F.

Figure 7–71   Construction of hydraulic hose. *(NRP-Jones, Inc., LaPorte, Ind.)*

Conference) at 37°, and the SAE 45 (Society of Automotive Engineers) at 45°, all of which seal by metal to metal contact (Figure 7–72). Other fittings include *quick-disconnect fittings* used when rapid or frequent connections are required (Figure 7–73).

A variety of methods can be used to attach fittings to hose. For high-pressure applications, fittings are permanently crimped onto the hose ends (Figure 7–74) using a hydraulic or pneumatic crimping machine. Similar results can be obtained with reusable fittings, called *skive type* fittings, which are hand mounted as shown in Figure 7–75.

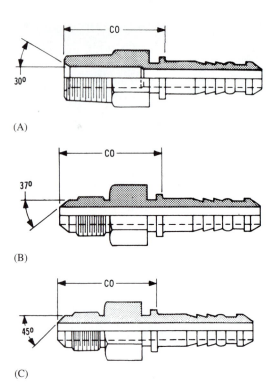

Figure 7–72 Common hose fittings: (A) NPTF solid male stem; (B) JIC 37° solid male stem; and (C) SAE 45° solid male stem. *(NRP-Jones, Inc., LaPorte, Ind.)*

These connectors require that the outer layer of hose be removed before the fitting is attached. Lines subject to low pressure, including plastic tubing, generally use *push-on* fittings (Figure 7–76), which are simply pushed onto the hose end and secured with a hose clamp.

Once a fluid power system has been assembled and becomes operational, its health or well-being is normally measured by line pressures taken at various locations in the hydraulic circuit. Any gauge used to measure pressure can be represented by the symbol of Figure 7–77(A); if a gauge connection exists but is not in use, its availability can be indicated by the symbol shown in Figure 7–77(B).

Most mechanical pressure gauges are of the Bourdon type, named after Eugene Bourdon, the French engineer who developed this instrument in 1849. The gauge movement centers around a hollow tube that is sealed at one end and bent into a circular shape (Figure 7–78). Attached to a source of pressurized liquid, this tube tends to straighten; its movement is transmitted through a linkage and gear to the needle indicator as shown in Figure 7–79. For applications where significant vibration occurs or fluctuations in pressure are frequent and rapid, a liquid-filled gauge is often used (Figure 7–80). The sealed housing of the gauge is filled with a clear liquid, usually glycerin, which acts as a shock absorber

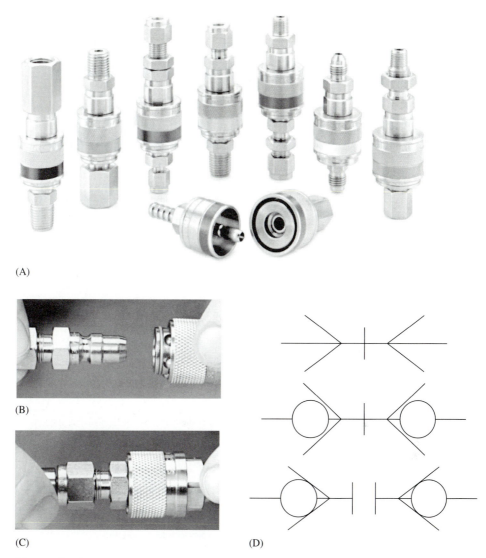

Figure 7–73   Quick disconnect fittings: (A) appearance, (B) and (C) operation, and (D) schematics. To couple fittings, as shown in (B), align stem with body and *slide* spring-loaded knurled sleeve back. Then, as shown in (C), insert stem into body until it bottoms. Return sleeve to its original position, making sure sleeve is completely forward. Schematics in (D) are: 1, connected—no checks; 2, connected—checks both sides; and 3, disconnected—checks both sides. *((A), (B), and (C) courtesy SWAGELOK®* *Co., Solon, Ohio)*

(A)

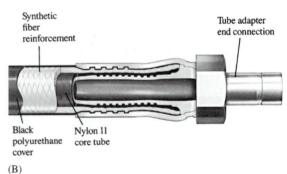

Synthetic
fiber
reinforcement

Tube adapter
end connection

Figure 7–74 Crimped hose
fittings (A) and assembly (B).
(SWAGELOK® Co., Solon, Ohio)

Black
polyurethane
cover

Nylon 11
core tube

(B)

by preventing rapid motion of the needle and possible damage to the gauge movement itself. Gauges can be either stem mounted or rear mounted (Figure 7–81) to the source of pressurized liquid.

Mechanical gauges have several advantages. They are easy to install and maintain, require no peripheral equipment, perform reliably in harsh environments, and are surprisingly economical. On the other hand, a system operator must physically visit the gauge to make an on-the-spot reading. In addition, mechanical dial gauges are not noted for being highly

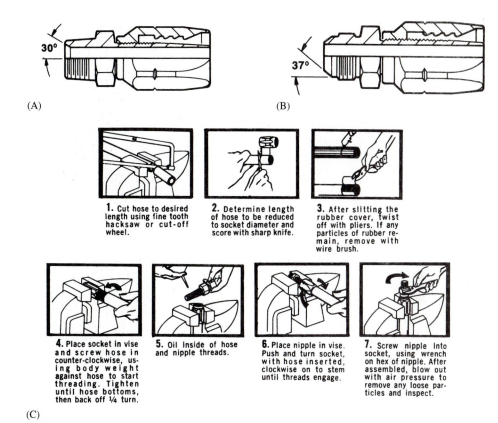

(A)

(B)

1. Cut hose to desired length using fine tooth hacksaw or cut-off wheel.

2. Determine length of hose to be reduced to socket diameter and score with sharp knife.

3. After slitting the rubber cover, twist off with pliers. If any particles of rubber remain, remove with wire brush.

4. Place socket in vise and screw hose in counter-clockwise, using body weight against hose to start threading. Tighten until hose bottoms, then back off ¼ turn.

5. Oil inside of hose and nipple threads.

6. Place nipple in vise. Push and turn socket, with hose inserted, clockwise on to stem until threads engage.

7. Screw nipple into socket, using wrench on hex of nipple. After assembled, blow out with air pressure to remove any loose particles and inspect.

(C)

Figure 7–75 Reusable skive type fittings and method of assembly. (A) NPTF solid male reusable skive type. (B) JIC 37° solid male reusable skive type. (C) Assembly. *(NRP-Jones, Inc., LaPorte, Ind.)*

Figure 7–76 Push-on type tubing fittings. *(SWAGELOK® Co., Solon, Ohio)*

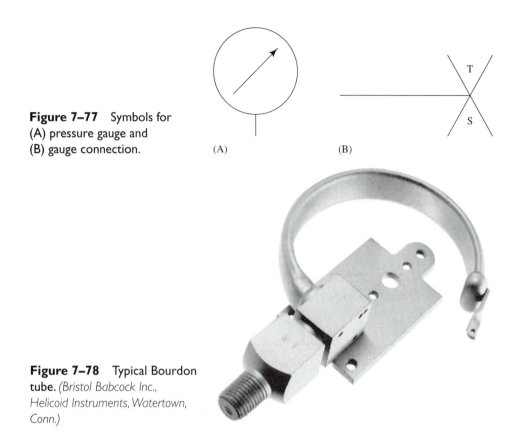

Figure 7–77 Symbols for
(A) pressure gauge and
(B) gauge connection.

(A)   (B)

Figure 7–78 Typical Bourdon
tube. *(Bristol Babcock Inc.,
Helicoid Instruments, Watertown,
Conn.)*

accurate, and as analog devices are often misread. Finally, mechanical gauges do not provide a record of the pressures that they measure and are not easily integrated into process control systems where pressure-related sensing and actuating functions might be required.

To increase the accuracy and utility of measured pressures, a wide variety of electrical and electronic pressure sensors have been developed. Those instruments that produce an output *voltage* proportional to applied pressure are designated as *pressure transducers;* devices whose output *current* is proportional to the applied pressure are known as *pressure transmitters.*

The pressure-sensing elements themselves operate on a variety of principles. *Diaphragm type* sensors contain a metal or foil diaphragm that deflects under an applied pressure. This deflection may be measured optically or electrically. If the diaphragm forms one side of a capacitor as shown in Figure 7–82(A), any deflection causes a change in capacitance that can be detected electrically. If a resistance-type strain gauge is bonded to the diaphragm, any deflections produce corresponding changes in resistance that can be measured using simple Wheatstone bridge circuitry. Sensors that operate on this latter principle are shown in Figure 7–83.

Other types of sensors include *linear variable differential transformers* (LVDTs), in which a metal rod moving under pressure within a wound coil produces a measurable change in the coil's inductance. Piezoelectric sensors (Figure 7–84) contain a quartz or tourmaline crystal that generates an electrical charge when the element is acted upon by an

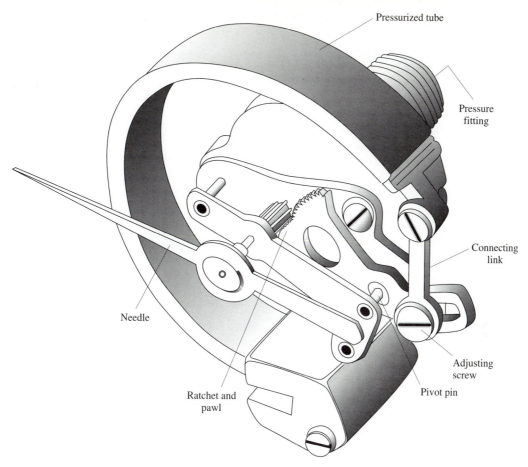

Figure 7–79 Bourdon gauge movement. When pressurized, the bent tube straightens slightly. This movement is carried through the connecting link to a ratchet and pawl assembly, whose rotary output causes the needle to pivot in an arc across the dial face (not shown).

Figure 7–80 Glycerin-filled gauge—stem mount. Glycerin, a clear liquid similar to vegetable oil, fills the gauge and acts as a shock absorber, cushioning the needle and gauge movement from abrupt changes in pressure. Because such gauges must be completely sealed to prevent leakage of the glycerin, they are considerably more expensive than a comparable dry gauge (see Figure 7–81). *(Voss, Inc., Valley View, Ohio)*

Figure 7–81 Dry gauge—rear mount. This is the standard analog pressure gauge used in industry today. Durable, inexpensive, and reliable, it is well adapted to applications where pressures remain fairly constant or change slowly over a narrow range. The biggest disadvantages to this mechanical gauge are that its readings must be observed and recorded manually. *(Voss, Inc., Valley View, Ohio)*

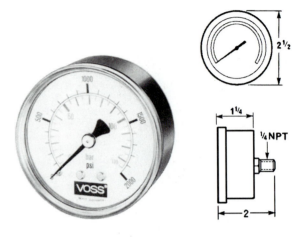

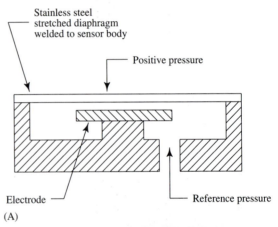

Stainless steel stretched diaphragm welded to sensor body

Positive pressure

Electrode

Reference pressure

(A)

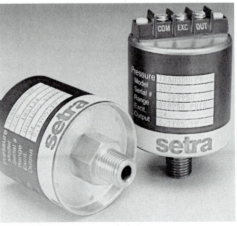

Figure 7–82 Capacitance-sensitive diaphragm gauges. (A) Together, the steel diaphragm and electrode form the plates of a capacitor. When positive pressure deflects this diaphragm, the air gap (and capacitance) change by a measurable amount, which may be converted directly into pressure. (B) Typical units. *(Setra Systems, Inc., Acton, Mass.)*

(B)

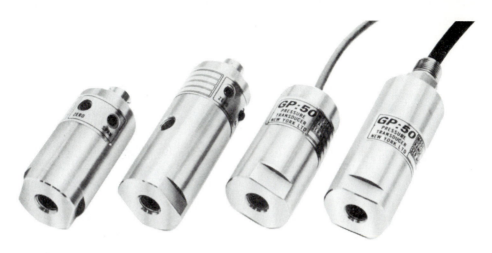

Figure 7–83 Sensors with strain gauge bonded diaphragms. Similar to capacitance gauges, such units each have a small rosette of fine wire (known as a strain gauge) bonded to their diaphragms. Under pressure, the diaphragm deflects and then stretches the wire, changing its resistance (rather than capacitance) by a measurable amount. *(GP:50 New York Ltd., Grand Island, N.Y.)*

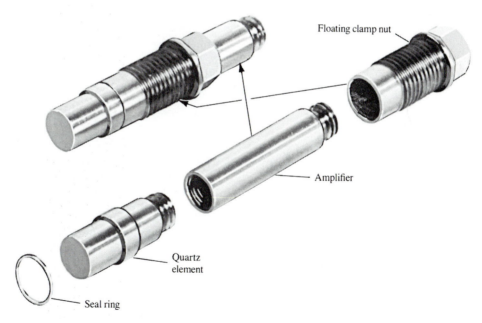

Figure 7–84 Complete quartz sensor assembly. When force (or pressure) is applied to a quartz crystal, the mineral reacts by generating a small voltage, which may be amplified and measured. These units are well suited to high-pressure environments and rapidly fluctuating dynamic pressures. *(PCB Piezotronics, Inc., Depew, N.Y.)*

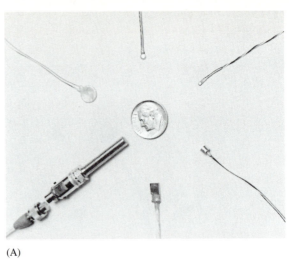

(A)

Figure 7–85 Subminiature pressure sensors. *((A) Precision Measurement Company, Ann Arbor, Mich. (B) Reproduced with permission of OMEGA Engineering, Inc., Stamford, Conn.)*

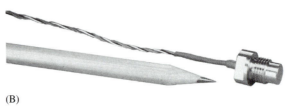

(B)

Figure 7–86 Digital pressure gauge. Not only are digital gauges easier to read than their analog (dial) counterparts, the electrical readings can often be carried over significant distances or to remote locations for ease in monitoring and recording pressure. *(HBM, Inc., Marlboro, Mass.)*

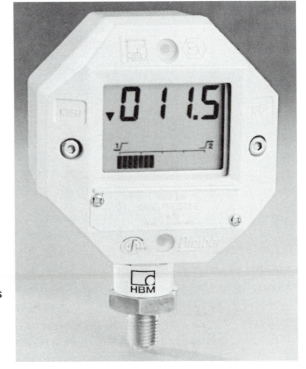

270

Figure 7–87   Digital pressure indicator. *(Setra Systems, Inc., Acton, Mass.)*

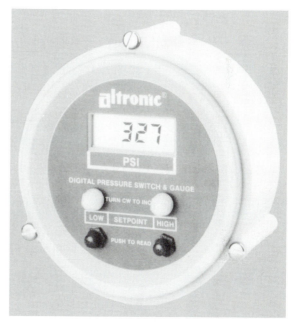

Figure 7–88   Digital pressure switch and gauge. *(Altronic Inc., Girard, Ohio)*

external pressure. These quartz sensors are particularly suited for the measurement of both dynamic and elevated pressures. Many of the sensors described above are also obtainable in subminiature form (Figure 7–85).

Sensors can be combined with digital indicators in a common housing (Figure 7–86) or sensor output may be fed to one or more indicators at remote locations (Figure 7–87). Instruments that integrate sensing, display, and some control functions are also available (Figure 7–88), as well as data acquisition systems to monitor and record pressures, store performance data in the memory bank of a computer, or actuate some form of process control.

Figure 7–89 presents a summary of schematic symbols for the most common types of hydraulic components.

Lines, fittings, and gauges

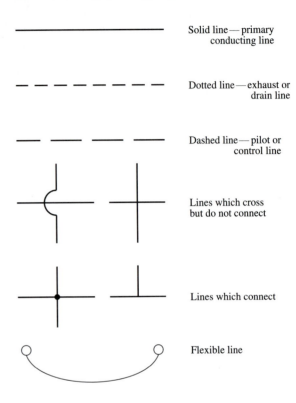

Solid line—primary conducting line

Dotted line—exhaust or drain line

Dashed line—pilot or control line

Lines which cross but do not connect

Lines which connect

Flexible line

Figure 7–89 Summary of schematic symbols.

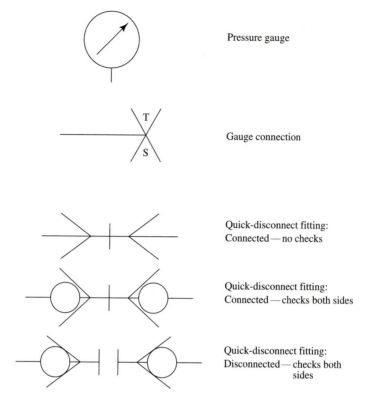

Pressure gauge

Gauge connection

Quick-disconnect fitting:
Connected — no checks

Quick-disconnect fitting:
Connected — checks both sides

Quick-disconnect fitting:
Disconnected — checks both
sides

Figure 7–89 (continued)

Reservoirs, filters, and strainers

Reservoir — vented tank, line enters tank below liquid level

Reservoir — vented tank, line enters tank above liquid level

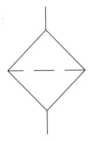

Strainer or filter

Pumps and motors

Pump — fixed capacity, nonreversible

Pump — fixed capacity, reversible

Pump — variable capacity, nonreversible

Pump — variable capacity, reversible

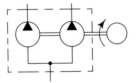

Pump — two-stage

Motor — reversible, fixed delivery

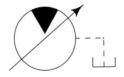

Motor — nonreversible, variable delivery, internal drain

Motor — oscillating

Figure 7–89 (continued)

274

Cylinders

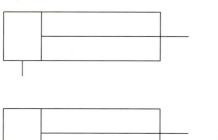 Single acting

 Double acting

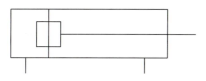 Double acting — fixed cushions both directions

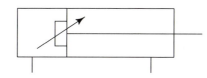 Double acting — adjustable cushion one direction

Directional control valves

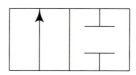 2-way, 2-position

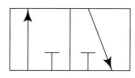 3-way, 2-position

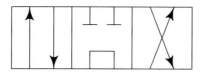 4-way, 3-position

Valve actuators

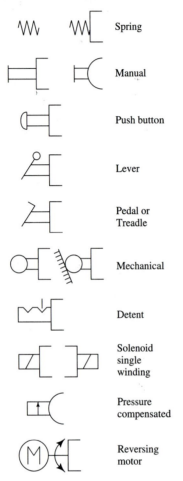

Spring

Manual

Push button

Lever

Pedal or
Treadle

Mechanical

Detent

Solenoid
single
winding

Pressure
compensated

Reversing
motor

Figure 7–89   (continued)

Flow control valves

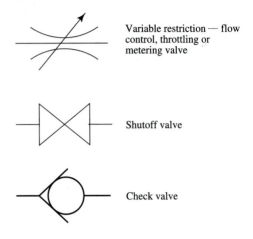

Variable restriction — flow
control, throttling or
metering valve

Shutoff valve

Check valve

276

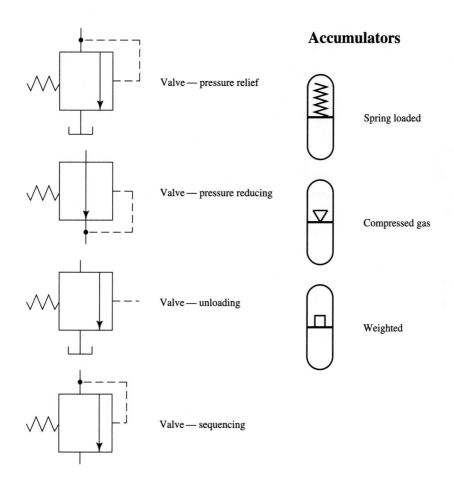

Valve — pressure relief

Valve — pressure reducing

Valve — unloading

Valve — sequencing

Accumulators

Spring loaded

Compressed gas

Weighted

SAFETY SIDEBAR

Avoid Costly and Dangerous Leaks

Leakage of pressurized fluid caused by improper use of tubing, lines, and fittings can result in a threat to the safety of nearby personnel. Always follow these guidelines:

1. Do not loosen or tighten fittings of a pressurized system.
2. Avoid mixing or combining ferrules, nuts, and fitting bodies from different manufacturers.
3. Always use tubing that is softer than the fitting material; never use stainless steel tubing with brass fittings.
4. Avoid scratches, cuts, dents, or gouges on tubing surfaces. Tubing should not be dragged out of a tubing rack or across concrete, asphalt, gravel, or other surfaces that could scratch it.
5. Route tubing and hose carefully to avoid kinking, abrasion, cutting, or excessive temperatures.
6. Never pull on tubing or hose to move portable components from one location to another.

Questions

1. List the three primary functions of a hydraulic reservoir.
2. What is the rule of thumb for determining reservoir capacity?
3. Occasionally, hydraulic reservoirs are sealed rather than vented. Such units generally contain a pressurized gas above the surface of the liquid instead of air at atmospheric pressure. How might a reservoir of this type affect component selection or system operation? Are there advantages or disadvantages in using such a reservoir?
4. What is the approximate size of the smallest particle that is visible to the naked eye?
   (a) 0.20 in.
   (b) 0.02 in.
   (c) 0.002 in.
   (d) 0.0002 in.

   Is this limit of visibility higher or lower than the clearances in most fluid power components?
5. Why are filters not used on the suction lines of pumps?
6. What type of filter is used in the lubrication system of an automobile? In the fuel system (both air and gasoline)?
7. Draw the schematic symbol for a variable-capacity, nonreversible pump.
8. List the advantages and disadvantages of centrifugal pumps.
9. Name the two most common types of positive displacement pumps found in fluid power systems.

10. At rest, a normal adult heart moves approximately 2100 gal of blood per day. What is this flow rate in gpm? (During strenuous activity, the heart may have a flow rate five to six times this amount.) Can you classify the heart as a pump in terms of its delivery (fixed/variable capacity, reversible/nonreversible) and displacement (positive/nonpositive)?

11. In an axial piston pump, would decreasing the angle between swash plate and pump shaft produce an increase or decrease in the flow rate? Explain.

12. On most dump trucks, the dump body is raised by a long-stroke hydraulic cylinder. In northern climates, when inclement weather is expected (especially a snowstorm) truck operators will often pressurize the cylinder and raise the dump body to its maximum height, leaving it in this position overnight. The practice is designed to reduce the amount of precipitation that might accumulate in the dump body during a storm. Can you see any disadvantages or dangers to this procedure?

13. Name the three operational functions used to classify valves found in fluid power systems.

14. Draw the schematic symbols for these directional control valves:
    (a) Two-way, two-position.
    (b) Three-way, two-position.
    (c) Four-way, two-position.

15. What is the difference between a check valve and a pressure relief valve?

16. When swimming just below the surface of a body of water, snorkelers are able to breath using an inverted U-tube such as the one shown in Figure 7–90. This device, known as a snorkel tube, contains a spherical float, which is lighter than water. How does the snorkel work? Can you see any similarities between its operation and that of a simple check valve?

17. List the three most common types of hydraulic lines.

18. Why are copper tubing and galvanized pipe not recommended for use in fluid power systems?

19. Describe the operation of a Bourdon gauge.

20. When should glycerin-filled gauges be used?

21. List several types of electrical pressure sensors. Describe their principles of operation, advantages, and disadvantages.

22. Based upon your own experience with clocks and automobile speedometers, what are some of the comparative advantages and disadvantages between analog and digital displays? Is there any reason why "idiot lights" rather than real gauges are used to indicate low oil pressure in automobiles?

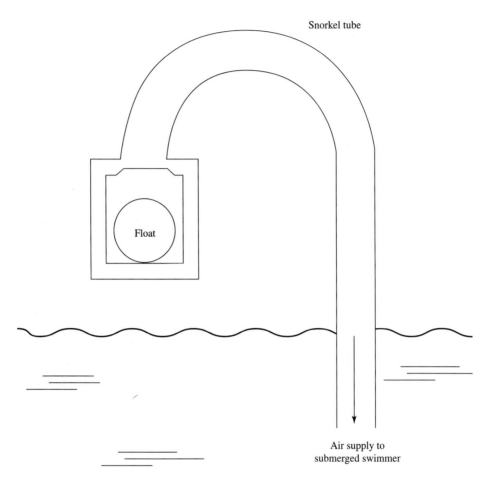

Figure 7–90   In some ways, operation of a snorkel tube resembles that of a simple check valve. (See question 16.)

8

A Few Basic Circuits

Now that we are somewhat familiar with various hydraulic components and their schematic representations, let us see how this hardware may be assembled to form several common circuits. The systems presented here are in their simplest forms and most are manually operated; complex circuits are often obtained by combining several basic circuits and, as we shall see in Chapter 11, virtually all can be adapted to automatic control.

Besides increasing your ability to correlate real components with their graphic symbols, this chapter should improve your understanding of the behavior and use of individual circuit elements and help you to visualize the operation of an entire fluid power system.

8.1   A Single-Piston System

In Chapter 4 we determined the size and power requirements for a device that used several double-acting pistons. Similar systems built around a single piston and designed to extend or retract on command may well be the most common hydraulic circuits in use today. A wide variety of agricultural, construction, and industrial machinery employs such systems to lift, position, clamp, compress, punch, and shear both materials and finished products.

The schematic diagram for a typical double-acting circuit is shown in Figure 8–1. Liquid from the reservoir (A) is drawn through a tank-mounted strainer (B) by a fixed-capacity, nonreversible pump (C). The fluid then passes through a lever-operated three-position, four-way directional valve (D) and on to the double-acting cylinder (E). Liquid leaving the cylinder flows back through the valve, passing through filter (F) on its return to the reservoir. (The various valve positions and flow modes for extension and retraction of the cylinder were presented in Figure 7–50.) If the valve itself does not contain provisions for internal pressure relief, then an external pressure relief valve (G) must be included in the circuit. This valve prevents the damaging effects of excess pressure that can develop in the system if the flow path is blocked by a "jammed," or overloaded, piston. (Blockage can occur, for example, if an attempt is made to lift some weight that is heavier than the system capability, or if a valve is held in the extend/retract position by an inattentive operator even though the cylinder is already at an extreme position.) For safety reasons, most manually operated directional valves are of the self-centering, or "dead man's," variety, meaning that they are spring-loaded to return to the neutral position if the operator's hand leaves the actuating lever, knob, or button.

Figure 8–2 contains several views of a mobile log splitter whose hydraulic circuit follows the above configuration and has pressure relief provided within the directional control valve. From Figure 8–2(B), can you correlate the actual circuitry of this device with its schematic diagram?

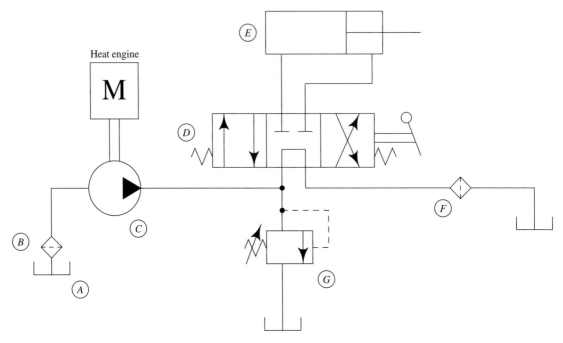

Figure 8–1 Typical circuit for manually operated, double-acting cylinder. Fixed-capacity, nonreversible pump *C* draws liquid from reservoir *A* through strainer *B*. Fluid flows through the manually operated, three-position, four-way valve *D* to activate double-acting cylinder *E* and then returns to the reservoir through filter *F*. Maximum system pressure is limited by adjustable pressure relief valve *G*.

(A)   (B)

Figure 8–2 This mobile log splitter utilizes a hydraulic circuit exactly like the one shown schematically in Figure 8–1: (A) rear view; (B) close-up from top front.

282

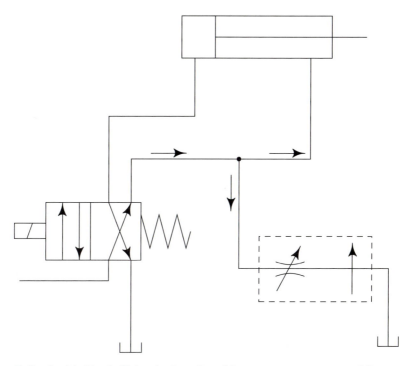

Figure 8–3   In this *bleed-off* circuit, the adjustable, pressure-compensated flow control valve shown within the dashed box controls cylinder speed on retraction by diverting a portion of the flow back to the reservoir. It presents an alternative path for the liquid and is said to be in *parallel* with the actuator.

It is often desirable to control the *speed* of a hydraulic cylinder during its work cycle. One of the simplest ways to do this is to install an adjustable flow control valve between the cylinder inlet line and reservoir (Figure 8–3). This configuration is known as a *bleed-off* circuit and operates by diverting a portion of the pump flow directly back to the reservoir at the system's normal operating pressure. Obviously, the amount of liquid bypassed to reservoir affects the amount of fluid delivered to the cylinder, thus controlling cylinder speed. Because the valve is in *parallel* with the cylinder, closing the valve reduces flow area and increases cylinder speed, while opening the valve increases flow area and decreases cylinder speed.

Bleed-off circuitry can be applied to a cylinder's extension or retraction feed lines (or both), but generally does not provide accurate speed control over the entire operating range of most fluid power systems. Additional control can be obtained if the flow control valves are *pressure-compensated* to provide uniform bypass flow despite variations in the system's

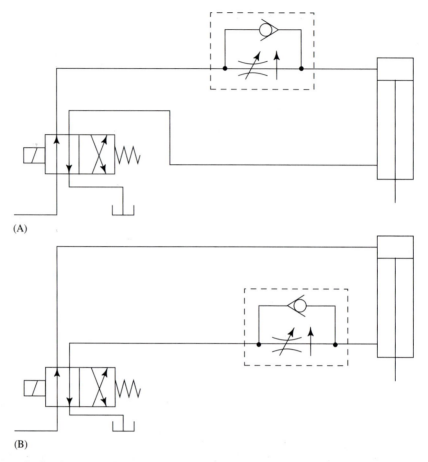

Figure 8–4   *Meter-in* (A) and *meter-out* (B) circuits are used to control cylinder speed by limiting the rate at which liquid enters or leaves the cylinder. Check valves allow for unrestricted flow in the opposite direction. Metering circuits are always in *series* with the actuator.

working pressure. This behavior is usually achieved through the use of a spring-loaded element such as a spool or piston that slides across the valve output port as inlet pressure increases. Since higher pressures tend to create higher flow rates through the valve, this decrease in flow area allows the valve to maintain a constant overall rate of delivery to the cylinder. As shown in Figure 8–3, pressure-compensated valves are denoted by an arrow located adjacent to the restrictor symbol and perpendicular to the direction of flow.

More accurate speed control can be obtained through the use of *metering* circuits. A *meter-in* circuit consists of a variable flow control valve located on the pressurized, or *in*let, line of the cylinder, while a *meter-out* circuit has the valve installed on the exhaust, or *out*let, line of the cylinder. Both types of circuits are shown in Figure 8–4.

Meter-in circuits are most effective in systems that act against *resistive* loads, where the reactionary force of the load is in the *opposite direction* to the cylinder movement. The wheel crusher that we analyzed in Chapter 4 operates against resistive loads, as do log splitters and any actuators that lift, clamp, shear, punch, or press. Since the reactionary force loads the cylinder *against* its source of pressurized fluid, positive speed control can be achieved simply by varying the flow rate of liquid into the cylinder.

Meter-out circuits are generally used with *overrunning* loads, which act in the *same direction* that the cylinder moves. Since these loads tend to pull the cylinder *away* from its source of liquid, any restriction in the output line acts as a shock absorber and prevents the load from "running away" with itself by limiting the rate at which fluid may leave the cylinder. One example of this is a hydraulic jack used to lift an automobile. As the vehicle is being lowered, its weight (downward) acts in the same direction as the movement of the jack and thus constitutes an overrunning load. Anyone who has used such a device knows that rapidly opening the jack's relief valve brings the vehicle back to earth abruptly, while "feathering" the relief valve allows for a slower and more controlled rate of descent.

Unlike bleed-off circuits that provide an alternative, or parallel, path for the working fluid, metering circuits are always in *series* with the actuator. Any liquid entering (meter-in) or leaving (meter-out) the cylinder must pass through the flow control valve, a characteristic that contributes to the accuracy of these circuits. Pressure-compensated valves are commonly used in metering circuits and are normally connected in parallel with a check valve. This check valve is often an integral part of the flow control valve itself and allows unrestricted flow in the nonmetered direction.

Since hydraulic systems can encounter both overrunning and resistive forces during the course of a complete work cycle, it is not uncommon to find meter-in and meter-out circuits within the same system. The vertically mounted cylinder of Figure 8–5, for example, raises a weight while extending and lowers that weight while retracting. During extension the weight is a resistive load, while on retraction it is overrunning. Reversing this cylinder end-to-end causes the resistive load to be applied during retraction and the overrunning load during extension. No matter what type of load occurs, however, appropriate use of the meter-in and meter-out circuits allows for continuous speed control of the cylinder during all segments of the work cycle.

Another type of circuit used to control cylinder speed is the *regenerative,* or *differential,* circuit in which the rod end of the cylinder is always pressurized (Figure 8–6). During extension of the cylinder, both the rod end and cap end are under the same pressure, but the *difference* in areas (see Figure 4–6) on opposite sides of the piston allows the cylinder to extend. Fluid leaving the rod end combines with the main pump flow entering the cap end, effectively doubling the flow rate available during extension. If the ratio of cap end area to rod end area for a particular cylinder is 2:1, then the cylinder will extend and retract at the same speed. Although the circuit of Figure 8–6 utilizes a three-way directional control valve, some four-way valves contain an extra spool position that allows regenerative flow (Figure 8–7). One disadvantage of a regenerative circuit is that the net force available during extension is reduced, since the effective area is limited to the difference between cap end and rod end areas. In other words:

$$F_{\text{ext}} = p \times A_{\text{rod}}$$

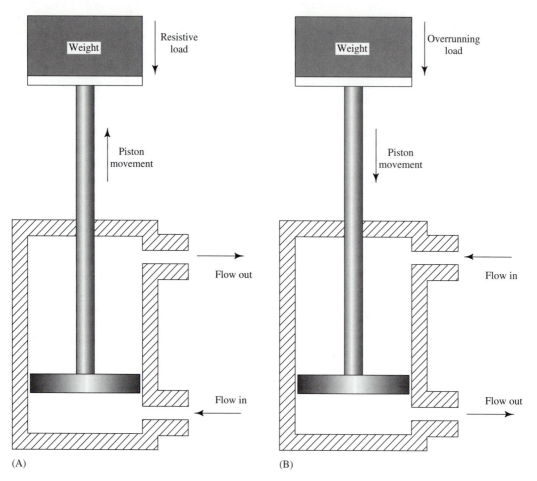

Figure 8–5   Even a constant load can change from resistive to overrunning as cylinder movement reverses direction. This weight is resistive as the cylinder extends (A) and overrunning as the cylinder retracts (B).

Our final circuit for controlling the speed of a cylinder is aimed at decelerating heavy loads near either end of the cylinder stroke. This behavior is achieved through the use of a *braking* or *deceleration* valve connected in series with the appropriate supply line as shown in Figure 8–8. These valves have several schematic representations, and generally contain a plunger, cam follower, or lever that is activated by the position of the cylinder output rod or by the load itself. Valve *A* is a normally open two-way valve that is gradually closed by the actuator as the cylinder reaches the end of its stroke. Closure of this valve allows pressure to build up in front of the moving piston, providing a liquid "cushion" that slowly

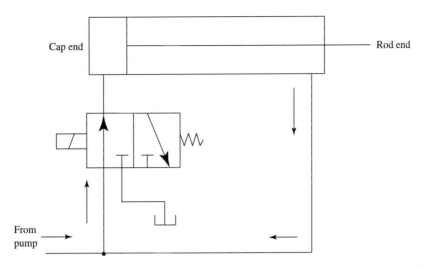

Cap end

Rod end

From pump

Figure 8–6 Typical *regenerative* circuit using a three-way valve. During extension of the cylinder (valve position shown), liquid exhausting from the rod end is fed directly back into the cap end. This addition to the normal pump flow increases cylinder speed on the extension stroke.

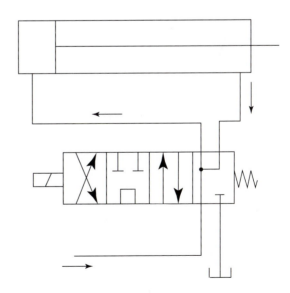

Figure 8–7 Some four-way valves contain a spool position (far right) that allows regenerative operation, as shown here.

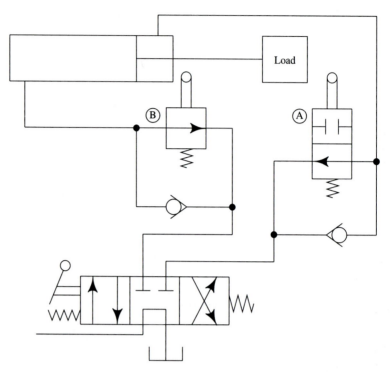

Figure 8–8 *Braking* or *deceleration* circuits contain normally open two-way (A) or relief (B) valves that are mechanically closed (by cylinder or load through a plunger, cam follower, or lever) before the cylinder reaches the end of its stroke.

brings the load to a halt. Valve *B* represents a normally open relief valve that operates in a similar manner and achieves the same result on retraction.

8.2 A Multipiston System

Consider the hydraulically operated snow plow assembly shown in Figure 8–9. A small, belt-driven gear pump housed inside the reservoir at *A* supplies liquid to a twin-spool directional control valve at *B;* one spool directs pressurized liquid to the single-acting cylinder at *C* used to lift the plow blade, while the other spool sends fluid to dual single-acting cylinders (*D*) used to change the blade angle about a vertical axis. The spools are activated individually by cables connected to a hand-operated dual-axis joystick (*E*) located inside the cab of the vehicle. Photographs of an actual plow frame assembly and its components are shown in Figure 8–10, while a schematic diagram for the hydraulic system is presented in Figure 8–11. The system shown here has several unique operational features:

- A small magnet located inside the reservoir and a 100 mesh screen on the pump inlet provide filtration for the liquid.

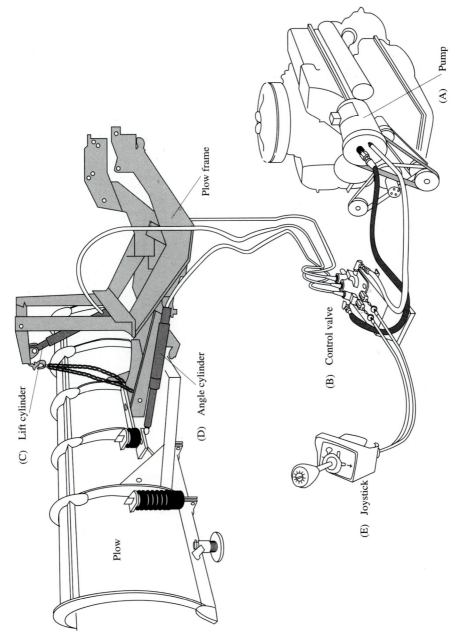

(C) Lift cylinder

Plow frame

Plow

(D) Angle cylinder

(A) Pump

(B) Control valve

(E) Joystick

Figure 8–9 Components of a typical snowplow assembly. (Text further identifies specific parts.)

(A)   (B)

(C)

Figure 8–10   Plow assembly components: (A) On most plow assemblies, two single-acting angling cylinders remain attached to the A-frame, as shown. (B) The single-acting lift cylinder is installed on a separate framework that is bolted or welded directly to the truck's undercarriage. Spring-loaded pins (visible at bottom center) allow the plow/frame of (A) to be quickly attached to or removed from the vehicle. (C) View of the under-hood hydraulic components, including the pump/reservoir (center) and cable-operated control valve (bottom center). Connections between the valve and cylinders are achieved using quick-disconnect fittings.

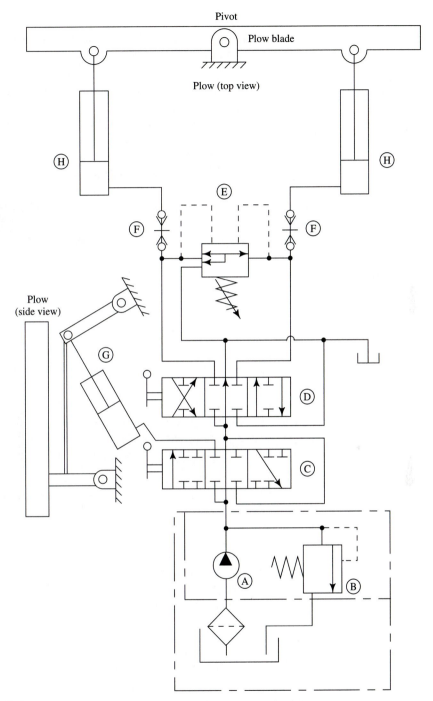

Figure 8–11 Hydraulic system for plow assembly of Figure 8–9 and Figure 8–10. (Text identifies parts.) *(From data provided by Fisher Engineering, Rockland, Maine.)*

291

- The system is protected by a 1500-psi ball-type pressure relief valve machined into the pump housing (element *B* in Figure 8–11).

- A 4000-psi cartridge-type crossover relief valve (*E* in Figure 8–11) housed in the directional control valve is used to protect the angling circuit. Since only one of the angling cylinders is under pressure at any given time, the crossover portion of the valve dumps minor pressure surges and differential oil volumes to the low-pressure cylinder as needed. For catastrophic pressure buildups caused by severe plow impact, relief to sump is provided sequential to crossover operation.

- The lift cylinder (*G*) is controlled by a three-position three-way valve (*C* in Figure 8–11), while the angling cylinders are controlled by a single three-position four-way valve (*D* in Figure 8–11). Note that the valve spools control six outlets in each position, rather than the four outlets common to most valves.

- The lift cylinder retracts due to gravity and the weight of the plow, but each angling cylinder (*H*) is driven to a retracted position by the action of its counterpart on the opposite side of the plow pivot point, as shown at the top of Figure 8–11. The plow blade itself simply acts like a giant lever. Note also that plowing normally takes place with the lift cylinder in its down, or retracted, position (right side of valve *C*). This locks the pressurized angling cylinder in place, yet allows the plow blade to "float" vertically over any variations in ground contour. To adjust the blade angle left or right, the lift control valve (*C*) must first be brought back to its mid-range (center position) in order to allow flow into the four-way valve (*D*).

- When the plow is removed from its host vehicle, the lift cylinder remains on the vehicle while both angling cylinders stay with the plow frame. Quick-release connections in the angling cylinder supply lines (*F* in Figure 8–11) facilitate the removal process.

You should take the time to trace the flow of liquid through this system for each position of the control valves and to match each flow condition to the mechanical behavior of the plow. Try to develop an understanding of how this simple system works, including the provisions for pressure relief and the interaction between cylinders. If you live in an area where snowplows are common, investigate the operation of a typical installation and compare it to the example presented here.

One important aspect of a snowplow is safety. To withstand the physical abuse received under normal operating conditions, most plows are ruggedly built and *heavy;* a typical 8-ft assembly popular for residential and light commercial use weighs well over 800 lb! Although plow blades are often released from their top-most position to shake off wet snow or crack through a thin layer of ice, there is also a possibility that the blade could drop down accidentally while personnel on foot are working near the plow or while the plow is being transported in its raised position at high speed over the road. Such an event could occur if the control cables were inadvertently jostled either at the valve itself or via the joystick inside the vehicle cab. Since the likelihood of such an event happening is small, most plow systems do not contain any safety provisions for hydraulically locking the plow in its raised position.

Many manufacturing operations, however, involve heavy elevated loads that pose a constant threat to workers and that therefore must be locked in position to minimize the possibility of a serious industrial accident. Typical examples include vertical presses and any devices used to lift or transfer heavy loads.

In a fluid power system, protection against such hazards is often achieved through the use of a *counterbalance circuit,* as shown in Figure 8–12. A normally closed two-way or pressure relief valve located between the cylinder and directional control valve maintains a preset back pressure in the cylinder output line, effectively locking the load in position. If this relief pressure setting is slightly higher than the pressure caused by

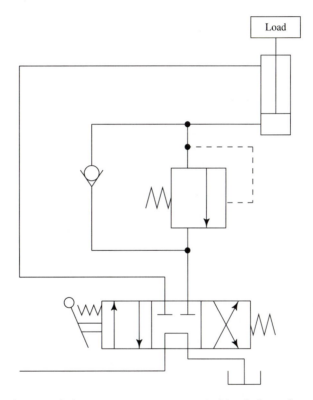

Figure 8–12 A *counterbalance* circuit prevents vertical loads from dropping unexpectedly and creating a safety hazard. During extension, liquid flows freely through the check valve into the cylinder; once this stroke has been completed, a normally closed two-way (or pressure relief) valve prevents the load itself from forcing liquid out of the cap end and thus causing the cylinder to retract. With the circuit shown, some retraction control is possible because the cylinder's rod (load) end must be pressurized until cap-end pressure increases sufficiently to open the two-way valve. Essentially, it is necessary to *drive* the cylinder downward. Because a broken or leaking hydraulic line between the two-way valve and cylinder will allow the load to fall, this valve is normally placed as close to the cylinder as possible.

the load itself, the valve will not open until the load is *driven* to retract by application of a positive pressure to the top of the cylinder. Since this normally requires a conscious, intentional act on the part of the operator, some degree of safety is introduced into the system.

Note that the check valve allows unrestricted flow during the extend, or lift, portion of the work cycle. In addition, the load will remain in an elevated position even if a rupture occurs downstream in the return line between counterbalance and directional control valves. Since the load will drop only if the line between cylinder and counterbalance valve breaks, the counterbalance valve itself is often installed as close to the cylinder as possible.

8.3 The Clamp-and-Work Circuit

One of the most useful hydraulic systems is the *clamp-and-work* circuit, in which a workpiece must first be securely clamped before any type of tool is applied to it. This simple two-step operation is often achieved through the use of a *sequence valve,* as shown in Figure 8–13. Liquid from supply flows into the primary circuit to activate the clamping mechanism. As pressure in this circuit reaches a preset level, the sequence valve opens, allowing liquid into a secondary circuit that controls the necessary tooling. These circuits are widely used both in manually operated and automated systems.

A typical device that employs sequential operation is the paper cutter shown in Figure 8–14. Used in printing shops to produce cleanly sheared edges on thick stacks of paper, these machines employ the same principles as many presses, shears, and stamping machines found in a variety of other industries. Figure 8–15 illustrates how a clamping bar on this paper cutter, actuated by hydraulic cylinder, first descends to hold the paper stack tightly in position before a razor-sharp cutting blade, driven down by a hydraulic motor and rotary crank, slices through the stack. To protect the operator's hands and fingers, two palm buttons spaced more than a foot apart on the front edge of the machine (well away from the clamping bar and cutter) must be depressed simultaneously during the entire work cycle.

The hydraulic circuit for this cutter is shown schematically in Figure 8–16. A vane pump at *A* delivers liquid to the system, with protection provided by the 1800-psi pressure

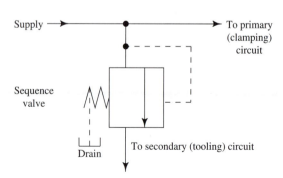

Figure 8–13 The sequence valve requires pressure to develop in a primary circuit before activation of the secondary circuit takes place.

relief valve at *B.* When the two palm buttons are activated, solenoids *d* on the directional control valves *D* and *F* are energized, shifting these valves to positions C_d and K_d respectively. This allows liquid to flow into the left-hand circuit, moving the clamping cylinder down. Note that clamping occurs on retraction, rather than extension, of cylinder *E,* and that pressure-reducing valve *C* limits the clamping pressure to 800 psi.

Once the primary (clamping) circuit has been fully activated, pressure begins to build up in the upper right-hand portion of the system. At a pressure of 1000 psi, sequence valve *G* opens, allowing liquid to reach a vane-type hydraulic motor at *H.* This actuator rotates through 180°, driving a crank mechanism that is attached to the cutting knife. When the knife has completed its cut, its position is detected by an electrical sensor called a *limit switch* that deenergizes knife solenoid *d* and energizes knife solenoid *u* on control valve *F.* This valve then shifts to the K_u position, reversing flow through the rotary actuator and raising the knife. Relief valve *J* returns liquid to reservoir if pressure in the motor's reverse flow line reaches 500 psi; this protects the hydraulic components should the cutter mechanism bind as the knife is being raised. Another limit switch detects the knife in its returned position, and deenergizes knife solenoid *u* on valve *F* while energizing clamp solenoid *u* on directional control valve *D.* This reverses flow to the cylinder and raises the clamping bar to its original position; a limit switch detects this condition and

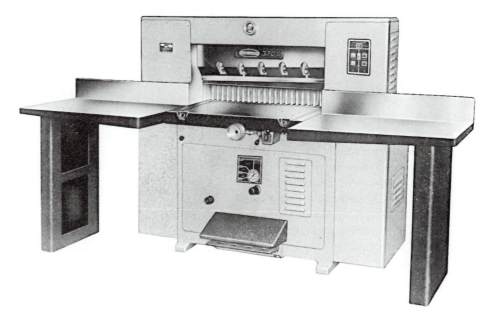

Figure 8–14 Commercial paper cutter. As with many industrial machines, the workpiece (stack of paper) must first be *clamped* in position before any *work* (shearing of edges) takes place. For operator safety, this shear cannot be activated until dual palm buttons on the front panel are depressed simultaneously. *(Challenge Machinery Company, Grand Haven, Mich.)*

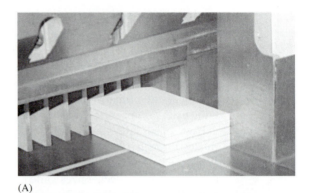

(A)

(B)

Figure 8–15 The paper cutter of Figure 8–14 has a typical clamp-and-work sequence, as follows: (A) A stack of paper pads is placed in the machine by operator. (B) The clamping bar descends to hold these pads firmly in position. (C) A razor-sharp cutting blade moves diagonally across the stack, shearing the pads to precise dimensions.

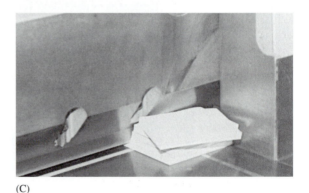

(C)

deenergizes clamp solenoid *u*. Both directional control valves are spring centered when neither solenoid is energized.

Note that this machine actually contains three separate yet interconnected systems: a *mechanical* system consisting of the clamping mechanism, cutting mechanism, machine base, and assorted pieces of hardware; a *hydraulic* system containing pump, liquid, valves, lines, and actuators; and an *electrical* system that includes solenoids, switches, and electric

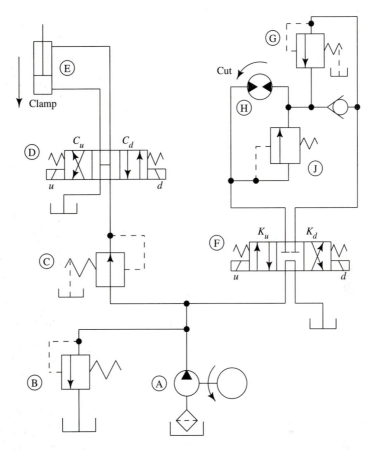

Figure 8–16 Hydraulic circuit diagram for the paper cutter shown in Figure 8–14 and Figure 8–15. (Text identifies components.) *(From data provided by Challenge Machinery Company, Grand Haven, Mich.)*

motor. Such a configuration is typical of that found on most industrial machines used in manufacturing processes; rarely does a production machine use control and actuation systems that are of the same type.

The paper cutter analyzed in this section operates on a simple two-step sequence. In Chapter 11 we shall see how electrical or pneumatic sensors and logic circuits may be used to provide sequential motion in more complicated systems, especially automated systems of the type shown in Figure 11–59. You should also note here that hydraulic systems that appear to be quite complicated are often constructed using combinations of several basic circuits. Such systems are best analyzed by examining the actual *function* performed by each portion of the total circuit.

8.4   Pump Unloading and Accumulator Circuits

Unloading circuits are used to bypass pump flow to reservoir during periods of low demand. We have already seen examples of how this can be accomplished. In its neutral position, for instance, a *tandem center* directional control valve simply routes the constant pump output back to tank (Figure 8–17(A)). Pressure relief valves perform the same function if a *closed center* valve is used (Figure 8–17(B)) or when a piston is jammed or locked

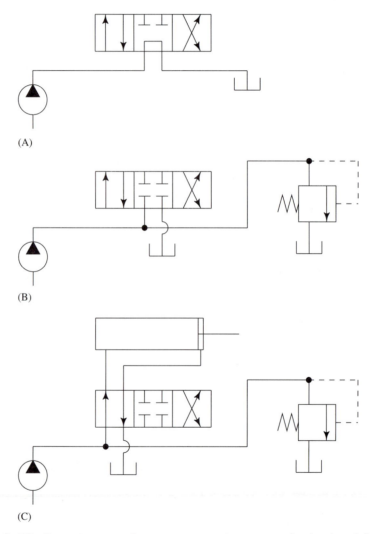

Figure 8–17   Bypassing pump flow to reservoir during periods of reduced demand is necessary to unload fixed delivery pumps: (A) tandem center valve; (B) closed center valve; and (C) locked piston.

at the end of its travel (Figure 8–17(C)). However, because relief valves are generally designed to open at high system pressures, flow through these devices generates considerable heat. Because of this, they are used for intermittent system protection only and are not recommended for continual, repetitive use during each work cycle.

Unloading valves are similar in construction, and, in fact, have the same schematic representation as pressure relief valves, but they are designed to open at lower pressures and transfer less heat to the working fluid. Their primary purpose is one of controlling pump loading, not of limiting maximum system pressure, but because they are pressure operated, unloading valves often provide this dual function.

A common application of unloading valves is in the control of two-stage pumps such as those discussed in Chapter 4 and Chapter 7. These actually consist of two pumps in one housing and are generally driven by a single input shaft. To accommodate work cycles that are not exactly repetitive, one pump is designed to provide high-volume–low-pressure flow, while the second pump has low-volume–high-pressure capabilities. Figure 8–18 shows how the two pumps operate in parallel and how each is unloaded at a different point in the work cycle. At the start of a cycle, both pumps contribute to the total flow that is directed by a control valve to the actuator. This combined flow allows the actuator to move quickly to the workpiece. Once contact has been made, system pressure rises beyond the range of the low-pressure pump, *L,* causing a pilot line to open unloading valve *A,* and returning the low-pressure flow to tank. This pressure buildup also closes check valve *C,* isolating the two pumps from each other and directing all of the output from high-pressure pump *H* to the actuator. Unloading valve *B* sends any excess flow from *H* back to reservoir as necessary. Circuits of this type are referred to as a *high-low* unloading circuits.

Pump unloading can also be accomplished using a solenoid-actuated two-way valve as shown in Figure 8–19. When pressure in the system has reached the desired level, a pressure-sensitive electrical switch energizes the solenoid, which shifts the valve to its open position, unloading the pump to tank. A check valve can be used to "lock" pressure in on the actuator side of the circuit while the pump is unloaded during periods of standby operation. Notice that pilot lines connected either directly to an unloading valve or to an electric switch/solenoid can sample pressures anywhere in the system and thus offer some degree of selective pressure control as well. Whatever their configuration, unloading circuits are used to provide an alternative, low-pressure flow path for fixed delivery pumps during periods of low demand and to reduce the workload on any pump wherever possible during the operating cycle.

Other circuits aimed at reducing direct demand on a pump are *accumulator circuits* such as the one shown in Figure 8–20. The accumulator is charged with pressurized liquid by the pump during slack periods in the work cycle and is isolated from the pump side of the system by a check valve. The pump may then be unloaded or taken out of service while system demand is met by the charge of accumulated liquid. A simple example of this is the residential water supply system shown in Figure 8–21. Water from an artesian well feeds the system and is pumped into the gas-charged accumulator until pressure in the accumulator reaches a preset upper limit, typically 60 psi. At this point, a pressure-sensitive electrical switch shuts the pump off. Liquid drawn from the system through faucets, bathroom

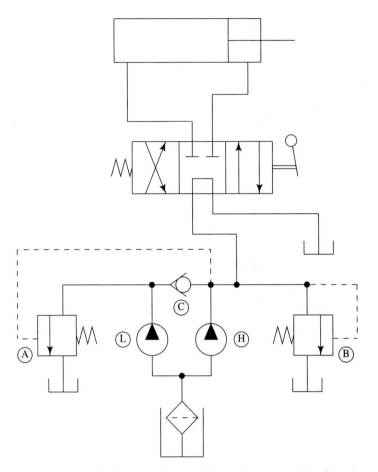

Figure 8–18 A typical *high-low* unloading circuit used to control the flow from a two-stage pump. At start-up, liquid from both low-pressure pump L and high-pressure pump H flows through the directional control valve to the cylinder, causing rapid extension toward the load, or workpiece. (Note that flow from pump L must pass through check valve C.) Once the load has been engaged, system pressure builds up to a level that is beyond the capability of pump L. At this point in the work cycle, check valve C closes and unloading valve A opens, thus isolating pump L from the high-pressure side of the circuit and allowing continuous flow from L to simply circulate back to reservoir. As pump H performs its high-pressure function, it is protected by unloading valve B, which diverts excess flow back to reservoir.

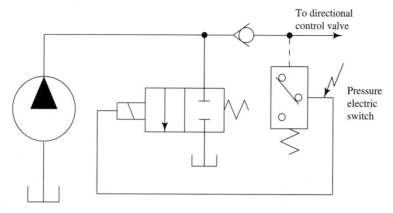

Figure 8–19 Pump unloading accomplished with a pressure-operated electric switch and solenoid-activated two-way valve. In the circuit shown, if pressure downstream (to the right) of the check valve is low, the pressure-operated electric switch is off. For this condition, the solenoid is not activated, so the two-way valve is held closed by the spring, and all the pump output flows through the check valve. When downstream pressure rises, the electric switch moves to the on position and activates the solenoid; this opens the two-way valve, dumping the pump flow to the reservoir.

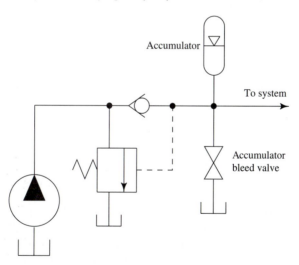

Figure 8–20 Accumulator circuits satisfy intermittent system requirements and thus reduce direct demand on pump. During initial pump operation, liquid flows into both the system and the accumulator. Once system pressure has reached a preset level, the check valve closes and the two-way valve opens, so that pump flow is diverted to the reservoir. Subsequent demand by the system is first satisfied by liquid stored in the accumulator. Eventually, as the accumulator empties, system pressure drops sufficiently to open the check valve and close the two-way valve, bringing the pump directly back on-line.

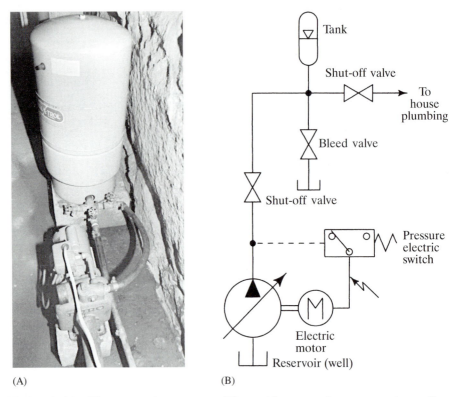

(A)   (B)

Figure 8–21   This accumulator system (A) provides water from an artesian well to a rural farmhouse. Intermittent demand is satisfied by the liquid accumulated in the tank, thus prolonging expected pump life considerably. The actual hydraulic circuit (B) for the system shown is quite simple.

fixtures, and kitchen appliances is provided by the accumulator until system pressure drops to a preset lower limit, usually near 40 psi, at which time the pressure-electric switch activates the pump. (Shut-off valves allow various segments of the system to be conveniently isolated for repair or replacement of components and lines.) If this system did not contain an accumulator, all demand would be directly on the pump and the unit would be forced to cycle on and off constantly to satisfy the requirements of a typical residence. Such operating conditions would drastically reduce the expected pump life; conversely, increasing accumulator size can be expected to prolong the pump life significantly.

Accumulator circuits are also used in emergency situations to provide pressurized liquid. Construction equipment, for example, is normally controlled using power steering systems that are supplied by pumps driven by the engines of these vehicles. In the event that an engine stalls, power steering is lost unless the vehicle is equipped with an accumulator that can provide enough fluid to help steer the vehicle to a safe stop. In fact, most mobile off-road vehicles contain an accumulator large enough to allow several complete steering cycles after engine power is lost.

SAFETY SIDEBAR

Beware of Residual Pressures

When working around accumulators, always remember that a charge of pressurized fluid represents a potential safety hazard. For this reason, accumulators should be equipped with either a manual or automatic bleed valve (see Figures 8–20, 8–21, and 8–22) to release pressure in the device when the system is shut down. Unexpected motion of an actuator driven by residual system pressure can have disastrous results!

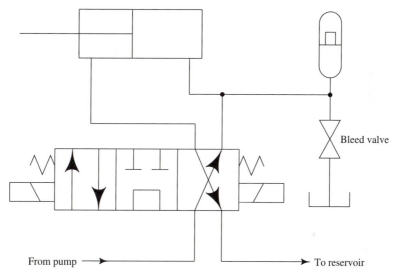

Figure 8–22 Accumulators are used in actuator supply lines to damp out pressure surges and pulsations caused by erratic loads.

Another important use of accumulators is in damping pressure surges and pulsations such as those caused by impacts or other variations in loading. Used in the actuator lines downstream of a directional control valve (Figure 8–22), the accumulator acts as a giant shock absorber or surge suppressor by accepting and releasing liquid as necessary to accommodate disturbances in the system.

8.5 Lowriders: A Case of Extreme Conditions

One general guideline for hydraulic circuit design is to try to protect the (expensive) major components under all anticipated operating conditions. However, even a system whose hardware has been carefully matched can occasionally experience severe overloads due to unforeseen circumstances or deliberately extreme demands.

Figure 8–23   Lowrider automobiles utilize a single-acting hydraulic cylinder at each wheel. Here, the front cylinders are extended to raise this end of the car, while the rear cylinders are retracted to lower the back end. *(Joan Eident Portrait Designs, Auburn, Mass.)*

An interesting example of the latter case may be found in lowrider automobiles. These customized vehicles contain modified suspension systems that utilize hydraulic cylinders on any two or on all four wheels. By extending and retracting the cylinders in various combinations, it is possible to raise the entire car or truck body high in the air, lower it completely to the ground, or tip it radically either from side to side or front to rear, as shown in Figure 8–23.

To achieve this behavior, a small-bore, short-stroke, single-acting hydraulic cylinder is installed at each wheel as desired; components are generally assembled as illustrated in Figure 8–24. Here, the specially designed cylinder (typically 1 to 2 in. in diameter, with a 6- to 12-in. stroke) passes through a load washer and on through the vehicle frame (or a bracket welded to the frame). The upper end of this cylinder is connected directly to a hydraulic line from the pump, while the large diameter end cap threaded onto the bottom of the cylinder forms a "step," upon which the load washer rests. An upper stepped bushing is attached to the cylinder's output rod; this bushing fits into the top of a coil spring, the lower end of which is held in alignment by a lower stepped bushing welded to either the vehicle's axle or leaf spring. If the cylinder now extends, it pushes apart the axle and frame, raising the car or truck body upward.

Although lowriders have been common in the western United States for many years, they are rapidly becoming popular in other areas around the world. Originally considered as simply a form of automotive expressionism, these vehicles and their owners may now compete in an extensive series of events (both amateur and professional) that subject their hydraulic suspension systems to extreme conditions.

In order to understand how severe these performance demands can be, let us examine a typical electrohydraulic circuit (Figure 8–25) used to activate one lift cylinder. When the

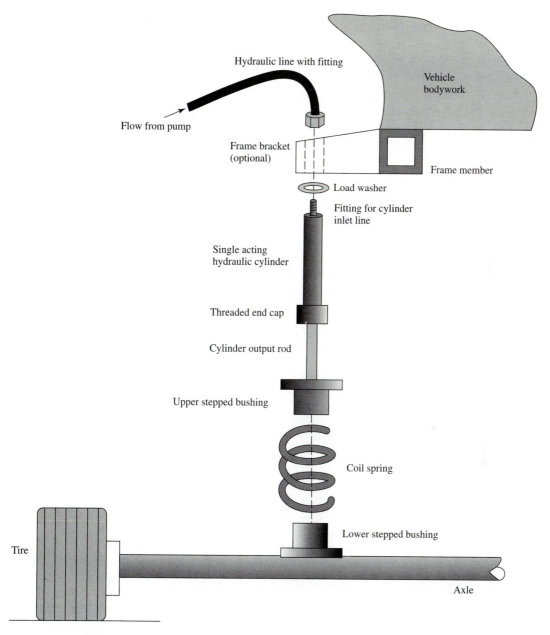

Figure 8–24 Assembly of components on a typical lowrider vehicle. As the cylinder output rod extends, it pushes down on the coil spring and axle. Simultaneously, the threaded end cap pushes upward against the load washer and frame bracket. The net effect is to raise the frame and body away from the axle.

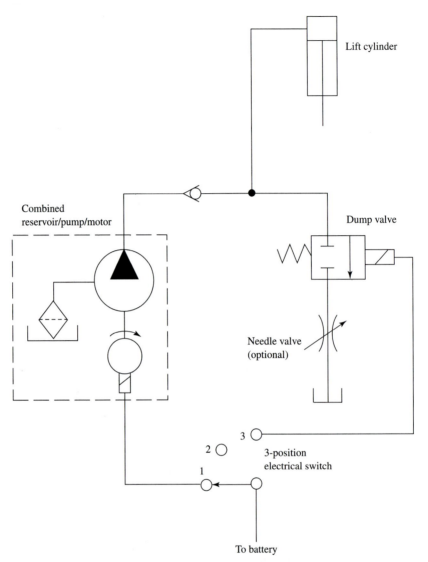

Lift cylinder

Combined
reservoir/pump/motor

Dump valve

Needle valve
(optional)

3

2

3-position
electrical switch

1

To battery

Figure 8–25 Electrohydraulic circuit used for a single lift cylinder. In switch position 1, pump flow extends the cylinder (raises body). Neutral position 2 allows the check valve to lock the cylinder in place. Switch position 2 allows the check valve to lock the cylinder in place. Switch position 3 activates the solenoid to open the "dump" valve, while the optional needle valve controls the rate of flow back to the reservoir; vehicle weight retracts the cylinder, causing the body to lower.

three-position electrical switch is in position 1, a 12-V electric motor (generally 3 hp or larger) activates a gear pump, whose average flow rate might consist of 0.25 in.^{3}/rev at a pressure of 2000 psi. Liquid flows through the check valve, causing the cylinder to extend (raise vehicle). If the switch is moved to position 2, the motor stops, and this cylinder is then locked in place by the check valve. When the switch is moved to position 3, an electric solenoid opens the normally closed two-way valve, "dumping" any liquid in the lines and cylinder back to the reservoir. The vehicle's own weight causes the cylinder to retract, while an optional needle valve controls the rate at which the body is lowered.

Consider several features about the operation of this system:

1. There is no provision for pressure relief. Because each cylinder is under the vehicle (and not visible), the operator must *guess* when full extension has been reached for switch position 1. If the motor/pump is not switched off (position 2) in time, severe damage can result to the positive displacement pump.

2. Even when the cylinders are only partially extended, they carry most of the frame, body, and passenger weight. Unless the vehicle is driven only on absolutely smooth roadways, pressure pulses generated in the hydraulic suspension system by bumps and potholes will create severe stresses on lines, fittings, and seals.

3. During lowrider competitions, owners routinely cycle their three-position switches between positions 1 and 3. If done in the proper rhythm, this causes the vehicle to lower itself onto its springs and then rise to progressively higher levels. Eventually, this bouncing can lift the front wheels of a vehicle off the ground by as much as 3 to 4 ft or cause the entire vehicle to hop *completely* off the ground to a height of 1 ft or more. (In fact, a skillful owner working from a panel of three-position switches can literally make his or her vehicle "dance" around within a prescribed area.) Although first-time viewers are amazed at the sight of an airborne car or truck, it is the hydraulic system components that pay the dearest price; when the weight of such a vehicle crashes back to earth, instantaneous pressures in the system approach 6000 psi.

4. In order to obtain the electrical system performance required to drive a lowrider hydraulic circuit, several automobile batteries are customarily wired in series to produce a 36- or 48-V supply. Because the solenoids and motor are designed for 12-V operation, this combination of high voltages and rapid switching can dramatically reduce the useful life of electrical components.

Because of the extremely severe demands made on a lowrider's electrohydraulic system, particularly in car-dancing or car-hopping competitions, virtually all components must be replaced or repaired on a regular basis, sometimes after every performance. For this reason, an entire industry segment has developed that specializes in the manufacture of components for this limited application. One example is an integrated motor/pump/reservoir of the type diagrammed in Figure 8–26(A) and shown in the typical installation of Figure 8–26(B). An electric motor is bolted to one face of a square mounting block, and a hydraulic gear pump is bolted to the opposite face. The motor and pump shafts are directly connected through a sealed coupling within the block itself. A metal cannister covers the pump and functions as a reservoir, with most units having capacities in the 3- to 4-qt range. These components are designed for frequent, rapid disassembly to allow reconditioning of the pump and motor as needed.

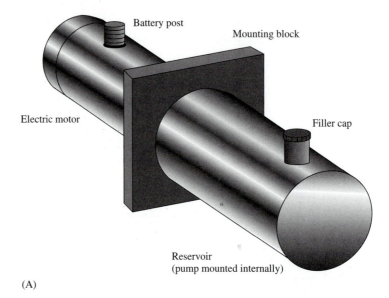

(A)

(B)

Figure 8–26 (A) One lowrider specialty component consists of an electric motor coupled to a gear pump inside the reservoir. The entire unit is easily disassembled for replacement or repair of components. (B) Two of these units installed in the trunk of the lowrider shown in Figure 8–23. Note the batteries wired in series to provide electric power for motors and solenoids. (*Joan Eident Portrait Designs, Auburn, Mass.*)

Questions

1. For the regenerative circuit of Figure 8–6, why does a 2:1 ratio of cap-end area to rod-end area produce equal extension and retraction speeds? What is the relationship between these speeds for an area ratio of 3:1?

2. In place of external circuitry of the type shown in Figure 8–8, what other option exists for deceleration of a piston at each end of its stroke?

3. Cylinder *E* in Figure 8–16 exerts a clamping force on its retract stroke. Would this be considered a resistive or overrunning load? What type of load is experienced by the cylinder during its extend stroke?

4. On a directional control valve, what is a tandem center?

5. What is the principal advantage of using a two-stage pump? Can you think of a potential disadvantage?

6. Why is a bleed-off valve used with accumulators?

7. For the lowrider circuit of Figure 8–25:
   (a) Which component replaces the standard directional control valve used to produce desired cylinder motion?
   (b) Name another major component normally found in a hydraulic system that is missing here.

8. Does the reservoir size of a unit such as that shown in Figure 8–26A satisfy the design guidelines for reservoirs given in Chapter 7? Explain.

9. For the partial circuit shown in Figure 8–27, which of the following is true:
   (a) As cylinder A extends, cylinder B extends.
   (b) As cylinder A extends, cylinder B retracts.
   (c) As cylinder A extends, cylinder B remains motionless.

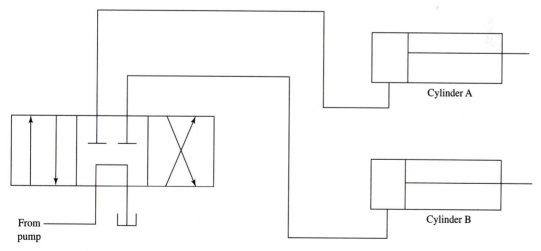

Figure 8–27   See question 6.

Pneumatics: The Behavior of Gases

Compressed gases find wide use in a variety of *process* and *production* industries. Typical process applications include the manufacture of chemicals, refining, food processing, waste water treatment, refrigeration, and power generation. Virtually all production operations use a compressed gas somewhere in their manufacturing process to power tools or automated transfer equipment. These gases may be of any type (Figure 9–1) and are often toxic, corrosive, or flammable; some gases or vapors, such as compressed steam, present an additional safety hazard because of their elevated temperatures.

By far the most widely used industrial gas is compressed air. Typically made up of 79% nitrogen and 21% oxygen, air also contains trace elements of gases such as argon, small amounts of water vapor, and oxides of nitrogen and carbon. This gas is nontoxic, nonflammable, inexpensive, and readily available, and generally can be released to the atmosphere without having a detrimental effect on the environment.

In this chapter we shall study the behavior of gases during the compression process and note the implications of this behavior for compressed gas supply systems and their use.

9.1 The Gas Laws

The condition, or *state,* of a gas may be completely specified by three variables: *pressure* of the gas; *volume* occupied by the gas; *temperature* of the gas. These three variables have been found to be related by the equation:

$$pV = WRT \tag{9–1}$$

where*: p* is the *absolute* pressure of the gas
      V is the volume occupied by the gas
      W is the amount (weight) of the gas
      R is a property of the gas called its *gas constant*
      T is the *absolute* temperature of the gas

Typical U.S. customary units for the quantities are p (psfa), V (ft^3), W (lb), R (ft/°R), and T(°R). Absolute temperatures on the Rankine scale (T_R) may be obtained by adding 460 to Fahrenheit temperatures: $T_R = T_F + 460$. Values of R for some common gases are presented in Table 9–1. Using these units, the product obtained on either side of equation (9–1) has units of ft-lb.

An *ideal gas* is defined as one that obeys the relationship stated by equation (9–1) *under all conditions.* For this reason, the equation is often called the *ideal gas law* and is frequently seen in other forms that use quantities such as the mass, density, and molecular

Figure 9–1   Portable tanks of compressed nitrogen are used by pit crews at most major stock car races to power impact wrenches such as the ones shown here.

Gas	R (ft/°R)	R(m/°K)
Air	53.3	29.2
Ammonia	89.5	49.1
Carbon dioxide	35.0	19.2
Hydrogen	766.5	419.9
Nitrogen	55.1	30.2
Oxygen	48.3	26.5

Table 9–1   Gas Constants for Common Gases

weight of the gas. The behavior of a real gas is affected somewhat by the size of its molecules and the cohesive forces developed between these molecules, but except at low temperatures or high pressures real gases closely follow the behavior of an ideal, or perfect, gas.

Although equation (9–1) presents the relationship between the properties and state of a gas, it is not the most useful form for calculating changes in the conditions of real gases. For a fixed quantity of a specific gas, equation (9–1) can be written as:

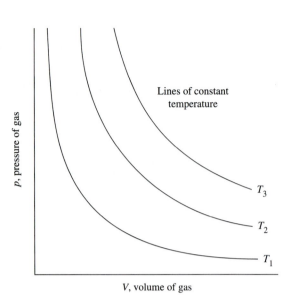

Figure 9–2 Typical *p-V-T* diagram showing behavior of a gas. Along each isotherm, pressure and volume of the gas are inversely related.

$$\frac{pV}{T} = \text{constant}$$

or:

$$\frac{p_1 V_1}{T_1} = \frac{p_2 V_2}{T_2} \qquad\qquad (9\text{–}2)$$

Notice that at constant temperature, pressure and volume are inversely related. In other words, as p increases, V decreases; as p decreases, V increases. This behavior is shown graphically in the *p-V-T* diagram of Figure 9–2, where each curve represents a line of constant temperature, called an *isotherm*, and $T_3 > T_2 > T_1$.

Over the last few centuries, researchers have investigated the behavior of gases under various conditions and have arrived at simplified forms of equation (9–2). Many useful problems can be solved using their results, while the *p-V-T* diagram enhances our understanding of the actual processes involved.

Consider, for example, the piston and cylinder containing a fixed amount of gas as shown in Figure 9–3. The gas can be maintained at constant temperature by immersing the cylinder in a large reservoir of constant-temperature fluid such as air or water. If the gas, initially at volume V_1, is compressed by the piston to a smaller volume V_2 while the temperature of the gas remains constant, then equation (9–2) becomes:

$$p_1 V_1 = p_2 V_2 \qquad\qquad (9\text{–}3)$$

This result, first observed by the English scientist Robert Boyle in 1660, is known as *Boyle's law*. As indicated on the *p-V-T* diagram of Figure 9–3, this *isothermal* compression produces an increase in the pressure of the gas from p_1 to p_2 as the gas volume decreases from V_1 to V_2.

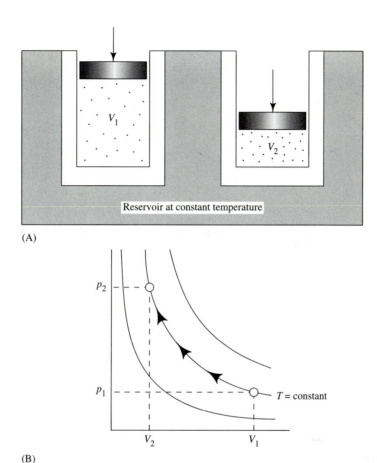

Figure 9–3 Isothermal compression of a gas. (A) During compression, the piston performs work *on* the gas. Although some of this work produces an increase in pressure energy of the gas, some work is also converted to heat. If compression occurs slowly in the presence of a large, constant-temperature reservoir, this heat is conducted away from the cylinder so that gas temperature remains unchanged. (B) The process is represented on a *p-V-T* diagram for a gas that is initially at pressure p_1 and volume V_1. Compression takes place along an isotherm, with pressure rising to p_2 as volume decreases to V_2.

EXAMPLE 1

A 2-in.-diameter pneumatic cylinder extends 14.0 in. to support a load of 250 lb, as shown in Figure 9–4(A).

(a) Find the pressure and volume of the compressed gas in the cylinder at the fully extended position.

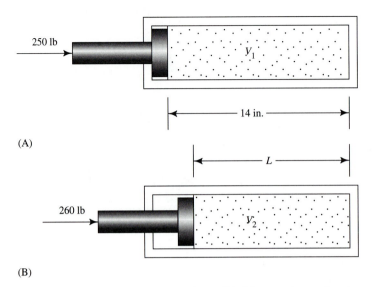

(A)

(B)

Figure 9–4   Load variations can cause significant "drift" in position of a piston due to the compressibility of the gas. For this reason, accurate positioning of a load can be difficult to achieve using pneumatic systems.

(b) If the load increases slightly to 260 lb, how far does the piston move back into its cylinder (Figure 9–4(B))?

Solution   (a) Area of the piston, A, may be computed as:

$$A = 0.7854 \times D^2 = 0.7854 \times (2 \text{ in.})^2 = 3.142 \text{ in.}^2$$

and cylinder volume, V_1, as:

$$V_1 = A \times \text{stroke} = 3.142 \text{ in.}^2 \times 14 \text{ in.} = \textbf{44.0 in.}^3$$

Gas pressure, p_1, is equal to:

$$p_1 = \frac{F}{A} = \frac{250 \text{ lb}}{3.142 \text{ in.}^2} = 79.6 \text{ psig}$$

$$= (79.6 + 14.7) \text{ psia} = \textbf{94.3 psia}$$

(b) If the load is increased to 260 lb, the new pressure is:

$$p_2 = \frac{260 \text{ lb}}{3.142 \text{ in.}^2} = 82.7 \text{ psig} = 97.4 \text{ psia}$$

Using Boyle's law, $V_2 = \left(\dfrac{p_1}{p_2}\right) \times V_1$, or:

$$V_2 = \frac{94.3 \text{ psia}}{97.4 \text{ psia}} \times 44.0 \text{ in.}^3 = 42.6 \text{ in.}^3$$

From Figure 9–4(B), $V_2 = A \times L$, or:

$$L = \frac{V_2}{A} = \frac{42.6 \text{ in.}^3}{3.142 \text{ in.}^2} = 13.6 \text{ in.}$$

In other words, a load variation of only 10 lb has moved the piston $(14.0 - 13.6)$ in., or **0.40 in.,** back into its cylinder! This example points out why hydraulic rather than pneumatic systems are used in applications that require accurate positioning of the load.

EXAMPLE 2

Free air is defined as air that is at whatever atmospheric conditions of pressure and temperature exist at any given time and place. How many cubic feet of free air at 14.7 psia and 68°F must be compressed to fill the cylinder of Example 1 at its pressure of 94.3 psia? Assume that the gas is compressed slowly so that its temperature remains constant (isothermal process).

Solution   Let the volume of free air outside the cylinder be V_1. Its pressure is then 14.7 psia. Inside the cylinder, the state of the gas is given by $p_2 = 94.3$ psia, and $V_2 = 44.0$ in^3. Since the temperature of the gas remains constant, Boyle's law applies, and:

$$p_1 \times V_1 = p_2 \times V_2$$
$$(14.7 \text{ psia}) \times V_1 = (94.3 \text{ psia}) \times 44.0 \text{ in.}^3$$

Solving this equation yields:

$$V_1 = \textbf{282 in.}^3$$

This is the amount of free air that must be gathered from the surroundings and squeezed into the pneumatic cylinder.

Another type of process affecting the state of a gas is the *isometric,* or *constant volume,* process shown in Figure 9–5. If the piston is locked in position, heating the gas increases both its temperature and pressure. Equation (9–2) then reduces to:

$$\frac{p_1}{p_2} = \frac{T_1}{T_2} \tag{9–4}$$

This result is known as *Gay-Lussac's law* after Joseph Gay-Lussac, the nineteenth-century French scientist who first noted the phenomenon.

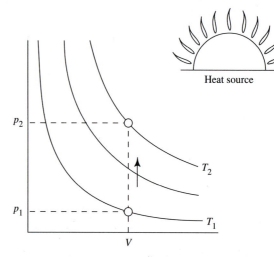

Figure 9–5   Adding heat to a fixed volume of gas increases both temperature and pressure of the gas. Any process that takes place at constant volume is called an *isometric* process.

EXAMPLE 3

When fully pressurized, commercial tanks of compressed nitrogen such as those shown in Figure 9–1 typically exhibit a gauge reading of 2600 psi at a temperature of 70°F. On a sunny day at the racetrack, gas temperatures in these tanks may easily top 120°F. At that temperature, what should the gauge read?

Solution   Since the volume of the gas remains constant, Gay-Lussac's law applies. Converting temperatures to the absolute, or Rankine, scale:

$$T_1 = (70 + 460)°R = 530°R$$

and:

$$T_2 = (120 + 460)°R = 580°R$$

Also:

$$p_1 = 2600 \text{ psig} = (2600 + 14.7) \text{ psia} = 2615 \text{ psia}$$

Then:

$$p_2 = p_1 \times \left(\frac{T_2}{T_1}\right)$$

$$= 2615 \text{ psia} \times \left(\frac{580°\text{R}}{530°\text{R}}\right)$$

$$= 2862 \text{ psia} = \textbf{2847 psig}$$

Does this example suggest how a confined gas and a pressure gauge might function as a simple thermometer?

Figure 9–6 illustrates an *isobaric,* or constant pressure, process. If the weighted piston is free to slide up and down within the cylinder, then the addition of heat causes the

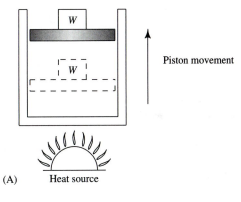

(A)   Heat source

Figure 9–6   (A) An *isobaric* process takes place at constant pressure, such as that created by fixed weight W. If the gas is heated, its volume increases, raising the piston. (B) On a *p-V-T* diagram, this process begins with the gas at an initial state (*p,* V_1, T_1) and ends at the final state (*p,* V_2, T_2).

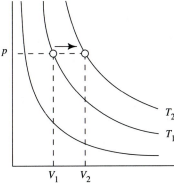

(B)

gas to expand while its pressure remains unchanged. Equation (9–2) can therefore be written as:

$$\frac{V_1}{T_1} = \frac{V_2}{T_2} \qquad (9\text{–}5)$$

This gas behavior was verified experimentally in 1787 by Jacques Charles and is known as *Charles's law.*

EXAMPLE 4

The air cylinder of Example 1 has an initial temperature of 70°F. If the 250-lb load remains constant but the gas temperature drops to 60°F, how far does the piston retract into its cylinder due to contraction of the gas?

Solution   From Example 1, $V_1 = 44.0$ in.^{3}. Converting temperatures to °R:

$$T_1 = 530°\text{R} \quad \text{and} \quad T_2 = 520°\text{R}$$

From Charles's law:

$$V_2 = V_1 \times \left(\frac{T_2}{T_1}\right)$$

$$= 44.0 \text{ in.}^3 \times \left(\frac{520°\text{R}}{530°\text{R}}\right)$$

or:

$$V_2 = 43.2 \text{ in.}^3$$

From Figure 9–4(B), $V_2 = A \times L$, where $A = 3.142$ in.^{2}. Then:

$$L = \frac{V_2}{A} = \frac{43.2 \text{ in.}^3}{3.142 \text{ in.}^2} = 13.7 \text{ in.}$$

The piston has retracted, then, by:

$$(14.0 - 13.7) \text{ in.} = \mathbf{0.30 \text{ in.}}$$

This problem illustrates that temperature as well as load variations can produce significant changes in position of loads supported by a compressed gas.

From the preceding examples, it is clear that the process of mechanically compressing a gas requires the movement of a force through some distance. By our definition in Chapter 3, this means that work must be done on the gas. Some of this work appears in the form of heat, which is a natural product of the compression process. Isothermal compression, for example, can be achieved only if the process takes place

Table 9–2 Adiabatic	Gas	Exponent, k
Exponents (Dimensionless) for Common Gases	Air	1.40
	Ammonia	1.32
	Carbon dioxide	1.30
	Ideal gas	1.67
	Nitrogen	1.40
	Oxygen	1.40

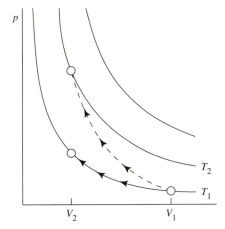

Figure 9–7 Comparison of adiabatic (dotted line) and isothermal (solid line) compression from V_1 to V_2.

slowly enough to allow the generated heat to be removed from the gas and conducted to the surroundings (see Figure 9–3).

Often, however, the compression process takes place in an insulated cylinder (a condition approximated if the cylinder is located deep inside a compressor and difficult to cool), or if compression takes place so quickly that heat does not have time to flow out of the gas to its surroundings. Such a process in which no heat is transferred is called an *adiabatic* process and satisfies the equation:

$$pV^k = \text{constant} \qquad (9\text{–}6)$$

where k is a dimensionless property of the gas, typical values of which are given in Table 9–2.

Figure 9–7 compares the adiabatic and isothermal processes. Because the gas retains its heat of compression in the adiabatic process, temperature of the gas increases. For a given change in volume, then, the final pressure achieved through an adiabatic compression is greater than that resulting from isothermal compression. Diesel engines, for example, depend upon an adiabatic process to reach the elevated temperatures that allow ignition of a fuel-air mixture without the use of a spark plug. These high temperatures, however, are accompanied by high pressures as well. For this reason, diesel engines are of much heavier construction than their spark-ignition counterparts.

EXAMPLE 5

Referring to Figure 9–3(A), the ratio of initial to final volumes, $\dfrac{V_1}{V_2}$, is called the *compression ratio,* a term often used to describe this characteristic of internal combustion engines. Compressors, however, are rated according to their *ratio of compression,* which is defined as the ratio of final pressure to initial pressure, $\dfrac{p_2}{p_1}$. Assume that air at 70°F and 14.7 psia is compressed to 1/12 of its original volume.

(a) Find the final pressure, p_2, final temperature, T_2, and ratio of compression if the process is isothermal.

(b) Repeat part (a) if the process is adiabatic.

Solution

(a) Since this process is isothermal, T_2 equals initial temperature T_1 at 70°F, or 530°R. The compression ratio is computed as:

$$\left(\frac{V_1}{V_2}\right) = \frac{(V_1)}{\left(\dfrac{1}{12}V_1\right)} = 12.0$$

From Boyle's law:

$$p_2 = p_1 \times \left(\frac{V_1}{V_2}\right) = 14.7 \text{ psia} \times 12.0$$

$$= \mathbf{176\ psia}$$

and the ratio of compression is:

$$\left(\frac{p_2}{p_1}\right) = \frac{176 \text{ psia}}{14.7 \text{ psia}} = \mathbf{12.0}$$

Notice that for an isothermal compression, the ratio of compression is always equal to the compression ratio:

$$\frac{p_2}{p_1} = \frac{V_1}{V_2}$$

(b) For the adiabatic process:

$$p_1 V_1{}^k = p_2 V_2{}^k$$

or:

$$p_2 = p_1 \times \left(\frac{V_1}{V_2}\right)^k$$

From Table 9–2, for air $k = 1.40$, so:

$$p_2 = 14.7 \text{ psia} \times (12.0)^{1.40} = \mathbf{477\ psia}$$

The ratio of compression becomes:

$$\left(\frac{p_2}{p_1}\right) = \frac{477 \text{ psia}}{14.7 \text{ psia}} = \textbf{32.4}$$

Since the ideal gas law is still valid during an adiabatic process, equations (9–2) and (9–6) may be combined and p eliminated to yield:

$$TV^{k-1} = \text{constant}$$

Then:

$$T_1 V_1^{k-1} = T_2 V_2^{k-1}$$

or:

$$T_2 = T_1 \times \left(\frac{V_1}{V_2}\right)^{k-1}$$

$$= 530°R \times (12.0)^{0.4}$$

$$= 1432°R = \textbf{972°F}$$

Problem Set I

1. Hydraulic oil in the cylindrical reservoir of Figure 9–8 is pressurized by a charge of compressed air at 120 psig and 68°F. If system demand draws the oil level down 4 in., what is the new air pressure inside the tank? Assume the process to be isothermal.

2. In a general compression process, all of the gas conditions may change. Such processes may be analyzed using equation (9–2) directly or by breaking the compression procedure down into several processes. Assume that 12,000 in.^{3} of free air at 68°F is compressed to 1/20 of its original volume while simultaneously being heated to 120°F.

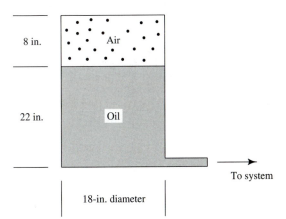

Figure 9–8 Cylindrical
reservoir for problem 1.

(a) Find the final pressure using equation (9–2).

(b) Find the final pressure by assuming that the compression is a two-step process: the air is first compressed isothermally and then heated isometrically.

3. If a 5000-cm³ volume of carbon dioxide at 138 kPa and 10°C is compressed adiabatically to a new volume of 2000 cm³, find the final pressure and temperature. (*Note:* The relationship between Fahrenheit and Celsius temperatures is defined as $T_F = (1.8 \times T_C) + 32$.)

4. A single-acting pneumatic cylinder whose diameter is 1.50 in. has a stroke of 10 in. This cylinder is extended by compressed air at the rate of 40 times per minute, retracting to its original position each time through the action of a spring. The compressed air that actuates this cylinder is at 90 psi and 80°F. How much free air must be compressed each minute to operate this cylinder if the intake for the compressor is located in a room at atmospheric pressure where the temperature is 70°F?

9.2 Properties of Gases

The preceding examples showed us that the volume occupied by a gas can be greatly affected by its temperature and pressure. In a typical isobaric expansion, for example, heating a fixed amount (mass or weight) of gas increases its volume and temperature (Figure 9–6), so that volume-related properties of the gas, namely its *density* and *specific weight,* both decrease. It is this behavior that allows a hot air balloon to rise from the earth (Figure 9–9). The density and specific weight of a gas, then, are generally specified for a particular temperature (typically 60°F or 68°F) and pressure (usually 14.7 psia).

Although the specific gravity of a liquid is determined by comparing its density or specific weight to that of water, the *specific gravity* of a gas uses "standard" air as a reference. The American Society of Mechanical Engineers (ASME) defines the properties of *standard air* to be those at 68°F and 14.7 psia, but most of the gas industries use 60°F and 14.7 psia as standard conditions.

EXAMPLE 6

(a) Compute the specific weight of ASME standard air.

(b) Calculate the specific weight and specific gravity of nitrogen at standard conditions.

(c) Determine the specific gravities of carbon dioxide and oxygen at standard conditions.

Solution   (a) The specific weight of air, γ_a, may be computed using equation (9–1):

$$\gamma_a = \frac{W}{V} = \frac{p}{RT}$$

At standard conditions $T = (68 + 460)°R = 528°R$, and $p = (14.7 \times 144)\text{lb/ft}^2$ $= 2120 \text{ lb/ft}^2$. From Table 9–1, R for air is 53.3 ft/°R. Then:

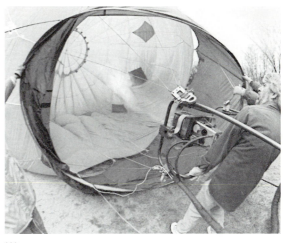

(A)

Figure 9–9 (A) Propane burners are commonly used to heat the air inside a hot-air balloon. Heating the air causes it to expand, decreasing its density below that of the surrounding atmosphere and resulting in a buoyant force that lifts balloon, basket, and passengers off the ground. (B) Because air is of such low density, the balloon volume must be quite large in order to generate the required buoyant force. *(Jean Dennett, Gardner, Mass.)*

(B)

$$\gamma_a = \left(\frac{2120 \text{ lb/ft}^2}{53.3 \text{ ft/}^\circ R \times 528^\circ R} \right)$$

$$\gamma_a = \mathbf{0.0753 \text{ lb/ft}^3}$$

(b) For nitrogen, $R = 55.1$ ft/$^\circ$R, so at standard conditions:

$$\gamma_n = \left(\frac{2120 \text{ lb/ft}^2}{55.1 \text{ ft/}^\circ R \times 528^\circ R} \right)$$

$$= \mathbf{0.0729 \text{ lb/ft}^3}$$

The specific gravity of nitrogen, S_n, is then:

$$S_n = \frac{\gamma_n}{\gamma_a} = \frac{0.0729 \text{ lb/ft}^3}{0.0753 \text{ lb/ft}^3}$$

$$= \mathbf{0.968}$$

(c) From parts (a) and (b), notice that when a gas is compared to air at the same conditions, the specific gravity of the gas can be computed as the ratio of gas constants. In other words:

$$S_{CO_2} = \frac{R_a}{R_{CO_2}} = \left(\frac{53.3 \text{ ft/}^\circ R}{35.0 \text{ ft/}^\circ R} \right)$$

$$= \mathbf{1.52}$$

and:

$$S_{oxy} = \frac{R_a}{R_{oxy}} = \left(\frac{53.3 \text{ ft/}^\circ R}{48.3 \text{ ft/}^\circ R} \right)$$

$$= \mathbf{1.10}$$

Table 9–3 presents specific gravities for some common gases.

Another gas property that differs from its liquid counterpart is that of *viscosity*. In Chapter 2 we saw that increasing the temperature of a liquid weakened the bonds between molecules, producing a *decrease* in viscosity. Raising the temperature of a gas, however, increases the velocity, internal energy, and effective volume of the molecules. Each of these changes is believed to increase the rate of interaction between molecules, resulting in a *higher* viscosity for the gas.

At any given time, a gas generally contains some amount of *water vapor*, which it can acquire directly from the atmosphere or from condensation that often forms on the inside surfaces of the lines, tanks, fittings, and actuators of a system. The moisture content of a gas is called its *humidity*, and several terms are used to describe this condition or state. Consider, for example, the behavior of air at atmospheric pressure and a given ambient temperature. When this gas contains the maximum amount of water that it can hold at that tem-

Table 9–3 Specific Gravities of Some Common Gases at 60°F and 14.7 psia (From Compressed Air and Gas Handbook of the Compressed Air and Gas Institute, 5th ed. Table 13.39. Englewood Cliffs, N.J.: Prentice Hall, 1989)	Gas	Specific Gravity
	Acetylene	0.9073
	Air	1.0000
	Ammonia	0.5888
	Carbon dioxide	1.529
	Carbon monoxide	0.9672
	Freon (F-12)	4.520
	Helium	0.1381
	Hydrogen	0.06952
	Methane	0.5544
	Nitrogen	0.9672
	Oxygen	1.105
	Propane	1.562
	Water vapor (steam)	0.6217

perature, the gas is said to be *saturated.* In this state, a slight increase in moisture content or a slight decrease in temperature will cause liquid to condense out of the gas. As we all know, saturated air in the environment generally produces rain.

The ability of a gas to hold moisture, however, is directly related to gas temperature: heating the gas increases the amount of moisture it can hold before reaching saturation, while cooling the gas decreases this allowable moisture content. In fact, any gas, regardless of its original humidity level, can be brought to saturation conditions simply by cooling it to a temperature known as the *dew point.* At this reduced temperature, the ability of the gas to hold moisture has decreased to a humidity level that equals the original moisture content of the gas, and condensation begins. It is this phenomenon that causes droplets of moisture to form on the outside of a lemonade pitcher that has been set outdoors on a warm summer day, as air directly adjacent to the cold surface of the pitcher is cooled to its dew point.

At any temperature, the water vapor contained in a gas exists at a pressure known as its *vapor pressure,* a property of the liquid that varies with temperature, as shown in Table 9–4. Vapor pressures are used to compute the *relative humidity* of a gas according to the formula:

$$H = \frac{p_a}{p_s} \tag{9–7}$$

where H is the relative humidity, p_a is the existing vapor pressure (equal to the saturation vapor pressure at the dew point temperature of the gas), and p_s is the vapor pressure required for saturation at the actual temperature of the gas. Relative humidity, usually expressed as a percentage, is used as an indicator of the humidity level or moisture content of a gas and is routinely included in most regional weather forecasts. Saturated air, for example, has a relative humidity of 100%, while ASME standard air has, by definition, a relative humidity of 36%.

Table 9–4 Vapor Pressure of Water Under Saturated Conditions	Temperature		Pressure (psi)
	°C	°F	
	0	32	0.0893
	10	50	0.1776
	20	68	0.3378
	30	86	0.6139
	40	104	1.068
	50	122	1.786
	60	140	2.884
	70	158	4.512
	80	176	6.855
	90	194	10.15
	100	212	14.7

EXAMPLE 7

On a particular day, it is determined that air at 68°F has a dew point of 50°F. Find the relative humidity of this air.

Solution   From Table 9–4, the vapor pressure required for *saturation* at the existing temperature of 68°F is $p_s = 0.3378$ psi. The *actual* vapor pressure in the air is found using the dew point, for we know that at this temperature the vapor pressure is the saturation value. Therefore, at a dew point of 50°F, Table 9–4 yields $p_a = 0.1776$ psi. The relative humidity is then:

$$H = \frac{p_a}{p_s} = \frac{0.1776 \text{ psi}}{0.3378 \text{ psi}}$$

$$= 0.5258 = \textbf{52.6\%}$$

The moisture content of a compressed gas is of considerable importance in a pneumatic system. We have seen that the heat of compression can raise the temperature of a gas and thus its ability to capture and retain moisture. Conversely, expansions of the gas that occur during use can drop temperatures to the dew point, causing condensation to form within the system. The long-term corrosive effects of this liquid are detrimental to components, and, in the case of gas storage tanks, can represent a serious threat to the safety of personnel working in the area. Other effects of water vapor on pneumatic components can include erratic operation of valves and cylinders, increased wear as lubricants are washed away, and malfunctioning of air-logic devices used for system control.

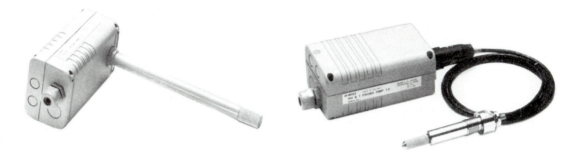

Figure 9–10   The sensor at left has a 9-in. probe and is used to measure relative humidity of gas flowing in a duct, while the unit at right performs the same function in pressurized spaces. *(Vaisala Inc., Woburn, Mass.)*

For this reason, most compressed gas systems contain some type of conditioning equipment to dry the gas before it reaches any major components in the circuit. These driers typically function much like a home dehumidifier that cools the air and separates out the resulting condensate. Humidity sensors such as those shown in Figure 9–10 can be used to measure/monitor the relative humidity of a gas.

Problem Set 2

1. In the SI system, absolute temperatures are represented on the Kelvin scale. These temperatures, T_K, are related to Celsius temperatures, T_C, by:

$$T_K = T_C + 273$$

   (a) Find the specific weight in N/m^3 of nitrogen at a pressure of 13,800 kPaa and a temperature of 50°C.
   (b) What is the specific gravity of the gas under these conditions?
   (c) Recompute (a) and (b) if the gas temperature drops to 25°C.

2. A television weather channel gives the ambient temperature as 86°F and the dew point temperature as 50°F. What is the corresponding relative humidity?

3. If ASME standard air has a relative humidity of 36% at a temperature of 68°F, what is the dew point? Estimate your answer from Table 9–4.

4. Air at 50°F has a relative humidity of 40%. To approximately what temperature must this air be heated so that its relative humidity drops to 20%?

9.3   The Compression Process

In Chapter 10 we will discuss several different types of compressors, including the popular reciprocating piston compressor. During each cycle, this type of machine draws in a fixed volume of gas, compresses the gas, and discharges this pressurized fluid to a pneumatic

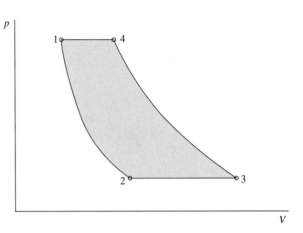

Figure 9–11 A typical single-stage compression cycle. (Numbers correspond to gas volumes shown in Figure 9–12.) The shaded area represents net work done on the gas during one cycle.

system. As shown in Figure 9–11, an entire compression cycle consists of four separate processes that correspond to the piston-valve positions of Figure 9–12.

In position 1, the piston is at the top of its stroke and both intake and exhaust valves are closed. The space above the piston, V_1, is called its *clearance volume* and generally contains some residual gas at the discharge pressure p_1. As the piston moves downward, pressure in the cylinder drops, but once the piston has reached position 2 the intake valve opens and gas flows into the cylinder at constant pressure. This process continues until the piston reaches the bottom of its stroke at position 3. Here the intake valve closes, and compression of the gas takes place as the piston moves upward to position 4. At this point the exhaust valve opens, and pressurized gas is discharged into the system as the piston continues to move back to its original position 1.

In an ideal compressor, processes 1-2 and 3-4 are considered to be adiabatic, while processes 2-3 and 4-1 are isobaric. It may be shown mathematically that the net work required for an entire compression cycle is the shaded area bounded by the process lines 1-2-3-4. Notice that this area may be broken down into small rectangles whose areas would each be computed as the product of pressure times volume and therefore would have units of work (lb/ft^2 × ft^3 = ft-lb). Although real compressors experience losses due to gas turbulence, heating of the gas, and flow through the intake and exhaust valves, machines of modern design usually achieve compression efficiencies in the range of 85% to 95%. However, the frictional effects of mechanical elements such as bearings and piston rings generally reduce these figures by about 10%, resulting in overall efficiencies of from 75% to 85%.

The compression cycle represented in Figure 9–11 is known as a single-stage process since the gas is compressed from initial pressure to final pressure in one step. Many modern compressors, however, are of multistage design in which the gas is removed from the cylinder before compression is complete, then cooled using a device known as an *intercooler* before being fed into the next stage where the compression process is continued. Figure 9–13 illustrates the compression cycle for a two-stage machine in which gas is removed from the first-stage cylinder at point *A*, cooled at constant pressure to the reduced volume at point *B*, and compressed by a second-stage cylinder to final pressure at

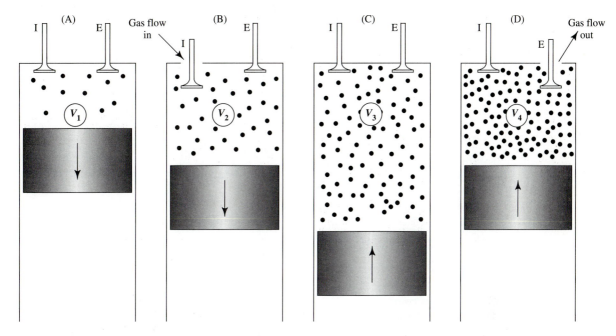

Figure 9–12 Piston and valve movements in a reciprocating piston compressor (I = intake valve; E = exhaust valve). (A) The piston is at the top of its stroke and begins to move downward. Because both valves are closed, gas pressure drops. (B) Partway through the piston stroke, the intake valve opens and gas begins to flow into cylinder at constant (low) pressure. (C) At the bottom of the stroke, the intake valve closes and the piston begins to move upward, compressing the gas. (D) Partway through the compression stroke, the exhaust valve opens and gas flows out of the cylinder at constant (high) pressure, as the piston returns to condition (A).

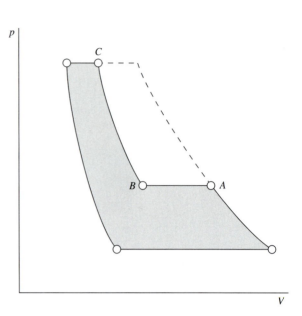

Figure 9–13   Compression cycle for a two-stage machine. Partially compressed gas at *A* is cooled in an *intercooler* to the lower temperature and volume at *B* before final compression to *C* takes place. Both the net work required (shaded area) and volume of gas delivered in one cycle are less than that of a single-stage compressor (compare to Figure 9–11).

C. Note that the total work required as indicated by the shaded area within the process lines is less than that required for a single-stage compression. You can also notice that because gas volume is reduced by the cooling process, the volume of gas delivered at final pressure is also reduced. Despite this fact, multistage compressors generally provide more economical operation than their single-stage counterparts. In addition, machines that supply gases at extremely high pressures invariably require several stages, often as many as six or seven, to reach the desired operating level.

Questions

1. Which two gases are the major components of air? List several minor constituents generally found in air.

2. What three variables are used to describe the state of a gas?

3. What is an ideal gas? How do real gases differ from an ideal gas?

4. What is free air? What is standard air?

5. Match each type of compression process with its description:

Process	**Description**
(a) Adiabatic	(i) Constant temperature
(b) Isobaric	(ii) Constant volume
(c) Isothermal	(iii) No heat transfer
(d) Isometric	(iv) Constant pressure

6. Match each compression process with the physical law that describes it:

Process	Physical Law
(a) Isobaric	(i) Boyle's law
(b) Isothermal	(ii) Gay-Lussac's law
(c) Isometric	(iii) Charles's law

7. If air is released from an inflated automobile tire, the moving fluid feels cool. Why? How would you classify this process? (*Hint:* See Figure 9–7 and Example 5(b).)

8. Explain the difference between compression ratio and ratio of compression.

9. Does an increase in temperature produce an increase or decrease in the viscosity of a gas? Explain.

10. What is a frangible disc? A fusible plug? Explain how these devices operate and where they are commonly used.

11. What is the relative humidity if dew-point temperature and actual air temperature are the same?

12. On a winter day, is the relative humidity inside a house the same as the relative humidity outside? Are the dew points inside and outside the house equal?

Pneumatic Components and Systems

Pneumatic systems using compressed air have three major applications: *actuation, processing,* and *control.* A variety of linear, rotary, and reciprocating motions are available through the use of pneumatic *actuators,* which are generally faster than their hydraulic counterparts. For this reason, they are often used in systems where speed is important, such as automated systems of the type shown in Figure 1–6. In addition, pneumatic equipment offers a low weight-to-power ratio and is well suited to mobile operation; rugged, durable tools such as rock drills (Figure 10–1) and pavement breakers are

Figure 10–1 Pneumatic rock drills are widely used in site preparation for new construction. The holes drilled by this worker were filled with explosives to blast out a rock formation on a lot where a service station was to be built.

commonplace in the construction industry, while fixed and hand tools such as riveters (Figure 10–2), drills, wrenches, grinders, and saws (Figure 10–3) find application in many manufacturing and service operations.

Compressed air can also be used in chemical and thermal applications where it becomes an actual *process component.* When mounted on an automobile engine, for example, the supercharger shown in Figure 10–4 draws in gasoline and air, compresses the mixture, and feeds it to the combustion process taking place inside the engine.

Finally, as we shall see in Chapter 11, there exists a large assortment of pneumatic sensors, logic elements, and switches for use in the *control* of fluid power systems; the many functions they perform include timing, counting, measuring, and sequencing.

Figure 10–2 Industrial pneumatic tools such as this riveter can provide continuously variable squeezing forces from less than 1 oz to more than 7 tons. They are well suited to a variety of operations, including hole punching, riveting, and press assembly. *(General Pneumatic Tools, Post Falls, Idaho)*

(B)

(A)

(C)

Figure 10–3 Pneumatic tools such as the grinder (A), wrench (B), and circular saw (C) offer lower weight-to-power ratios than their electrical counterparts and are not prone to motor burnouts if the tools stall. *(Ingersoll-Rand Company, Power Tool Division, Liberty Corner, N.J.)*

Figure 10–4 This automotive supercharger furnishes a mixture of compressed air and gasoline for the combustion *process.* Although such devices are driven by belt and pulley from the engine, many modern automobiles utilize *turbochargers* that perform the same function but are driven by the engine's exhaust gases.

10.1 A Compressed Air Supply System

Unlike hydraulic fluids, air is available anywhere and is virtually free for the asking. To be used in a conventional pneumatic system, this fluid must first be pressurized to a level typically in the range of 90 psi to 150 psi (620 kPa to 1035 kPa). The compressed air is then stored or accumulated in tanks called *receivers,* from which it is drawn on demand for use in various processes and actuators. Before its use, however, most compressed air first passes through one or more devices such as *filters, driers, regulators,* and *lubricators* that *condition* the fluid for its end use.

Although many small compressed air systems supply a single application, industrial systems such as the one shown in Figure 10–5 often provide working fluid for a variety of diverse uses throughout a manufacturing facility. Stationary systems are usually engineered for a specific installation, while commercially available mobile or portable units are selected based upon the power requirements for a particular application. For these reasons, compressed air supply systems—power source, compressor, and receiver—are rarely assembled from components as are their hydraulic counterparts; it is much more likely for a technologist to become involved in simply establishing a supply line of conditioned air from some central source. Since this does require a familiarity with both source and conditioning components, let us examine a supply system from compressor to end use.

10.2 Compressors

Compressors represent the pneumatic counterparts of hydraulic pumps, moving specified volumes of gas against some working pressure. Although machines are available that can operate on a variety of gases and vapors, including steam, the majority of industrial units deliver compressed air for the operation of tools and actuators. Although compressors generally take free air at atmospheric conditions and raise the pressure of the gas significantly, there are several exceptions. *Blowers* are compressors that move large volumes of air at low pressures, generally less than 40 psi, for use in cleaning, drying, ventilation, aeration, media agitation, and combustion processes. *Vacuum pumps* operate at inlet pressures considerably less than atmospheric and have discharge pressures at, or slightly above, atmospheric. Applications for these units include evaporators, power plant condensers, packaging machinery, and freeze dryers used in the processing of food.

Compressors are generally classified either as *positive displacement* or *dynamic* machines. Positive displacement compressors can be of the *reciprocating* or *rotary* type; with each stroke or rotation they deliver a specific volume of pressurized gas. Dynamic compressors, often called *turbomachines,* produce a significant increase both in pressure and velocity of the gas and deliver a continuous stream of fluid. Depending upon the direction of flow within these machines, they can be defined as *axial, centrifugal* (radial), or *mixed* flow compressors.

One of the most common types of positive displacement compressors is the reciprocating piston machine, an example of which is shown in Figures 10–6 and 10–7. Because they are compact and completely self-contained, these units are popular in the

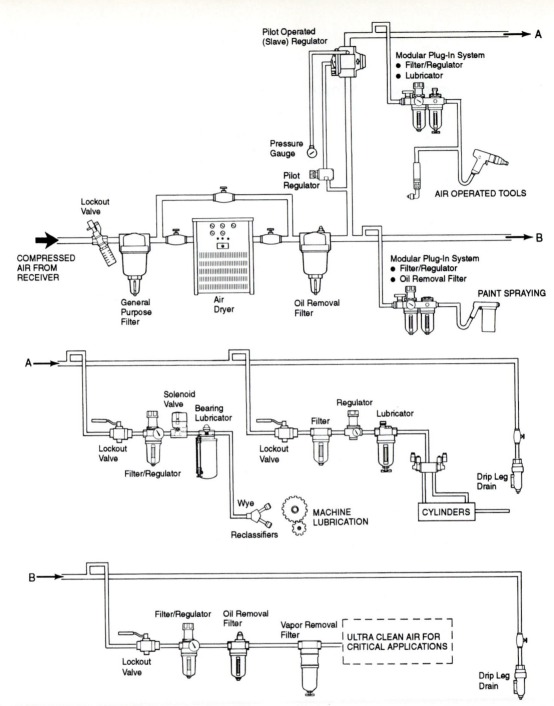

Figure 10–5 Typical conditioners for an industrial compressed air supply system. In this diagram, the main supply line (top left) is split into two legs, which feed a total of five outlets used for different applications. Line A (top right) supplies fluid for air-operated tools, machine lubrication, and pneumatic cylinders (middle). Line B provides air for paint spraying (top) and critical applications (bottom). Note that each leg of the system uses a different combination of hardware to condition the fluid for its specific use. *(Norgren Co., Littleton, Colo.)*

Figure 10–6 An electrically driven reciprocating compressor. *(Champion Pneumatic Machinery Co., Princeton, Ill.)*

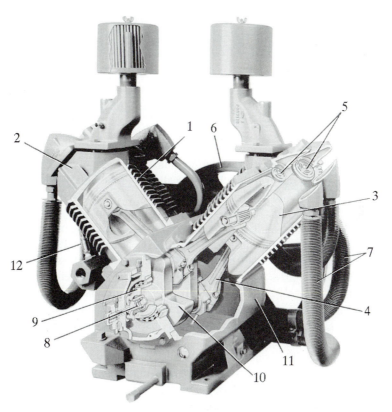

Figure 10–7 Internal construction of a reciprocating compressor: 1, finned cylinder; 2, cylinder/head; 3, piston; 4, connecting rod; 5, valves; 6, flywheel; 7, intercoolers; 8, centrifugal unloader; and 9, main bearings. The compression process for such a machine is illustrated in Figure 9–12. *(Champion Pneumatic Machinery Co., Princeton, Ill.)*

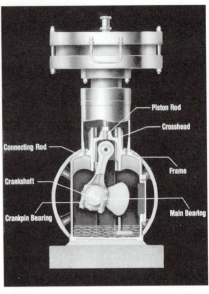

(A)

(B)

Figure 10–8 Devices used to supply oil-free air include (A) the oil-less compressor *(JUN-AIR (USA) Inc., Buffalo Grove, Ill.)* and (B) the diaphragm compressor *(Pressure Products Industries, Warminster, Pa.).*

service station and construction industries where portable or mobile machines are used for intermittent service. They can also be found in large, stationary installations used to supply a complete manufacturing facility.

Since hydraulic pumps experience a continuous flow of the working liquid (oil), no special system is required for lubrication of internal moving parts. A gas, however, does not possess this lubricating quality, so care must be taken in the design of compressors to ensure that the bearings and compression pistons receive an adequate supply of oil. Some of this lubricant invariably mixes with the compressed gas itself and travels throughout the system. Although this oil mist also helps lubricate any tools and actuators that operate on the compressed gas, certain applications require that the oil be removed by filtration; some uses, in fact, demand that the gas be *oil-free.* (One example of this is the compressed air provided in aqualungs for use by scuba divers.) Such gas purity can be achieved by constructing the cylinder and piston of carbon-based or synthetic materials that operate without lubrication, or by using a *diaphragm compressor* (Figure 10–8). In diaphragm machines, reciprocation of a mechanically driven hydraulic piston compresses a pocket of liquid against a flexible diaphragm, which in turn compresses a fixed volume of gas on the backside of the diaphragm (Figure 10–9). This process isolates the gas from possible sources of contamination during the compression process.

The *sliding vane* compressor is a rotary type positive displacement machine whose construction (Figure 10–10) and operation are similar to that of a vane type hydraulic pump. (See Figure 7–25.) *Rotary screw* compressors (Figure 10–11 and Figure 10–12)

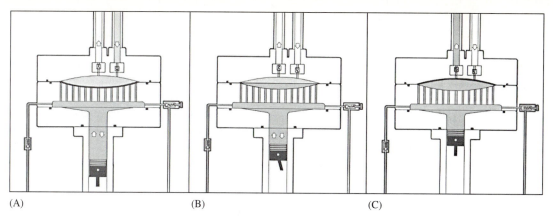

Figure 10–9 Operation of a diaphragm compressor. (A) Hydraulic piston is at bottom dead center. Diaphragm group is fully deflected to bottom of cavity by process gas that has entered through the inlet check valve. (B) As crankshaft rotates, hydraulic piston moves upwards. Hydraulic pressure deflects diaphragm group, compressing the process gas. (C) Gas in cavity is compressed to system discharge pressure. Discharge check valve opens; process gas is discharged. Hydraulic piston reaches top dead center and diaphragm group is fully deflected. Hydraulic fluid is overpumped through the hydraulic relief valve, and compression cycle is complete. *(Pressure Products Industries, Warminster, Pa.)*

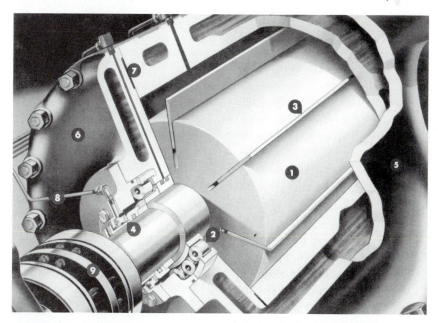

Figure 10–10 Internal construction of sliding vane rotary compressor: 1, rotor and shaft; 2, cylindrical roller bearings; 3, vanes (made of laminated cloth impregnated with phenolic resin and heat treated); 4, mechanical seals; 5, cylinder housing (Note cooling jacket cast into housing.); 6, cylinder head; 7, gaskets; 8, lubrication system; and 9, motor coupling. The appearance and principal of operation resemble those of a vane-type hydraulic pump such as the one pictured in Figure 7–25. *(A-C Compressor Corporation, Elm Grove, Wis.)*

339

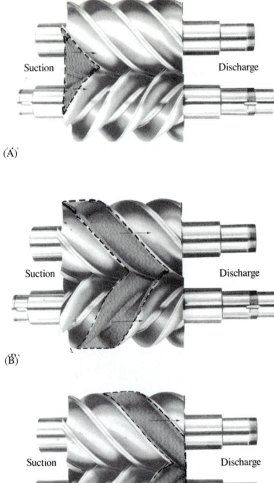

(A)

(B)

(C)

Figure 10–11 Movement of gas (rippled area) caused by meshing action of lobes on male and female rotors in a rotary screw compressor. Such machines produce a smooth, continuous supply of gas that is relatively pulsation-free.
(Dunham-Bush, Inc., West Hartford, Conn.)

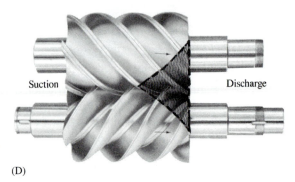

(D)

(A)

(B)

(C)

Figure 10–12 Like these rotary screw machines, most small compressors may be tank mounted (A) or base mounted (B). Cabinet-enclosed installations (C) are becoming more popular for their safety and sound-deadening qualities. *(Ingersoll-Rand Company, Rotary-Reciprocating Compressor Division, Davidson, N.C.)*

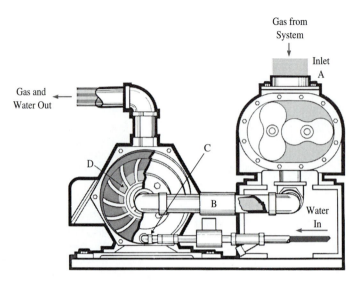

Figure 10–13 Liquid ring compressor used as a vacuum pump. The system is evacuated as rotating lobes draw gas into the machine at A, and convey this fluid through pipe B to the main compressor at D. Under the influence of centrifugal force, water entering at C forms a liquid ring around the rotor, trapping gas between the vanes. As radial clearance decreases, this liquid compresses and cools the gas, with the entire mixture discharged at the compressor outlet. If inlet A is open to atmosphere, the machine operates in the same manner but serves as a compressor rather than a vacuum pump. *(Atlantic Fluidics, Inc., So. Norwalk, Conn.)*

use the meshing action of lobes on male and female rotors to deliver a relatively pulsation-free supply of compressed gas.

An unusual type of rotary, positive displacement compressor is the *liquid ring* unit shown operating as a vacuum pump in Figure 10–13. Gas enters at inlet A, passing through the counterrotating lobes of a blower, and then on to the compressor through line B. Water enters the compressor at C and is thrown to the outside of the compression chamber by centrifugal force. The water actually forms a liquid ring that traps pockets of gas (D) between the blades of the impeller. As radial clearance decreases, this liquid is forced into the spaces between blades, compressing the trapped gas. The entire water-gas mixture is then discharged and separated as necessary. Because of the water's cooling effect, compression in a liquid ring machine is closer to an isothermal process rather than an adiabatic one. Remember also that vacuum pumps are simply compressors whose *inlets* are connected to the system, typically as shown in Figure 10–14. An air compressor, for example, draws free air in through its inlet and discharges pressurized air into the system, while a vacuum pump pulls air *from* the system and discharges this air to atmosphere.

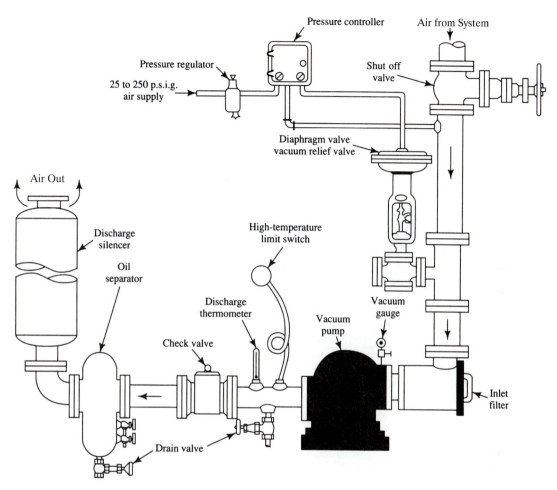

Figure 10–14 Typical piping arrangement for vacuum pump used to evacuate air from a system. *(A-C Compressor Corporation, Elm Grove, Wis.)*

Dynamic compressors find their greatest use in continuous flow processes such as those common to refineries, power plants, chemical production facilities, steel mills, and some food-processing industries. These units, often called *turbocompressors,* may be quite large, produce high flow rates (often at low pressures), and, like the machine shown in Figure 10–15, generally operate without a *receiver* or storage tank. Flow in a dynamic compressor can be axial, radial (often called centrifugal), or a combination of the two known as *mixed* flow (Figure 10–16). For a given impeller diameter, mixed-flow machines produce flows two to three times greater than those of a radial machine, but

Figure 10–15 Typical mixed-flow dynamic compressor unit. Also called turbocompressors, dynamic machines can produce high flow rates and often operate without a receiver. *(Ingersoll-Rand Company, Centrifugal Compressor Division, Mayfield, Ky.)*

generally at lower pressures. Construction of a typical turbocompressor is detailed in Figure 10–17.

Schematic representations of compressors and pneumatic motors are similar to those of hydraulic pumps and motors except that compressors are never reversible and arrows indicating directions of flow are in outline rather than solid form. These differences are illustrated in Figure 10–18.

As with hydraulic pumps, compressor selection is based upon delivery of a given flow rate at a specified working pressure. Although many of the machines discussed here have different operating characteristics that suit them for particular applications, there is considerable overlap on the flow rates and pressures obtainable from the various types of compressors (see Figure 10–19). Economic factors that may influence the selection process include initial cost of the machine, cost of installation, maintenance costs, and energy costs for operation. Interestingly, over the life of a typical compressor, initial, installation, and maintenance costs are normally quite small compared to the total energy costs.

Required compressor capacity for various tools and actuators can be estimated using Tables 10–1 and 10–2. However, care must be taken to distinguish between the

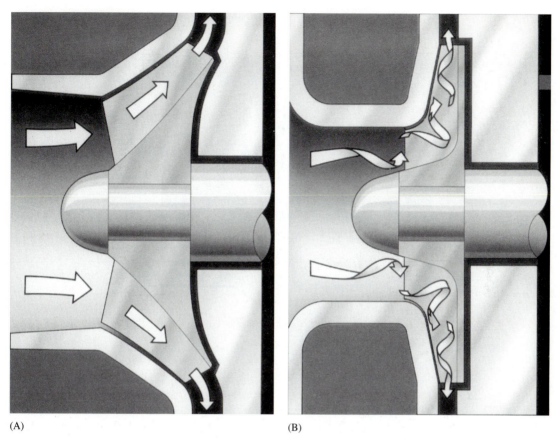

(A)   (B)

Figure 10–16   Flow characteristics of turbocompressors. In the mixed-flow machine (A), movement of fluid is a combination of axial and radial flow, while that of a centrifugal machine (B) is primarily in the radial direction. For a given impeller diameter, mixed-flow machines can produce flow rates up to three times greater than those of radial machines but generally at lower pressures. *(Ingersoll-Rand Company, Centrifugal Compressor Division, Mayfield, Ky.)*

various flow rates commonly used to describe gas production and usage. Note that while the consumption rates of Table 10–1 are given in *cubic feet per minute of free air*, both the compressor flow rates of Figure 10–19 and the cylinder consumption rates of Table 10–2 are given in *actual cubic feet per minute* (acfm) delivered at the specified or required pressure. Consumption rates should also be adjusted through the use of a *load factor*, which represents a combination of the fractional time for which a tool is in use and the percentage of its maximum performance capacity actually used. A tool that is operated continuously at its maximum capability has a load factor of 1.00 or 100%.

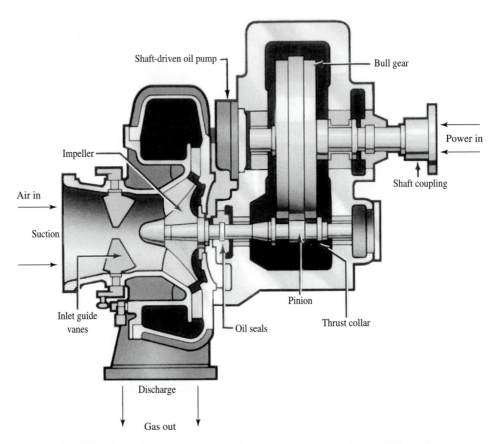

Figure 10–17 Assembly of mixed-flow dynamic compressor. *(Ingersoll-Rand Company, Centrifugal Compressor Division, Mayfield, Ky.)*

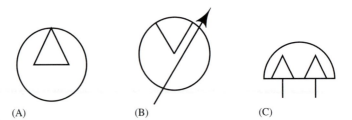

(A)   (B)   (C)

Figure 10–18 Schematic representations of several common pneumatic devices: (A) fixed capacity compressor; (B) variable capacity, nonreversible pneumatic motor; (C) oscillating pneumatic motor.

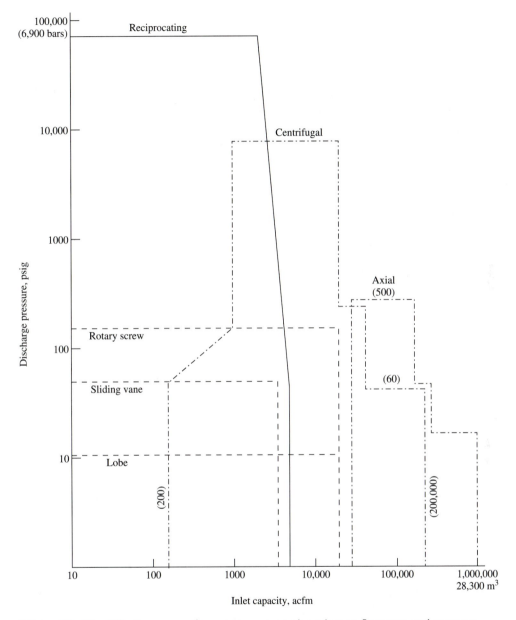

Figure 10–19 Effective range of compressor types based upon flow rate and pressure. Note that the scales are logarithmic, and so divisions along each axis increase by *factors of ten.* For a typical sliding-vane compressor, then, this figure yields approximate values of 3500 acfm for maximum flow rate and 50 psig for maximum discharge pressure. *(From* Compressed Air and Gas Handbook *of the Compressed Air and Gas Institute, 5th ed., Figure 4.6. Upper Saddle River, N.J.: Prentice Hall, 1989.)*

Table 10–1 Consumption Rates of Various Tools*

(From Compressed Air and Gas Handbook, 4th ed., Table 4.1. New York: Compressed Air and Gas Institute, 1973)

Tool	Free Air, cfm at 90 psig, 100% Load Factor	Tool	Free Air, cfm at 90 psig, 100% Load Factor
Grinders, 6- and 8-in. wheels	50	Plug drills	40–50
Grinders, 2- and $2^1/2$-in. wheels	14–20	Riveters, $^3/_{32}$–1-in. rivets	12
File and burr machines	18	larger weighing 18–22 lb	35
Rotary sanders, 9-in. pads	53	Rivet busters	35–39
Rotary sanders, 7-in. pads	30	Wood borers to 1 in. diameter weighing 4 lb.	40
Sand rammers and tampers		2 in. diameter weighing 26 lb	80
1 × 4-in. cylinder	25	Steel drills, rotary motors	
$1^1/4$ × 5-in. cylinder	28	Capacity up to $^1/4$ in. weighing $1^1/4$–4 lb	18–20
$1^1/2$ × 6-in. cylinder	39	Capacity $^1/4$ to $^3/8$ in. weighing 6–8 lb	20–40
Chipping hammers, weighing 10–13 lb	28–30	Capacity $^1/2$ to $^3/4$ in. weighing 9–14 lb	70
heavy	39	Capacity $^7/8$ to 1 in. weighing 25 lb	80
weighing 2–4 lb	12	Capacity $1^1/4$ in. weighing 30 lb	95
Nut setters to $^5/_{16}$ in. weighing 8 lb	20	Steel drills, piston type	
Nut setters $^1/2$ to $^3/4$ in. weighing 18 lb	30	Capacity $^1/2$ to $^3/4$ in. weighing 13–15 lb	45
Sump pumps, 145 gal (a 50 ft head)	70	Capacity $^7/8$ to $1^1/4$ weighing 25–30 lb	75–80
Paint spray, average	7	Capacity $1^1/4$ to 2 in. weighing 40–50 lb	80–90
varies from	2–20	Capacity 2 to 3 in. weighing 55–75 lb	100–110
Bushing tools (monument)	15–25		
Carving tools (monument)	10–15		

Cubic Feet of Air per Minute Required by Sand Blast

Nozzle Diameter, in.	Compressed Air Gage Pressure			
	60 lb	70 lb	80 lb	100 lb
$^1/_{16}$	4	5	5.5	6.5
$^3/_{32}$	9	11	12	15
$^1/8$	17	19	21	26
$^3/_{16}$	38	43	47	58
$^1/4$	67	76	85	103
$^5/_{16}$	105	119	133	161
$^3/8$	151	171	191	232
$^1/2$	268	304	340	412

*Note that these values are approximate; tools from different manufacturers may vary by more than 10% from the given consumption rates.

348

Table 10–2 Volume of Compressed Air (ft^3 per Stroke) Required to Operate Various Pneumatic Cylinders*

(From Compressed Air and Gas Handbook, 4th ed., Table 4.2, New York: Compressed Air and Gas Institute, 1973)

Piston Dia. (in.)	Length of Stroke (in.)											
	1	2	3	4	5	6	7	8	9	10	11	12
1¼	.00139	.00278	.00416	.00555	.00694	.00832	.00972	.0111	.0125	.0139	.0153	.01665
1⅞	.00158	.00316	.00474	.00632	.0079	.00948	.01105	.01262	.0142	.0158	.0174	.01895
2	.00182	.00364	.00545	.00727	.0091	.0109	.0127	.0145	.01636	.0182	.020	.0218
2⅛	.00205	.0041	.00615	.0082	.0103	.0123	.0144	.0164	.0185	.0205	.0226	.0244
2¼	.0023	.0046	.0069	.0092	.0115	.0138	.0161	.0184	.0207	.0230	.0253	.0276
2⅜	.00256	.00512	.00768	.01025	.0128	.01535	.01792	.02044	.0230	.0256	.0282	.0308
2½	.00284	.00568	.00852	.01137	.0142	.0171	.0199	.0228	.0256	.0284	.0312	.0343
2⅝	.00313	.00626	.0094	.01254	.01568	.0188	.0219	.0251	.0282	.0313	.0345	.0376
2¾	.00343	.00686	.0106	.0137	.0171	.0206	.0240	.0272	.0308	.0343	.0378	.0412
2⅞	.00376	.00752	.0113	.01503	.01877	.0226	.0263	.0301	.0338	.0376	.0413	.045
3	.00409	.00818	.0123	.0164	.0204	.0246	.0286	.0327	.0368	.0409	.0450	.049
3⅛	.00443	.00886	.0133	.0177	.0222	.0266	.0310	.0354	.0399	.0443	.0488	.0532
3¼	.0048	.0096	.0144	.0192	.024	.0288	.0336	.0384	.0432	.0480	.0529	.0575
3⅜	.00518	.01036	.0155	.0207	.0259	.031	.0362	.0415	.0465	.0518	.037	.062
3½	.00555	.0112	.0167	.0222	.0278	.0333	.0389	.0445	.050	.0556	.061	.0644
3⅝	.00595	.0119	.0179	.0238	.0298	.0357	.0416	.0477	.0536	.0595	.0655	.0715
3¾	.0064	.0128	.0192	.0256	.032	.0384	.0447	.0512	.0575	.064	.0702	.0766
3⅞	.0068	.01362	.0205	.0273	.0341	.041	.0477	.0545	.0614	.068	.075	.082
4	.00725	.0145	.0218	.029	.0363	.0435	.0508	.058	.0653	.0725	.0798	.087
4⅛	.00773	.01547	.0232	.0309	.0386	.0464	.0541	.0618	.0695	.0773	.0851	.092
4¼	.0082	.0164	.0246	.0328	.041	.0492	.0574	.0655	.0738	.082	.0903	.0985
4⅜	.0087	.0174	.0261	.0348	.0435	.0522	.0608	.0694	.0782	.087	.0958	.1042
4½	.0092	.0184	.0276	.0368	.046	.0552	.0643	.0735	.0828	.092	.101	.1105
4⅝	.0097	.0194	.0291	.0388	.0485	.0582	.0679	.0775	.0873	.097	.1068	.1163
4¾	.01025	.0205	.0308	.041	.0512	.0615	.0717	.0818	.0922	.1025	.1125	.123
4⅞	.0108	.0216	.0324	.0431	.054	.0647	.0755	.0862	.097	.108	.1185	.1295
5	.0114	.0228	.0341	.0455	.0568	.0681	.0795	.091	.1023	.114	.125	.136
5⅛	.01193	.0239	.0358	.0479	.0598	.0716	.0837	.0955	.1073	.1193	.1315	.1435
5¼	.0125	.0251	.0376	.0502	.0627	.0753	.0878	.100	.1128	.125	.138	.151
5⅜	.0131	.0263	.0394	.0525	.0656	.0788	.092	.105	.118	.131	.144	.158
5½	.01375	.0275	.0412	.055	.0687	.0825	.0962	.110	.1235	.1375	.151	.165
5⅝	.0144	.0288	.0432	.0575	.072	.0865	.101	.115	.1295	.144	.1585	.173
5¾	.015	.030	.045	.060	.075	.090	.105	.120	.135	.150	.165	.180
5⅞	.0157	.0314	.047	.0628	.0785	.094	.110	.1254	.142	.157	.1725	.188
6	.0164	.032	.0492	.0655	.082	.0983	.1145	.131	.147	.164	.180	.197

*These volumes are for single-acting cylinders. For double-acting cylinders, multiply by 2 and subtract the volume of the piston rod.

EXAMPLE I

A manufacturing facility used for light assembly work employs 20 pneumatic riveting machines requiring compressed air at 90 psi. Because of their intermittent operation and the fact that each riveter functions at only a fraction of its maximum potential, the overall load factor is 38%.

   (a) Find the total acfm required at this pressure to supply the manufacturing facility.

   (b) Repeat part (a) for an operating pressure of 120 psi.

Solution   (a) From Table 10–1, light riveters operated at 90 psi and 100% capacity each require 12 cfm of *free air.* Assuming that the air is compressed adiabatically from atmospheric pressure to the working pressure, then equation (9-6) can be used with $p_1 = 14.7$ psia, $p_2 = 104.7$ psia, and $k = 1.40$ (see Table 9–2). Equation (9-6) becomes:

$$\left(\frac{p_1}{p_2}\right)^{1/k} = \frac{V_2}{V_1}$$

and:

$$\left(\frac{14.7 \text{ psia}}{104.7 \text{ psia}}\right)^{1/1.40} = \frac{V_2}{V_1}$$

or:

$$V_2 = 0.2460 V_1$$

Note that if both sides of this equation are divided by time, t, then:

$$\frac{V_2}{t} = 0.2460 \frac{V_1}{t}$$

or:

$$q_2 = 0.2460 q_1$$

where q_1 and q_2 are the volume flow rates. Thus:

$$q_2 = 0.2460 \times 12 \text{ cfm} = 2.952 \text{ acfm}$$

Then each riveter consumes 2.952 acfm, and the total flow required at 90 psi, q_{tot}, by 20 machines having a 38% load factor is:

$$q_{tot} = 0.38 \times 20 \text{ machines} \times 2.952 \text{ acfm/machine}$$

$$= \textbf{22.4 acfm}$$

   (b) Although increasing the pressure to 120 psi may increase the riveting *forces* available from each machine, *speed* of operation is controlled by available flow rate. In order to maintain the original speed available at 90 psi, each machine must still receive 2.952 acfm and the compressor must still supply a total flow of 22.4 acfm. What does change is the consumption rate of free air for each machine, for if $p_2 = 120$ psi $= 134.7$ psia, then:

$$\left(\frac{14.7 \text{ psia}}{134.7 \text{ psia}}\right)^{1/1.40} = \frac{V_2}{V_1}$$

or:

$$V_2 = 0.2055V_1$$

and:

$$q_2 = 0.2055q_1$$

Since q_2 remains at 2.952 acfm per machine, then the flow rate of free air, q_1, required for *each* riveter becomes:

$$q_1 = \frac{2.952 \text{ acfm}}{0.2055} = \textbf{14.4 cfm}$$

Pressure capability of a compressor must be determined by taking into account friction losses that occur through hoses, pipes, and fittings between the compressor itself and the tools or actuators that it services. These losses can be considerable, since pneumatic supply lines are usually much longer than their hydraulic counterparts. Table 10–3, Table 10–4, and Table 10–5 give typical values for pressure drops in hose, pipe, and pipe fittings respectively.

SAFETY SIDEBAR

Don't Misuse Compressed Air

Wherever compressed air is available in a manufacturing facility, workers commonly use jets of air to blow debris from their machines and even to clean themselves by directing these jets of air against their own clothes or those of their fellow workers. This is an extremely dangerous practice! Metal particles and dirt can easily travel 30 ft or more through the air to lodge in the eyes of unsuspecting personnel standing nearby. In addition, documented fatalities have shown that compressed air can be deadly if it enters the bloodstream through cuts or abrasions in the skin. Treat any pressurized fluid with respect and avoid this practice of "blowing down." Always use compressed air in a safe and responsible manner.

Table 10–3 Friction Losses for Pulsating Flow in Air Hose.*
(From Compressed Air and Gas Handbook of the Compressed Air and Gas Institute, *5th ed., Table 13.27. Upper Saddle River, N.J.: Prentice Hall, 1989)*

Size of Hose, Coupled Each End (in.)	Gage Pressure at Line (lb)	Cu ft Free Air per min Passing Through 50-ft Lengths of Hose													
		20	30	40	50	60	70	80	90	100	110	120	130	140	150
		Loss of Pressure (psi) in 50-ft Lengths of Hose													
	50	1.8	5.0	10.1	18.1										
	60	1.3	4.0	8.4	14.8	23.4									
1/2	70	1.0	3.4	7.0	12.4	20.0	28.4								
	80	0.9	2.8	6.0	10.8	17.4	25.2	34.6							
	90	0.8	2.4	5.4	9.5	14.8	22.0	30.5	41.0						
	100	0.7	2.3	4.8	8.4	13.3	19.3	27.2	36.6						
	110	0.6	2.0	4.3	7.6	12.0	17.6	24.6	33.3	44.5					
	50	0.4	0.8	1.5	2.4	3.5	4.4	6.5	8.5	11.4	14.2				
	60	0.3	0.6	1.2	1.9	2.8	3.8	5.2	6.8	8.6	11.2				
3/4	70	0.2	0.5	0.9	1.5	2.3	3.2	4.2	5.5	7.0	8.8	11.0			
	80	0.2	0.5	0.8	1.3	1.9	2.8	3.6	4.7	5.8	7.2	8.8	10.6		
	90	0.2	0.4	0.7	1.1	1.6	2.3	3.1	4.0	5.0	6.2	7.5	9.0		
	100	0.2	0.4	0.6	1.0	1.4	2.0	2.7	3.5	4.4	5.4	6.6	7.9	9.4	11.1
	110	0.1	0.3	0.5	0.9	1.3	1.8	2.4	3.1	3.9	4.9	5.9	7.1	8.4	9.9
	50	0.1	0.2	0.3	0.5	0.8	1.1	1.5	2.0	2.6	3.5	4.8	7.0		
	60	1.1	0.2	0.3	0.4	0.6	0.8	1.2	1.5	2.0	2.6	3.3	4.2	5.5	7.2
	70	—	0.1	0.2	0.4	0.5	0.7	1.0	1.3	1.6	2.0	2.5	3.1	3.8	4.7
	80	—	0.1	0.2	0.3	0.5	0.7	0.8	1.1	1.4	1.7	2.0	2.4	2.7	3.5
	90	—	1.1	0.2	0.3	0.4	0.6	0.7	0.9	1.2	1.4	1.7	2.0	2.4	2.8
	100	—	1.1	0.2	0.2	0.4	0.5	0.6	0.8	1.0	1.2	1.5	1.8	2.1	2.4
	110	—	0.1	0.2	0.2	0.3	0.4	0.6	0.7	0.9	1.1	1.3	1.5	1.8	2.1
	50	—	—	0.1	0.2	0.2	0.3	0.4	0.5	0.7	1.1				
	60	—	—	—	0.1	0.2	0.3	0.3	0.5	0.6	0.8	1.0	1.2	1.5	
	70	—	—	—	0.1	0.2	0.2	0.3	0.4	0.4	0.5	0.7	0.8	1.0	1.3
1 1/4	80	—	—	—	—	0.1	0.2	0.2	0.3	0.4	0.5	0.6	0.7	0.8	1.0
	90	—	—	—	—	0.1	0.2	0.2	0.3	0.3	0.4	0.5	0.6	0.7	0.8
	100	—	—	—	—	—	0.1	0.2	0.2	0.3	0.4	0.4	0.5	0.6	0.7
	110	—	—	—	—	—	0.1	0.2	0.2	0.3	0.3	0.4	0.5	0.5	0.6
	50	—	—	—	—	—	0.1	0.2	0.2	0.2	0.3	0.3	0.4	0.5	0.6
	60	—	—	—	—	—	—	0.1	0.2	0.2	0.2	0.3	0.3	0.4	0.5
1 1/2	70	—	—	—	—	—	—	—	0.1	0.2	0.2	0.2	0.3	0.3	0.4
	80	—	—	—	—	—	—	—	—	0.1	0.2	0.2	0.2	0.3	0.4
	90	—	—	—	—	—	—	—	—	—	0.1	0.2	0.2	0.2	0.3
	100	—	—	—	—	—	—	—	—	—	—	0.1	0.2	0.2	0.2
	110	—	—	—	—	—	—	—	—	—	—	0.1	0.2	0.2	0.2

*For longer or shorter lengths of hose the friction loss is proportional to the length, i.e., for 25 ft one-half of the above; for 150 ft. three times the above, etc.

Table 10–4 Pipe Friction Losses in psi per 1000 ft of Pipe at an Initial Pressure of 100 psi* *(From Compressed Air and Gas Handbook of the Compressed Air and Gas Institute, 5th ed., Table 13.23. Upper Saddle River, N.J.: Prentice Hall, 1989)*

Cu ft Free Air per min	Cu ft Equivalent Compressed Air per min	Nominal Diameter (in.)															
		1/2	3/4	1	1 1/4	1 1/2	2	2 1/2	3	3 1/2	4	4 1/2	5	6	8	10	12
10	1.28	6.50	.99	0.28													
20	2.56	25.9	3.90	1.11	0.25	0.11											
30	3.84	58.5	9.01	2.51	0.57	0.26											
40	5.12	—	16.0	4.45	1.03	0.46											
50	6.41	—	25.1	6.96	1.61	0.71	0.19										
60	7.68	—	36.2	10.0	2.32	1.02	0.28										
70	8.96	—	49.3	13.7	3.16	1.40	0.37										
80	10.24	—	64.5	17.8	4.14	1.83	0.49	0.19									
90	11.52	—	82.8	22.6	5.23	2.32	0.62	0.24									
100	12.81	—	—	27.9	6.47	2.86	0.77	0.30									
125	15.82	—	—	48.6	10.2	4.49	1.19	0.46									
150	19.23	—	—	62.8	14.6	6.43	1.72	0.66	0.21								
175	22.40	—	—	—	19.8	8.72	2.36	0.91	0.28								
200	25.62	—	—	—	25.9	11.4	3.06	1.19	0.37	0.17							
250	31.64	—	—	—	40.4	17.9	4.78	1.85	0.58	0.27							
300	38.44	—	—	—	58.2	25.8	6.85	2.67	0.84	0.39	0.20						
350	44.80	—	—	—	—	35.1	9.36	3.64	1.14	0.53	0.27						
400	51.24	—	—	—	—	45.8	12.1	4.75	1.50	0.69	0.35	0.19					
450	57.65	—	—	—	—	58.0	15.4	5.98	1.89	0.88	0.46	0.25					
500	63.28	—	—	—	—	71.6	19.2	7.42	2.34	1.09	0.55	0.30					
600	76.88	—	—	—	—	—	27.6	10.7	3.36	1.56	0.79	0.44					
700	89.60	—	—	—	—	—	37.7	14.5	4.55	2.13	1.09	0.59					
800	102.5	—	—	—	—	—	49.0	19.0	5.89	2.77	1.42	0.78					
900	115.3	—	—	—	—	—	62.3	24.1	7.6	3.51	1.80	0.99					

Table 10–4 continued

Flow																	
1,000	128.1	—	—	—	—	—	76.9	29.8	9.3	4.35	2.21	1.22	—	—	—	—	—
1,500	192.3	—	—	—	—	—	—	67.0	21.0	9.8	4.9	2.73	1.51	0.57	—	—	—
2,000	256.2	—	—	—	—	—	—	—	37.4	17.3	8.8	4.9	2.72	0.99	0.24	—	—
2,500	316.4	—	—	—	—	—	—	—	58.4	27.2	13.8	8.3	4.2	1.57	0.37	—	—
3,000	384.6	—	—	—	—	—	—	—	84.1	39.1	20.0	10.9	6.0	2.26	0.53	—	—
3,500	447.8	—	—	—	—	—	—	—	—	58.2	27.2	14.7	8.2	3.04	0.70	0.22	—
4,000	512.4	—	—	—	—	—	—	—	—	69.4	35.5	19.4	10.7	4.01	0.94	0.28	—
4,500	576.5	—	—	—	—	—	—	—	—	—	45.0	24.5	13.5	5.10	1.19	0.36	—
5,000	632.8	—	—	—	—	—	—	—	—	—	55.6	30.2	16.8	6.3	1.47	0.44	0.17
6,000	768.8	—	—	—	—	—	—	—	—	—	80.0	43.7	24.1	9.1	2.11	0.64	0.24
7,000	896.0	—	—	—	—	—	—	—	—	—	—	59.5	32.8	12.2	2.88	0.87	0.33
8,000	1,025	—	—	—	—	—	—	—	—	—	—	77.5	42.9	16.1	3.77	1.12	0.46
9,000	1,153	—	—	—	—	—	—	—	—	—	—	—	54.3	20.4	4.77	1.43	0.57
10,000	1,280	—	—	—	—	—	—	—	—	—	—	—	67.1	25.1	5.88	1.77	0.69
11,000	1,410	—	—	—	—	—	—	—	—	—	—	—	—	30.4	7.10	2.14	0.83
12,000	1,540	—	—	—	—	—	—	—	—	—	—	—	—	36.2	8.5	2.54	0.98
13,000	1,668	—	—	—	—	—	—	—	—	—	—	—	—	42.6	9.8	2.98	1.15
14,000	1,795	—	—	—	—	—	—	—	—	—	—	—	—	49.2	11.5	3.46	1.35
15,000	1,926	—	—	—	—	—	—	—	—	—	—	—	—	56.6	13.2	3.97	1.53
16,000	2,050	—	—	—	—	—	—	—	—	—	—	—	—	64.5	15.0	4.52	1.75
18,000	2,310	—	—	—	—	—	—	—	—	—	—	—	—	81.5	19.0	5.72	2.22
20,000	2,560	—	—	—	—	—	—	—	—	—	—	—	—	—	23.6	7.0	2.74
22,000	2,820	—	—	—	—	—	—	—	—	—	—	—	—	—	28.5	8.5	3.33
24,000	3,080	—	—	—	—	—	—	—	—	—	—	—	—	—	33.8	10.0	3.85
26,000	3,338	—	—	—	—	—	—	—	—	—	—	—	—	—	39.7	11.9	4.65
28,000	3,590	—	—	—	—	—	—	—	—	—	—	—	—	—	46.2	13.8	5.40
30,000	3,850	—	—	—	—	—	—	—	—	—	—	—	—	—	53.0	15.9	6.17

*For longer or shorter lengths of pipe the friction loss is proportional to the length, i.e., for 500 ft, one-half of the above; for 4,000 ft, four times the above, etc.

Table 10–5   Pressure Loss Through Screw Type Pipe Fittings, Given in Equivalent Feet of Schedule 40 Straight Pipe
(From Compressed Air and Gas Handbook of the Compressed Air and Gas Institute, 5th ed., Table 13.26. Upper Saddle River, N.J.: Prentice Hall, 1989)

Nominal Pipe Size (in.)	Actual Inside Diameter (in.)	Gate Valve	Long Radius Ell or on Run of Standard Tee	Standard Ell or on Run of Tee Reduced in Size 50%	Angle Valve	Close Return Bend	Tee Through Side Outlet	Globe Valve
$1/2$	0.622	0.36	0.62	1.55	8.65	3.47	3.10	17.3
$3/4$	0.824	0.48	0.82	2.06	11.4	4.60	4.12	22.9
1	1.049	0.61	1.05	2.62	14.6	5.82	5.24	29.1
$1 1/4$	1.380	0.81	1.38	3.45	19.1	7.66	6.90	38.3
$1 1/2$	1.610	0.94	1.61	4.02	22.4	8.95	8.04	44.7
2	2.067	1.21	2.07	5.17	28.7	11.5	10.3	57.4
$2 1/2$	2.469	1.44	2.47	6.16	34.3	13.7	12.3	68.5
3	3.068	1.79	3.07	6.16	42.6	17.1	15.3	85.2
4	4.026	2.35	4.03	7.67	56.0	22.4	20.2	112
5	5.047	2.94	5.05	10.1	70.0	28.0	25.2	140
6	6.065	3.54	6.07	15.2	84.1	33.8	30.4	168
8	7.981	4.65	7.98	20.0	111	44.6	40.0	222
10	10.020	5.85	10.00	25.0	139	55.7	50.0	278
12	11.940	6.96	11.0	29.8	166	66.3	59.6	332

Problem Set 1

1. At a manufacturing facility that produces large metal castings, pneumatic grinders are used to eliminate sharp edges and corners from the newly cast pieces.
   (a) If 10 grinders, each using a 6-in.-diameter wheel, operate at 90 psi and a 50% load factor, find the total acfm required.
   (b) How many cfm of free air are required if the grinders operate at 100 psi?

2. Small electronic components are packaged by a machine that utilizes one pneumatic cylinder. This cylinder has a stroke of 10 in., a diameter of 1 1/4 in., and an output-rod diameter of 3/8 in. If the cylinder operates at 75 psi and completes 20 cycles/min:
   (a) Find the acfm required.
   (b) Determine the cfm of free air needed to supply this machine.

3. Using spreadsheets, construct a table that lists tool consumption rates, in acfm, for various free-air rates and operating pressures at a 100% load factor. Your completed results should resemble the partial table shown at the top of the following page. (How well do your entries compare with the values obtained for Example 10–1?)

CFM of Free Air	Operating Pressure (psig)					
	40	**50**	**60**	**70**	**80**	**90**
10		3.47				
15					3.96	
20			6.26			

10.3   Receivers

Receivers are the pneumatic equivalents of hydraulic reservoirs and generally consist of cylindrical steel tanks that may be installed either vertically or horizontally. On small portable, permanent, or mobile units, the compressor is frequently mounted directly on the receiver itself (Figures 10–6 and 10–12(A)). By storing compressed gas for use during periods of high demand, the receiver acts as an accumulator, reducing direct loading on the compressor. This accumulation of gas also helps dampen any pulsations that occur in the system and serves to precipitate moisture out of the pressurized fluid.

The schematic symbol for a receiver (Figure 10–20) further reflects its similarity to an accumulator. However, whereas the accumulator is essentially a dead-ended or closed device, the receiver may allow through-flow of fluid. Recommended receiver sizes are presented in Table 10–6.

Table 10–6   Recommended Receiver Sizes*
(From Compressed Air and Gas Handbook of the Compressed Air and Gas Institute, 5th ed., Table 4.4, Upper Saddle River, N.J.: Prentice Hall, 1989)

Tank Dimensions			Compressor Capacities in cfm of Free Air at 40–125 psig	
Diameter (in.)	Length (ft)	Volume (ft^3)	Constant-Speed Regulation	Automatic Start-and-Stop Service
14	4	4.5	45	22.5
18	6	11	110	55
24	6	19	190	95
30	7	34	340	170
36	8	57	570	285
42	10	96	960	480
48	12	151	2115	1058
54	14	223	3120	1560
60	16	314	4400	2200
66	18	428	6000	3000

*For example, a receiver 30 in. in diameter and 7 ft. long can be used safely with a compressor that operates intermittently at flow rates up to 170 cfm of free air and discharge pressures between 40 and 125 psig.

Beware of the Unseen Danger

Because the moisture in compressed air tends to precipitate and collect in receivers, these tanks should be checked frequently for defects that could endanger lives. Several organizations, including ASME and OSHA, have established guidelines for the inspection of pressurized tanks with regard to damage caused by corrosion and pitting. Many states and municipalities have incorporated these guidelines into regulations requiring that tanks located in public buildings, as well as units found in business or industry, can be used by permit only and are subject to periodic examination by certified inspectors. Since most corrosion occurs *inside* such tanks, units must often be dismantled for a thorough visual inspection. Unfortunately, many people, especially home mechanics and do-it-yourselfers, do not appreciate the dangers posed by a tank of pressurized gas. They tend to use anything available—from empty freon or propane tanks to air tanks from the brake systems of tractor trailers—as makeshift receivers. Unless you like playing with bombs, *never* pressurize any tank that has not been specifically designed for use as a receiver. When in doubt as to the condition of a receiver, have the unit inspected and hydrostatically tested by qualified personnel. Don't take chances with your own life or those of your family or fellow workers!

Figure 10–20 Schematic symbol for a pneumatic receiver.

   In addition to inlet and outlet lines, receivers are generally equipped with fittings for relief valves, pressure gauges, drain valves, and inspection ports. Most tanks are also stamped with a maximum allowable pressure for which the unit has been designed and fabricated. Working pressures for systems using such tanks are usually held 5% to 10% below this allowable level.

10.4 Fluid Conditioners

A number of devices may be used to condition or prepare air or other gases for various applications. Solid contaminants are removed by *filters* at the compressor inlet and by individual air line filters located in the supply lines feeding each pneumatic tool or actuator. These filters can also eliminate a small percentage of the water and oil particles that exist in most compressed gases.

Aftercoolers are water-cooled heat exchangers generally located between the compressor and receiver and are used to cool the compressed gas. In a typical air compressor, these devices are able to condense approximately 70% of the moisture content from the gas. Normally, an aftercooler also contains a *separator* that, as the name implies, helps to separate vapor particles from the gas. Separators can be located at other locations in a pneumatic circuit as well, typically at the point of use, and can operate on several different principles. *Gravity* separators are simply large volumes, such as a receiver, which reduce the gas velocity and allow water and oil vapors to settle out of the mixture. *Impingement* separators discharge the mix against a surface such as a corrugated vane element that deflects the gas but collects the vapor particles in its corrugations. *Centrifugal* or *cyclone* separators create a swirl or vortex in the gas, causing any vapor particles to be thrown to the outside wall by centrifugal force. Separators can be of the *manual drain* or *automatic drain* type.

To help in removing condensate from pneumatic systems, lines are generally sloped away from the compressor, typically at a rate of $^1/_8$ in. to $^1/_4$ in. per foot, with *traps* and *drains* installed at low points. Traps may consist of a simple U-tube or tee that collects condensate; a drain plug or small valve at the bottom of the trap allows this captured liquid to be withdrawn from the system. Commercial traps are also available with either manual or automatic drains.

Some systems require drier air or gases than those obtained through the use of filters, separators, and aftercoolers. Figure 10–21, for example, shows that the amount of water vapor contained in atmospheric air at various temperatures and relative humidities can be substantial. Within a system, excessive moisture can inhibit lubrication, thus increasing wear and maintenance of components. It can also produce sluggish or erratic operation of valves and cylinders, damage sensitive instruments, and cause exposed lines to freeze in cold weather. To avoid these problems, particularly in compressed air systems, additional moisture is often removed using *air driers*. *Chemical* driers contain beads or pellets that absorb moisture from the air and then turn into a liquid that is drained from the system. These beads or pellets must be replaced periodically. *Adsorption* driers use a desiccant that attracts and holds water vapor in its pores. A packet of this same material, typically activated alumina or silica gel, is often packaged with delicate electronic or photographic equipment to prevent moisture damage. Commercial adsorption driers consist of desiccant beds that are periodically replenished or reactivated by the application of heat and exposure to a drier gas. *Refrigeration* driers are essentially refrigerators that operate using commercial refrigerants. These driers achieve much lower dew points, and correspondingly lower moisture contents, than af-

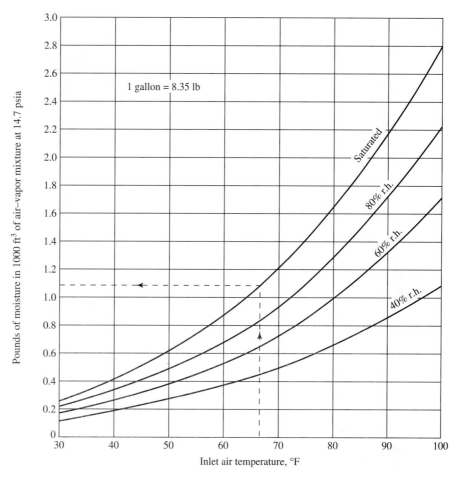

Figure 10–21   Water vapor per 1000 ft^3 of atmospheric air. *(From* Compressed Air and Gas Handbook, *4th ed., Figure 4.15. New York: Compressed Air and Gas Institute, 1973.)*

tercoolers. Small- and medium-sized driers of this type generally pass the air directly across the refrigeration coils. Larger units, called *chiller driers,* first cool water, which then circulates through coils to cool the flow of air.

    Although the above devices are quite successful in extracting moisture from a compressed gas, they are generally able to remove only a small fraction of the oil vapor contained in a gas. In fact, *oil-free* gas is best obtained through the use of either diaphragm or nonlubricated types of compressors (see Figure 10–8). On the other hand, compressed gas is often too dry to satisfactorily lubricate pneumatically operated actuators and tools. For this reason, *mist-type lubricators* are used for adding oil to gas supplied to high-speed tools, while *drop-type lubricators* are used with actuators that move slowly

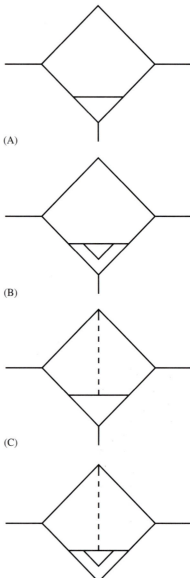

(A)

(B)

(C)

Figure 10–22 Schematic symbols for traps and filters: (A) liquid trap—manual control; (B) liquid trap—automatic drain; (C) filter with liquid trap—manual control; and (D) filter with liquid trap—automatic drain.

(D)

or infrequently. In addition to providing lubrication, these oils reduce corrosion in tools and actuators. Specially formulated winter grades of oil can also contain deicing additives to prevent freezing during cold weather operation. Schematic symbols for various types of filters, traps, driers, heat exchangers, and lubricators are shown in Figure 10–22 and Figure 10–23.

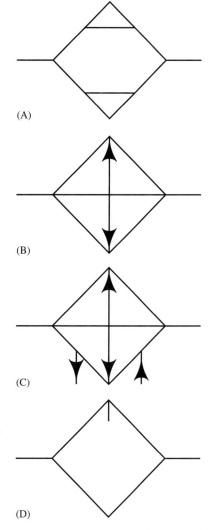

Figure 10–23 Schematic symbols for drier, coolers, and lubricator: (A) air drier (such as a chemical drier); (B) heat exchanger used to cool liquid or gas (arrows show direction of heat); (C) heat exchanger showing direction of flow of external coolant; and (D) lubricator.

(A)

(B)

(C)

(D)

Diaphragm-type *pressure regulators* are commonly used with individual tools and actuators to eliminate surges and variations in supply line pressure, guaranteeing consistent, predictable operation of these devices. Often, a filter, regulator, lubricator, and pressure gauge are combined into a single unit such as that shown in Figure 10–24. Sometimes referred to as *FRLs,* these units are assembled so that the direction of flow is filter-to-regulator-to-lubricator. Ideally, the gas supply line for each tool or actuator would have its own FRL. Detailed and simplified schematic representations for these combination devices are shown in Figure 10–25.

Figure 10–24 A filter, regulator, lubricator, gauge assembly. *(Michigan Pneumatic Tool, Inc., Detroit, Mich.)*

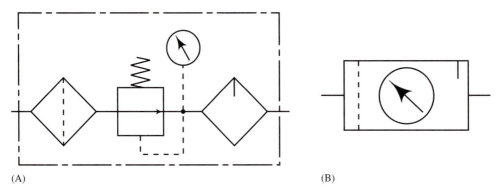

(A)   (B)

Figure 10–25 FRL (filter/regulator/lubricator) detailed (A) and simplified (B) symbols.

10.5 Other Pneumatic Components

Generally, pneumatic system components are similar, and in some cases identical, to their hydraulic counterparts. Differences, when they exist, are usually attributable to one of the following:

- Increased difficulty in *sealing* a gas rather than a liquid. (Typical pneumatic systems are designed at 10% over required compressor capacity to account for gas leakage.)
- *Corrosiveness* of some gases themselves, as well as the moisture that they contain.
- Completely different *operating ranges* of such conditions as pressure and temperature.
- Different *operating characteristics,* such as the speed of linear actuators or the speed and torque of rotary actuators.

Many valves, lines, and cylinders can be used interchangeably in pneumatic or hydraulic (oil) systems. Other pneumatic components such as pressure/temperature switches, filters, fittings, gauges, and rotary actuators have significant differences in construction, materials, or operating characteristics from equivalent hydraulic devices. Whenever the suitability of a particular component is in question, check with your local sales representative or contact the manufacturer directly.

Questions

1. What is an FRL? Draw the schematic symbol for this device.

2. In order to operate the pneumatic gripper on an industrial robot (see cover photo), it is necessary to construct a supply line between the robot and the central source of compressed air in a building. What types of hardware would likely be installed in the supply line to condition the air for this use?

3. Explain the difference between positive displacement and dynamic compressors.

4. Name one advantage of a mixed-flow compressor.

5. Why might the compression process in a real machine be considered adiabatic rather than isothermal?

6. Describe the operational differences between a compressor and a vacuum pump.

7. What types of compressors produce oil-free air?

8. Approximately how many pounds of water are contained in 1000 ft^3 of saturated air at 68°F?

9. List several ways in which excessive moisture can affect the performance of a pneumatic system.

10. What type(s) of compressors would be satisfactory for the flow rates and operating pressures described in Example 10–1?

Control of Fluid Power Systems

Although many fluid power systems operate under manual control, the trend in recent years has been to an increased use of automated systems, most of which involve some form of electrical or electronic control. *Process control* systems are used to maintain conditions such as pressure, liquid level, temperature, or flow rate of fluids present in continuous or batch processes. Industrial operations that utilize such systems include food processing, generation of electric power, and the manufacture of chemicals or petroleum products. *Sequential control* systems are used to control a specific series of events that can involve a workpiece, various tools, and robot arms or other mechanical actuators. Many repetitive operations such as sorting, counting, loading/unloading, positioning, measuring, and packaging are well suited to sequential control, and can easily be combined with a variety of machining or other manufacturing processes in a single automated system. This often increases productivity, quality, and operator safety, while reducing material waste and energy usage. Examples of automated sequential control systems are shown in Figure 11–1 and Figure 11–2.

Although a detailed analysis of control systems is beyond the scope and objectives of this book, we will, in the sections that follow, examine some of the methods and components commonly used to achieve automatic control of fluid power systems.

11.1 Elements of a Typical Control System

The most common type of control system used in manufacturing today is the *feedback* control system, such as the one diagrammed in Figure 11–3. Such a system, also known as a *closed-loop* system, has three major functions: It first *senses,* or measures, the conditions existing at some point in the system; it then *compares* these conditions with a preselected value or set of conditions; finally, it *adjusts* the existing conditions to match the desired values. Changes in the existing conditions are "fed back" through the sensing elements, so that when the desired conditions have been met no further adjustments are made.

A simple example of closed-loop operation is a residential heating system. The temperature of any given room may be measured continuously by a thermostat, which compares this temperature to some preset value. If room temperature drops below the desired level, the thermostat activates a heating system, which adjusts conditions in the room by supplying some form of heat. When existing temperature reaches the preset value, this "match" is sensed by the thermostat, which then deactivates the source of heat but continues to monitor room conditions.

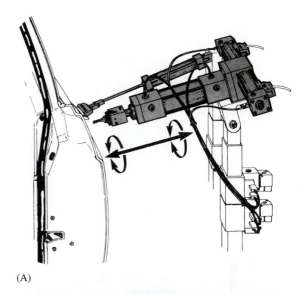

(A)

Figure 11–1 At Epcot Center's World of Motion, multimotion actuators are used to conduct a lock/unlock endurance test on a car door. *(A, courtesy of PHD, Inc., Fort Wayne, Ind.)* **Similar actuators mounted on the inside of the door are used to duplicate the reach and turn actions performed by car occupants on the door latch and door pull (B).**

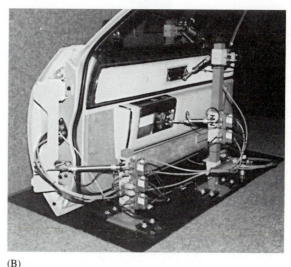

(B)

From this example, you should notice several important features of control systems. First, a single device can perform several functions; here the thermostat senses, compares, and activates. Second, control systems are often "multimedia." A typical thermostat senses temperatures *mechanically* by the action of a bimetallic strip. This controls an *electrical* circuit that activates some source of heat. If the heat source is a furnace, then the final portion of our control system may be *pneumatic* (forced hot air heating) or *hydraulic* (forced hot

Figure 11–2   The pneumatic cylinder shown extends down into an injection molding machine where vacuum grippers mounted on the cylinder rod remove an **x**-shaped piece of waste from the mold (arrow points to waste piece inside box). The cylinder then retracts, rotates through 90°, and extends beyond the edge of the machine, where the piece is released into a scrap bin.

water heating). These mixed components must interface smoothly and efficiently. Finally, no control system is perfect. An average thermostat may allow existing temperatures to vary by several degrees above and below the preset value before activating or deactivating the heat source. Such behavior exists to some degree in all control systems; it affects both accuracy and repeatability of the control function and must be held within specified limits determined by the particular application.

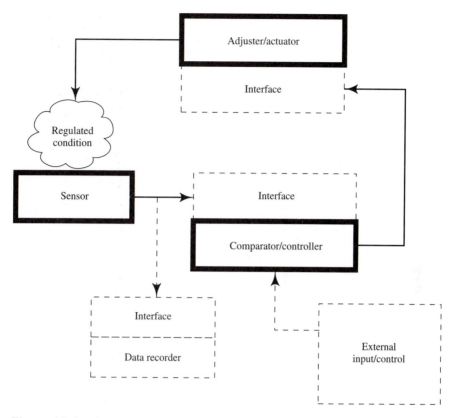

Figure 11–3   A typical feedback control system contains sensors, comparators, and adjustors. It can also include recorders (for data acquisition) and computer access via keyboard or control panel (for alteration of preset conditions). Interface devices are used to connect different media—mechanical, electrical, hydraulic, and pneumatic—that may be used in various parts of the system.

Figure 11–4 shows a typical automated system that contains pneumatic, hydraulic, and electrical actuators being activated by various interface devices through control circuitry connected to pneumatic sensors. Until recently, such systems generally used either pneumatic or electronic *analog controllers,* circuitry whose output signal was proportional to some measured position, force, velocity, or acceleration. Today, microprocessors allow the use of *digital controllers* in which signals are represented in computer language and offer the ability to calculate anticipated changes in operating conditions before they actually occur.

In the next few sections we will examine some of the hardware used in the automated control of fluid power systems. Since virtually all control functions can be performed using pneumatic, hydraulic, electrical, or mechanical methods either alone or in combination,

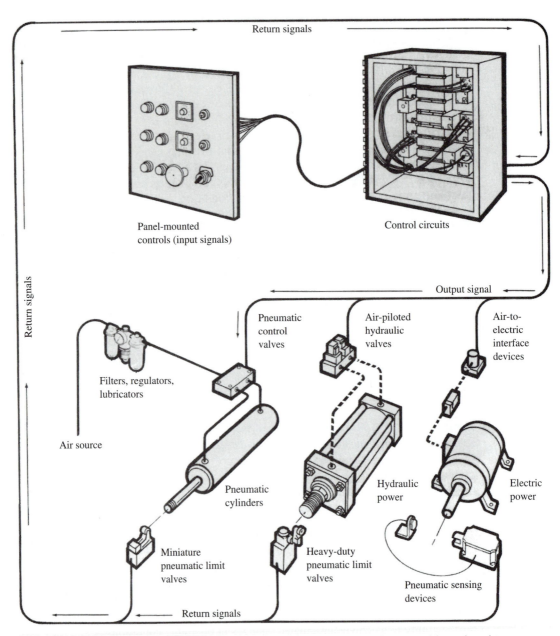

Figure 11–4 Elements of a typical fluid power control system. Note that this system contains many of the elements—sensors, actuators, interface devices, and comparators/controllers—shown in the feedback control circuit of Figure 11–3. *(The Aro Corporation, Bryan, Ohio)*

the number of commercial devices available today is staggering! As much as possible, components have been selected that operate on hydraulic or pneumatic phenomena, including both sensors and logic devices. Although these components may not be the best choice for every application, their operation is simple and easy to understand. You should remember, however, that each of the fluid power devices discussed *probably* has at least one electrical, mechanical, or electronic counterpart that performs the same function.

11.2   Sensors

The most commonly monitored quantities in process control systems that use liquids or gases are pressure, temperature, velocity/flow rate, and liquid level. Any sensors used to detect and measure these conditions must be capable of producing some type of active output signal—usually electrical, mechanical, or pneumatic—to activate those system components that evaluate and adjust the existing condition.

In Chapter 7, we saw that pressure sensors containing either capacitance sensitive or strain gauge bonded diaphragms (Figure 7–82 and Figure 7–83) produce electrical outputs that vary in response to applied pressure. Sensing elements of piezoelectric materials (Figure 7–84) also produce direct electrical outputs and are well suited to the measurement of dynamic pressures. In other sensors, triggering of switches and external circuitry can be accomplished mechanically by either a plunger or rod directly connected to some movable element such as a diaphragm, bellows, or piston.

Temperature measurements are generally obtained using devices whose output is electrical; the most popular of these include the *thermocouple, resistance temperature detector* (RTD), and *thermistor.* Thermocouples consist of two dissimilar metals such as copper and constantan (a copper-nickel alloy), which are joined together as shown in Figure 11–5. When this junction is heated, a small voltage and current are produced, the magnitude of which increases with increasing temperature. The typical RTD is constructed using either platinum wire or a thin film of the same metal. Variations in temperature produce measurable changes in electrical resistance of the platinum element, a phenomenon that may be used to produce corresponding voltage or current changes in a simple electric circuit.

Figure 11–5   In 1821, German physicist T. J. Seebeck discovered that a voltage exists across the junction of any two dissimilar metals. Sensors that operate using this principle are called *thermocouples* and generally consist of two wires made from different metals, as shown above. Such devices function as electrical thermometers by generating voltages and currents that vary with the junction temperature. Because of their simplicity, durability, and low cost, thermocouples are widely used in industry.

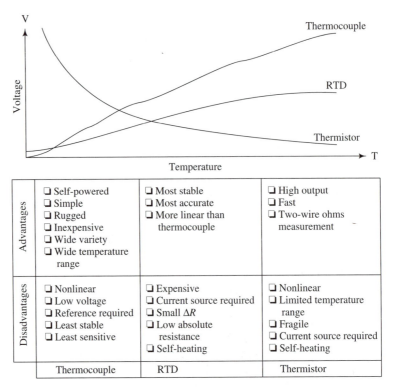

Advantages	❏ Self-powered
	❏ Simple
	❏ Rugged
	❏ Inexpensive
	❏ Wide variety
	❏ Wide temperature
	   range
		❏ Most stable
		❏ Most accurate
		❏ More linear than
		   thermocouple
			❏ High output
			❏ Fast
			❏ Two-wire ohms
			   measurement

Disadvantages	❏ Nonlinear
	❏ Low voltage
	❏ Reference required
	❏ Least stable
	❏ Least sensitive
		❏ Expensive
		❏ Current source required
		❏ Small ΔR
		❏ Low absolute
		   resistance
		❏ Self-heating
			❏ Nonlinear
			❏ Limited temperature
			   range
			❏ Fragile
			❏ Current source required
			❏ Self-heating

	Thermocouple	RTD	Thermistor

Figure 11–6 Characteristic operating curves for several common temperature sensors, with advantages and disadvantages of each type.

Similarly, the thermistor is a semiconductor device whose resistance is temperature sensitive. Characteristics, advantages, and disadvantages of these three types of temperature sensors are given in Figure 11–6.

A variety of techniques exist for measuring velocity or mass flow rates of fluids. The *pitot gauge* discussed in Chapter 6 may be used for both liquids and gases (see Figures 6–20, 6–21, 6–22), as may the variable-area flowmeter, or *rotameter,* illustrated in Figure 11–7. This device contains a float that moves up or down in a tapered tube at some distance proportional to the flow rate. A magnet is often attached to the float and used to trigger electrical proximity switches near the tube wall when the flow rate nears preset maximum or minimum values. Gas velocity and mass flow rate are often measured using *anemometers;* the propellor type consists of a small propellor attached to a generator whose electrical output varies with windspeed; hot-wire anemometers employ an exposed platinum wire whose temperature and resistance are related to the flow velocity. Anemometers are insertion devices; they are inserted into a flowing stream that then passes *around* them. For the tube-type mass flowmeter shown in Figure 11–8, a portion of the moving gas passes

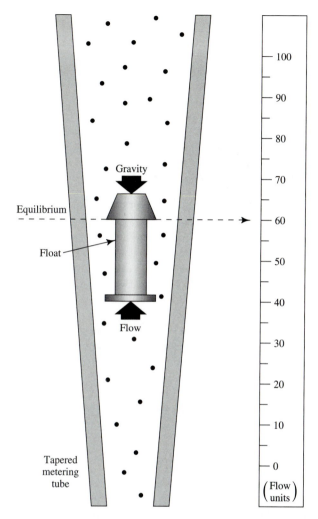

Figure 11–7 Operation of a variable-area flowmeter or rotameter. As flow rate increases, so does dynamic pressure on the bottom surface of the float, thus causing the float to rise in its metering tube. However, because tube area *increases* vertically, fluid velocity and dynamic pressure *decrease*. The float remains suspended at its new equilibrium position when the dynamic pressure force exactly equals float weight.

through a tube that is heated at a fixed rate. Sensors upstream and downstream of the heating element measure the increase in temperature of the gas; as mass flow rates increase, more gas particles pass through the tube in a given period of time, each particle absorbs less heat, and the net temperature increase of the gas is reduced. A commercial tube-type flowmeter is shown in Figures 11–9 and 11–10.

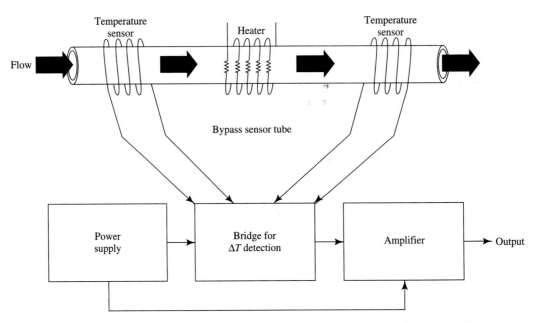

Figure 11–8   Operation of a heated-tube mass flowmeter. Heat flows from an external source into the moving fluid at a known rate. For low flows, each fluid particle absorbs a fixed amount of heat and temperature rises to some particular value. At higher flows, the same amount of heat is absorbed by a larger number of particles, so final fluid temperature is less than that for lower flows. This change in fluid temperature (obtained by sampling fluid temperatures upstream and downstream of the heat source) can be directly correlated to fluid velocity or mass flow rate.

Measurement of liquid flow rates can be accomplished in several ways. *Differential pressure* devices employ some type of flow obstruction such as a calibrated orifice or venturi (Figures 6–24, 6–25, and 6–26); the pressure drop required to force a given flow of liquid through the obstruction is determined using a manometer or differential pressure transmitter (Figure 6–23). *Ultrasonic Doppler* flowmeters transmit an ultrasonic signal into the flowing liquid and measure the frequency shift, or Doppler effect, as the signal is reflected from suspended particles or bubbles in the liquid. *Turbine* or *paddlewheel* flowmeters (Figure 11–11) use vaned rotors whose movement in a flow induces voltages in magnetic pickup coils located adjacent to the rotor. *Vortex* flowmeters operate on the principle that a nonstreamlined shape placed in a flow of moving fluid will "shed" vortices, or swirls of fluid, in a regular pattern (Figure 11–12) whose frequency is related to the flow velocity. (This phenomenon is what causes a flag to ripple in the wind downstream of a flagpole; it can also produce sounds such as those emitted by "singing" telephone or electrical transmission lines, sometimes called Aeolian harps.) The vortices produce pressure pulses that are detected using a sensing element such as

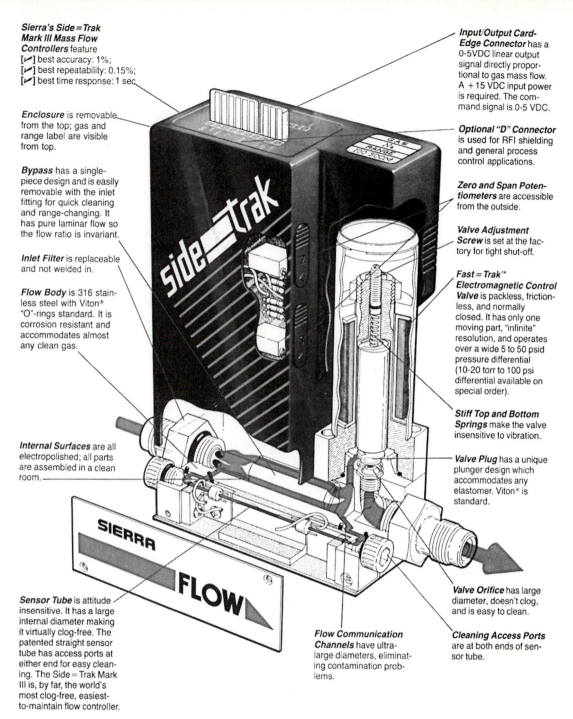

Sierra's Side=Trak Mark III Mass Flow Controllers feature
[✓] best accuracy: 1%;
[✓] best repeatability: 0.15%;
[✓] best time response: 1 sec.

Enclosure is removable from the top; gas and range label are visible from top.

Bypass has a single-piece design and is easily removable with the inlet fitting for quick cleaning and range-changing. It has pure laminar flow so the flow ratio is invariant.

Inlet Filter is replaceable and not welded in.

Flow Body is 316 stainless steel with Viton® "O"-rings standard. It is corrosion resistant and accommodates almost any clean gas.

Internal Surfaces are all electropolished; all parts are assembled in a clean room.

Sensor Tube is attitude insensitive. It has a large internal diameter making it virtually clog-free. The patented straight sensor tube has access ports at either end for easy cleaning. The Side=Trak Mark III is, by far, the world's most clog-free, easiest-to-maintain flow controller.

Input/Output Card-Edge Connector has a 0-5VDC linear output signal directly proportional to gas mass flow. A +15 VDC input power is required. The command signal is 0-5 VDC.

Optional "D" Connector is used for RFI shielding and general process control applications.

Zero and Span Potentiometers are accessible from the outside.

Valve Adjustment Screw is set at the factory for tight shut-off.

Fast=Trak™ Electromagnetic Control Valve is packless, frictionless, and normally closed. It has only one moving part, "infinite" resolution, and operates over a wide 5 to 50 psid pressure differential (10-20 torr to 100 psi differential available on special order).

Stiff Top and Bottom Springs make the valve insensitive to vibration.

Valve Plug has a unique plunger design which accommodates any elastomer. Viton® is standard.

Valve Orifice has large diameter, doesn't clog, and is easy to clean.

Flow Communication Channels have ultra-large diameters, eliminating contamination problems.

Cleaning Access Ports are at both ends of sensor tube.

Figure 11–9 A commercial tube-type mass flowmeter. Note that this device has the capability of both *measuring* and *controlling* flow rates using electrical output signals available at connectors on the top surface. Designed to regulate mass flow rates of gases, the unit has a response time of 1 s and an accuracy of 1%. *(Sierra Instruments, Inc., Monterey, Calif.)*

373

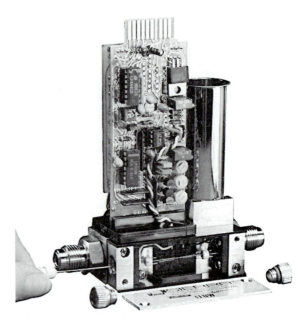

Figure 11–10 Internal electronic circuitry for meter of Figure 11–9. *(Sierra Instruments, Inc., Monterey, Calif.)*

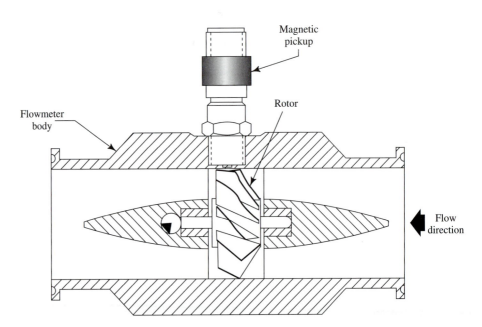

Figure 11–11 Typical turbine flowmeter. The vane-type rotor spins like a propeller in the moving fluid, causing a voltage to be induced in the magnetic pickup. Because this voltage varies with rotational speed, the signal can be correlated directly to fluid velocity.

Figure 11–12 Vortex shedding in the wake of a nonstreamlined or "bluff" object.

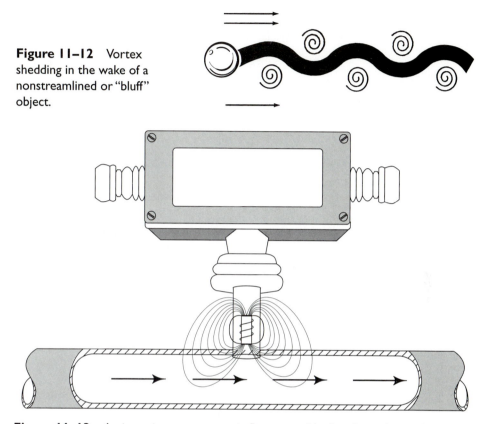

Figure 11–13   An insertion-type magnetic flowmeter. Used with conductive liquids, the device creates a magnetic field within the flow area and senses changes to this field caused by the moving fluid.

a piezoelectric crystal. As their name implies, *magnetic* flowmeters create a magnetic field within the flow area (Figure 11–13). When a conductive liquid flows through this field, a voltage is induced between sensing electrodes.

One of the most common sensors used to detect changes in liquid levels is a simple *float,* such as the type used to control the amount of water stored for flushing an everyday bathroom commode. Direct linear or levered motion provided by buoyant forces of the liquid can be used to actuate electrical switches or mechanical valves to replenish the spent fluid.

Another simple liquid level sensor is the *bubble tube* shown in Figure 11–14. A compressed gas such as air flows continuously through the tube into the regulated liquid; as the liquid level rises, back pressure in the tube increases, providing a pilot signal that may be used to actuate many types of logic or control hardware.

Other liquid level sensors operate using *ultrasonic* waves. The noncontacting type shown in Figure 11–15 transmits a signal to the liquid surface and measures the reflection

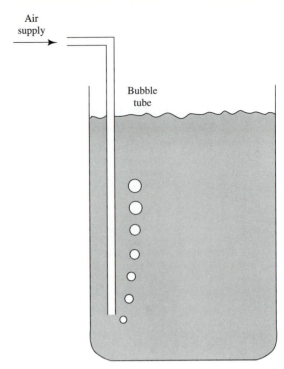

Figure 11–14 Back pressure developed within bubble tube is proportional to liquid level in tank.

Bubble tube

Air supply

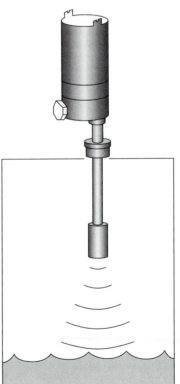

Figure 11–15 A noncontact-type ultrasonic liquid level sensor. This unit acts as a small radar system, measuring the time required for a transmitted signal to reach the liquid surface, reflect back, and return to the sensor.

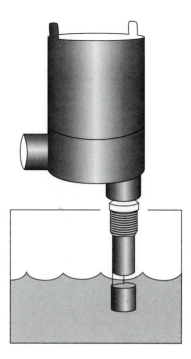

Figure 11–16 Contact-type ultrasonic liquid level sensor. The strength of a signal transmitted across an air gap indicates whether the liquid level lies within or above this gap.

of this wave; the contact sensor of Figure 11–16 transmits a signal across its fixed gap distance, with signal strength indicating whether the liquid level lies within this gap or whether the gap is totally submerged.

By far the most common property measured in sequential control systems is *position*. The presence of an object at a precise location is often a necessary condition for continuation of many manufacturing operations such as machining, measuring, counting, and packaging. Position may be measured using contact-type sensors such as the plunger-operated LVDT (linear variable differential transformer) or noncontact devices such as photoelectric "eyes" that use visible or infrared light transmitted either directly or via fiber-optic cables.

One other type of noncontact sensor commonly used in fluid power systems is the *Hall-effect* proximity switch, whose principle of operation is shown in Figure 11–17. When an electric charge is applied across the ends of a flat conductor, no measurable voltage exists at the side edges of this conductor (Figure 11–17(A)). However, if a magnetic field approaches the charged conductor, a voltage appears at the side edges (Figure 11–17(B)). This phenomenon, discovered in 1879 by American physicist Edwin Hall, has several variations in practice: a magnet can be placed on the moving object (such as a cylinder output rod) whose position is to be detected, or a slight magnetic field can be created within the Hall-effect device itself in order to detect voltage changes caused by the approach of any object made from a ferrous (iron-bearing) material.

Another type of detector that finds wide usage in proximity sensing is the *fluidic* sensor. As we shall see in the following section, fluidic devices employ a moving fluid, usually

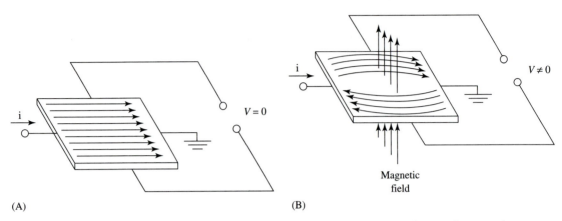

(A)   (B)

Figure 11–17 (A) When an electrical charge is applied to a flat conductor, voltage measured across the side edges is zero. (B) If a magnetic field is brought into proximity with the charged conductor, a voltage appears across these side edges. Discovered in 1879 and known as the *Hall effect,* this phenomenon forms the basis of operation for many of the position sensors used to control today's fluid power systems.

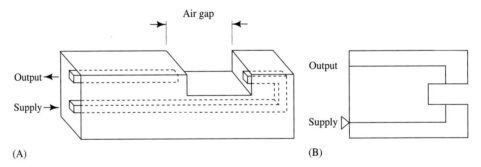

(A)   (B)

Figure 11–18 (A) Interruptable jet sensor and (B) schematic symbol. If the air gap is unobstructed, flow from the supply line traverses this gap and creates a positive pressure at the output. When an object enters the air gap, fluid flow is interrupted, and output pressure drops.

air, to sense, process information, and actuate various components in a control system. Two of the most common fluidic sensors are the *interruptable jet sensor* and the *back pressure switch,* both of which are noncontacting devices. Figure 11–18 shows a typical interruptable jet sensor and its schematic representation. Compressed air, normally at pressures less than 10 psi, is applied to the supply port and flows through the sensor, across the air gap, to the output port. Any object passing through the air gap interrupts this flow and eliminates the output signal. Most commercial sensors of this type (see Figure 11–19) can operate at speeds up to several hundred cycles per second across air gaps of approximately 0.5 in. However, if the device is used as a cross-jet detector as shown in Figure 11–20, sensing

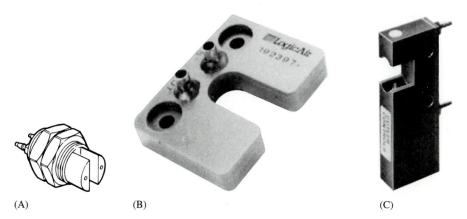

(A)   (B)   (C)

Figure 11–19   Interruptable jet sensors. *((A) Norgren Co., Littleton, Colo.; (B) Logic Air, a Division of TEKNOCRAFT, Inc., Palm Bay, Fla.; and (C) Component Engineering & Sales Co., Inc., Fort Wayne, Ind.)*

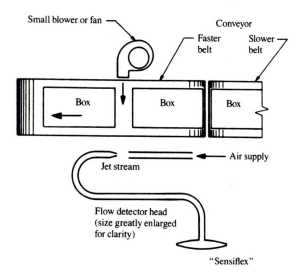

Figure 11–20   Interruptable jet sensor used as a cross jet detector. For the condition shown, air from the fan disrupts the jet stream between air supply and flow-detector head so that no dynamic pressure is felt by the detector. If this fan flow is blocked by the presence of a box, air supply impinges directly on the detector head, which then senses a positive pressure. *(Gagne Associates, Inc., Binghamton, N.Y.)*

gaps of several feet are possible. Here a cross flow of air provided by the blower or fan disrupts air flow across the sensing gap of the jet sensor; when a box passes by on the conveyor, flow from the fan is blocked and flow across the jet sensor is resumed. In fact, the actual sensing gap for this application becomes the distance from the fan to the jet sensor, rather than the distance across the jet sensor's own air gap. Notice also that this configuration reverses the output signal; when an object is present, a positive pressure is obtained at the sensor's output port. This pressure can be applied to the input of a fluidic logic circuit

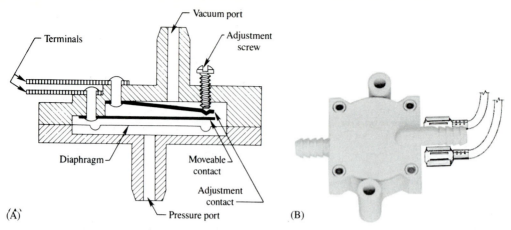

(A) (B)

Figure 11–21 (A) In this typical pneumatic-to-electric switch, air pressure pushes the movable contact until it touches the adjustment contact whose fixed position, and therefore the switching pressure, is controlled by the adjustment screw. (B) The actual device. *(Micro Pneumatic Logic, Inc., Fort Lauderdale, Fla.)*

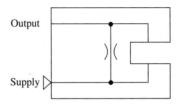

Figure 11–22 Schematic symbol for impacting jet sensor. Similar to the interruptable jet sensor of Figure 11–18, the device shown here allows a restricted internal flow that constantly purges, or cleans, the receiver port.

such as a binary counter, used to operate pneumatic-to-electric switches such as the type shown in Figure 11–21, or power many types of pneumatic actuators.

Although fluidic devices are generally slower than their electronic counterparts, they are extremely rugged and will perform reliably under conditions of high temperature, vibration, and shock; in the presence of magnetic fields; and in dusty or explosive environments. However, they do require a constant supply of *clean,* oil-free air and will clog easily in the presence of lubricating mists. To combat this problem, the impacting jet sensor, shown schematically in Figure 11–22, was developed. This device looks much like a regular interruptable jet sensor, but reduced flow through the internal restriction provides a constant purging of the receiver port.

A typical back pressure switch and its schematic symbol are shown in Figure 11–23. Under normal conditions, air supplied at the inlet port flows directly through to output port $O2$, with a secondary flow discharging from sensing port S. An object blocking port S creates a back pressure within the sensing line; this pressure then deflects the main air flow from port $O2$ to port $O1$.

Other proximity sensors (Figure 11–24) use focused air jets that either diverge or converge during normal operation. Figure 11–25 illustrates how the output of a diverging cone sensor increases as flow is blocked by a nearby object.

380

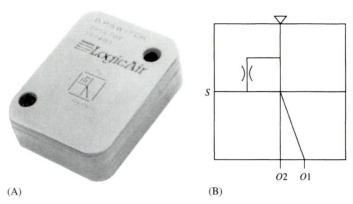

(A)   (B)

Figure 11–23 (A) Typical back pressure switch. *(Logic Air, a Division of TEKNOCRAFT, Inc., Palm Bay, Fla.)* (B) Its schematic symbol. For normal operation, air supplied at the inlet (top) flows directly to outlet $O2$, with a smaller secondary flow leaving the device at sensing port S. When an object approaches the sensing port, secondary flow is blocked, and sufficient back pressure develops in this line to switch the main flow to outlet $O1$. Internal geometry of the device allows the main flow to automatically switch back to outlet $O2$ when the object is no longer in proximity to outlet S.

(A)   (B)   (C)

Figure 11–24 Proximity sensors that use focused air jets. *((A) Component Engineering & Sales Co., Inc., Fort Wayne, Ind.; (B) Logic Air, a Division of TEKNOCRAFT, Inc., Palm Bay, Fla.; and (C) Norgren Co., Littleton, Colo.)*

381

Figure 11–25 (A) Output of a divergent cone air jet sensor increases (B) as supply flow is obstructed by presence of object. *(Component Engineering & Sales Co., Inc., Fort Wayne, Ind.)*

(A) Low output (vacuum)

(B) High output (ambient)

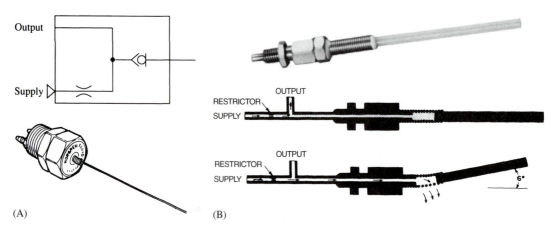

(A)

(B)

Figure 11–26 Contact sensors that operate on fluidic principles. (A) When the whisker is undisturbed, air from supply exits this device through an opening that functions as a check valve, so that no pressure is felt at the output. Upon contact with an object, the whisker closes off this opening, and supply flow passes directly to output. *(Norgren Co., Littleton, Colo.)* (B) For noncontact conditions, this spring element represents a closed flow path, so air moves directly from supply to output. Upon contact, the spring deflects, causing its coils to separate; these openings provide alternative exit paths for supply air, so that flow to output is virtually eliminated. *(Air Logic, Racine, Wis.)*

Contact type sensors that use fluidic principles are shown in Figure 11–26. The device at (A) normally has no output unless the "feather" or "whisker" sensor is disturbed. Sensor (B) has a positive output until deflection of the spring element allows supply air to escape through the open spring coils.

Another popular contact-type sensor is the limit switch or limit valve shown in Figure 11–27. Mechanically operated by a plunger, roller, or lever, these devices are normally used in automated systems to switch pneumatic flows at system pressures up to 150 psi.

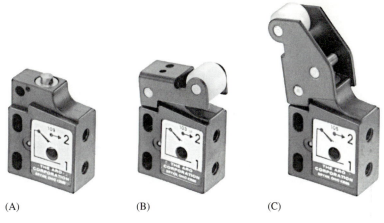

Figure 11–27 Various limit switch configurations: (A) straight plunger, (B) roller lever, (C) 90° roller lever. *(The Aro Corporation, Bryan, Ohio)*

11.3 Controllers and Logic Circuits

Sensor outputs are generally compared to a desired condition so that corrective adjustments or sequential operations can be initiated. Many control functions are of the "if/then" type: *if* certain conditions exist, *then* specific actions must be taken. A number of different electronic and pneumatic switches are available that, when assembled in various combinations, determine the existence of these conditions and trigger the necessary response. Combinations of switches are found in most automated controllers and are often referred to as *logic circuits*.

Pneumatic switches that operate on fluidic principles are commonly used in process and sequential motion control systems. These devices can be of the *wall attachment* or *moving parts* type. Consider the fluid jet, either liquid or gas, discharging from an opening in a wall as shown in Figure 11–28(A). Ambient air close to the edges of this jet is entrained in the moving stream and swept away, with air from the surroundings flowing in to take its place. However, if the wall on one side of the jet is moved as shown in Figure 11–28(B), make-up air cannot easily flow into region F, a partial vacuum is created in that area, and the jet attaches itself to the wall (Figure 11–28(C)). This phenomenon is known as the *Coanda effect* after Henri Coanda, who discovered its existence in the 1930s.

Wall attachment can be used in a switch whose internal flow configuration is as shown in Figure 11–29. If a fluid, typically compressed air, is applied at *Ps,* the moving stream will attach itself to one wall and produce a flow at the appropriate output port, in this case *O*1 (Figure 11–29(A)). Either a positive pressure at *C*2 or a partial vacuum at *C*1 will cause the flow to move (switch) to the opposite wall and produce an output at *O*2 (Figure 11–29(B)). This device is called a flip/flop and is the pneumatic equivalent of an electrical on/off switch. Control signals *C*1 and *C*2 are often provided by sensors such as a back pressure switch and need not be steady or continuous; during the operation of a *bistable* flip/flop, once the main flow has attached itself to a wall it remains attached even if the control signal is removed. The schematic representation for a flip/flop is given in Figure 11–30(A) and its operation is summarized in the *truth table* of Figure 11–30(B). The existence of pressure at

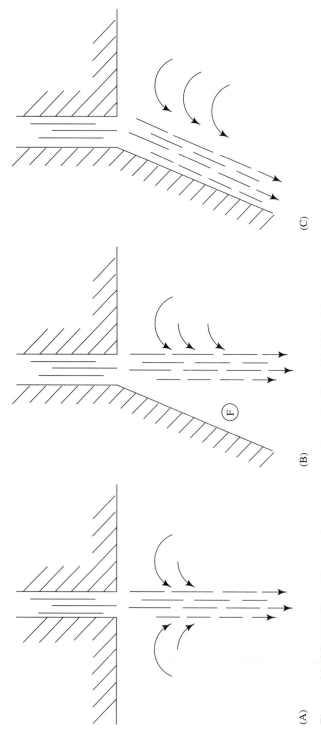

Figure 11–28 Wall attachment phenomenon or Coanda effect. (A) If fluid discharges from an opening, as shown, surrounding air is entrained and replenished at the same rates on both sides of the moving stream, so that flow is essentially undisturbed. (B) Angling one wall closer to this stream decreases the rate at which air entrained from region F can be replenished. Eventually a partial vacuum develops between the moving fluid and angled wall. (C) With atmospheric pressure acting on one side and a partial vacuum on the other, the stream of fluid attaches itself to the angled wall.

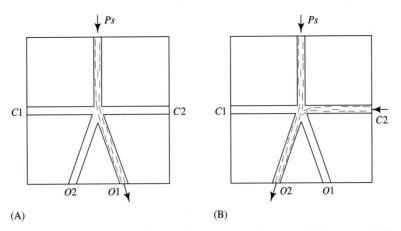

(A)   (B)

Figure 11–29   Pneumatic flip/flop equivalent to on-off switch. (A) Due to the Coanda effect, flow from supply attaches itself to the wall of one output path or the other. (B) When pressure is applied to the appropriate control port, supply flow switches to the opposite output. If flow remains in the switched position even after the control signal is removed, the device is said to be *bistable*.

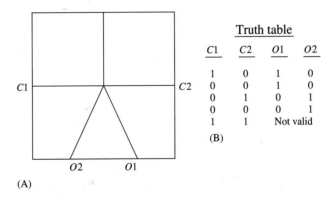

Truth table

C1	C2	O1	O2
1	0	1	0
0	0	1	0
0	1	0	1
0	0	0	1
1	1	Not valid	

(B)

(A)

Figure 11–30   Graphic symbol (A) and truth table (B) for bistable flip/flop. A 1 indicates the presence of pressure or flow at that port; 0 represents lack of pressure or flow. Thus, the first two lines of this table show that pressure at *C*1 produces an output at *O*1 and that this output continues even after control signal *C*1 is removed.

a given control or output port is denoted by a 1, while the absence of pressure is indicated by a 0. A commercially available flip/flop is shown in Figure 11–31.

Notice that the wall attachment device has no moving parts to wear out. Other fluidic control devices achieve the same switching functions with moving parts. Consider the device shown in Figure 11–32, which consists of supply, control and output ports, a ridge, and a flexible diaphragm. If no pressure is applied at the control port, the diaphragm remains clear of the ridge and flow entering the device passes between ridge and diaphragm to the output port (Figure 11–32(A)). When pressure is applied at the control port, the diaphragm deflects downward, where it comes in contact with the ridge and forms a seal that prevents flow through the element (Figure 11–32(B)). Multiple ridges and diaphragms can be combined within a single fluidic device to achieve various switching functions.

Figure 11–31 This commercial flip/flop, also available with fittings, can be assembled with other devices in modular form using O-rings. Device shown has dimensions of 1.3 in. L × 1.0 in. W × 0.33 in. T. Supply pressure is 3–10 psi; required control (switching) pressure is 10% to 30% of supply. *(TEKNOCRAFT, Inc., Palm Bay, Fla.)*

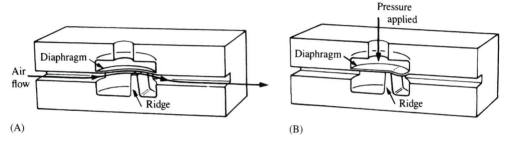

Figure 11–32 Switching in a typical moving-part fluidic logic device. (A) With no pressure applied to the control port (top), clearance exists between the diaphragm and the ridge, so flow occurs through device. (B) If control pressure is applied, the diaphragm is pressed against the ridge, thus blocking the flow path. This flow/no-flow device acts as an on-off switch but is not bistable. *(Component Engineering & Sales Co., Inc., Fort Wayne, Ind.)*

Of the many fluidic devices available, several find widespread application. Schematic representations and truth tables for these devices are given in Figure 11–33. The *OR/NOR* element normally produces an output at *O*2 if control pressure is applied at neither *C*1 nor *C*3; if pressure is applied at *C*1 *or* *C*3, output switches to *O*1. Note that when *C*1 and *C*3 are both removed the output does not remain at *O*1 but automatically returns to *O*2. Such behavior is achieved in wall attachment devices by altering the internal flow geometry. The *AND/NAND* gate produces an output at *O*2 unless signal pressures are present at both *C*1 *and* *C*3. Absence of pressures at *C*1 and *C*3 is known as a "not AND" condition (NAND). The *fixed one-shot* generates a pulse of pressure regardless of how long a control pressure is applied at *C*1. The flow geometry of this device is skewed so that output normally is present at *O*2. Pressure applied at *C*1 instantaneously switches the output to *O*1, but this control pressure builds up in the longer curved flow path as well, eventually balancing on both sides of the jet and causing it to switch back to port *O*2. In lieu of a truth table, the pressure characteristics of this device are presented. Finally, the *Schmitt trigger* converts an analog

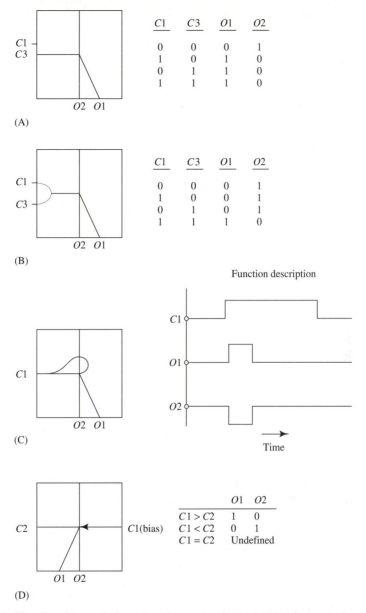

C1	C3	O1	O2
0	0	0	1
1	0	1	0
0	1	1	0
1	1	1	0

(A)

C1	C3	O1	O2
0	0	0	1
1	0	0	1
0	1	0	1
1	1	1	0

(B)

Function description

Time

(C)

	O1	O2
$C1 > C2$	1	0
$C1 < C2$	0	1
$C1 = C2$	Undefined	

(D)

Figure 11–33 Graphic symbols and performance characteristics for several common fluidic devices: (A) In the OR/NOR, positive control pressure at $C1$ *or* $C2$ switches output from $O2$ to $O1$, but output returns to $O2$ if control pressure is removed. (B) The AND/NAND requires pressures at $C1$ *and* $C2$ to obtain output at $O1$; this output returns to $O2$ if either $C1$ or $C2$ is removed. (C) Because of a curved internal flow loop, the fixed one-shot produces a single pulse of output pressure at $O1$ regardless of how long control pressure $C1$ is applied. (D) In the Schmitt trigger, control pressure $C1$ may vary over a range of values but must exceed bias pressure $C2$ in order to switch output flow to $O1$. This effectively converts an analog (continuously varying) input ($C1$) into a digital (on-off) output ($O1$–$O2$).

387

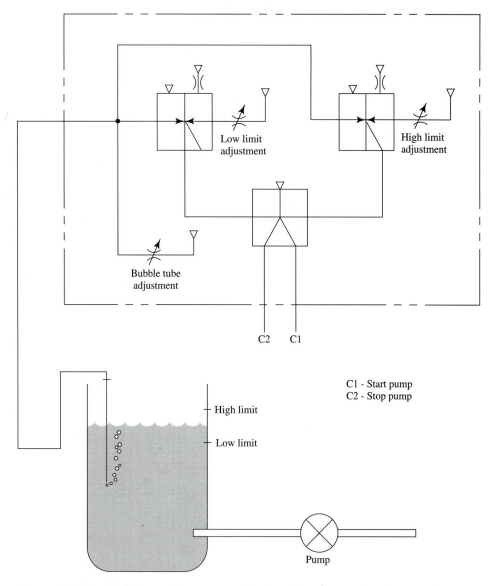

Figure 11–34 Fluidic logic elements used in a liquid level controller. (See text for a description of the system operation.)

signal into a digital signal. Under the influence of a preset bias pressure at $C1$, output occurs at $O1$ as long as control $C2$ is less than $C1$. When $C2$ exceeds $C1$, output switches to $O2$. The device, then, is either "on" or "off" (digital) even though $C2$ may vary (analog) in ranges above and below $C1$.

Let us see how these fluidic devices can be arranged to control several common systems. Figure 11–34 shows a liquid level controller consisting of two Schmitt triggers and one flip/flop that are activated by back pressure from a bubble tube sensor. Variable restrictors on the bubble tube supply and the bias lines for the "low-limit" and "high-limit" Schmitt triggers are used to control the sensitivity of the system and thus the allowable range of liquid levels. If the liquid level drops to its low limit, pressure in the bubble tube drops and bias control pressures on both Schmitt triggers dominate. This causes output from the high limit trigger to vent to atmosphere, while output from the low limit trigger acts as a control signal for the flip/flop, producing an output at $C1$. A simple pressure-to-electric switch (see Figure 11–21) connected between $C1$ and the pump will initiate the filling process. Once the liquid level reaches its high limit, pressure in the bubble tube is sufficient to overcome bias in both triggers; the low limit element then vents to atmosphere, while output from the high limit trigger switches the flip/flop to output $C2$, where it vents to atmosphere, thus opening the pressure-to-electric switch and shutting off the pump.

Another common logic system is shown in Figure 11–35. This is a safety control circuit for any type of machine that performs a stamping, shearing, pressing, or clamping operation. The circuit consists of two fixed one-shots, one OR/NOR device, and one flip/flop, along with three manually operated palm buttons located at *A, B,* and *C.* Workpieces (parts) are often fed into, or removed from, machines of this type by the hands of an operator whose production rate is reflected in his or her paycheck. This encourages the operator to work at the highest possible speed, thus increasing the danger of a serious accident to fingers, hands, and arms. For this reason, most machines in use today require that two hands be used to initiate a work cycle, thereby assuring that the operator's hands are safely outside the danger zone during operation of the machine. This safety precaution can often be circumvented by taping or otherwise fastening down one of the control buttons/switches; the operator can then initiate a work cycle with one hand while feeding and removing parts with his or her free hand. To avoid this dangerous situation, fixed one-shots are used. Until these devices are activated, both apply control signals to the OR/NOR, venting the OR/NOR output to atmosphere and leaving the flip/flop in the off position. When both one-shots are activated by hand(s), they momentarily switch their outputs to atmosphere, causing the OR/NOR output to trigger the flip/flop into the on position. Neither of the one-shots can be permanently tied down, since each element delivers only a single pulse no matter how long its palm button is depressed. In the circuit shown, the operator returns the machine to its original condition (off) by pressing the palm button at *C.*

Both of the preceding control systems are *event driven;* a particular condition or series of events is required to initiate the control function. Other types of control systems are *time driven;* these systems operate on a repetitive time pattern such as the mechanical oscillator shown in Figure 11–36. Here an air cylinder extends and retracts continuously under the control of a single flip/flop. Sensors $S1$ and $S2$, typically back pressure switches or limit switches, detect the allowable extreme positions of the cylinder output rod and switch the fluidic element into its appropriate control mode. Accumulators and/or restrictors can be placed in the flip/flop output lines to slow the buildup of pressure in the cylinder and thereby provide a time delay at the end of each stroke. Also, every time the flip/flop switches its output to one side of the driven piston, pressurized gas on the backside of the piston bleeds out through vents in the fluidic device itself.

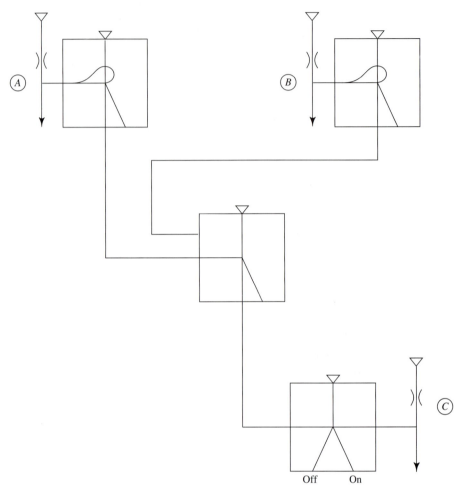

Figure 11–35 Typical press safety control circuit. Control pressures for the fixed one-shots and flip/flop are initiated manually by palm buttons at *A*, *B*, and *C*, respectively. (See the text for a description of circuit operation.)

Groups or modules of fluidic logic devices are often assembled in specific combinations to produce control systems known as *programmable air controllers* or PACs. These units generally utilize plug-in elements (Figure 11–37) that are mounted on circuit boards and interconnected by manifolds or tubing as shown in Figure 11–38. Besides the logic elements previously discussed, timers, counters, accumulators, amplifiers, and time delay and memory devices are available, as well as a variety of panel-mounted indicators and switches. Typically housed in a metal cabinet (Figure 11–39), the PAC represents the heart of many types of process and motion control systems (Figure 11–40).

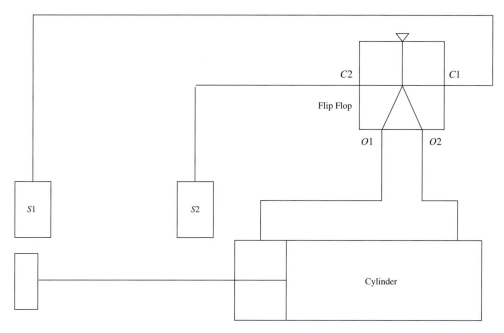

Figure 11–36 This double-acting air cylinder cycles continuously and acts as a time-driven mechanical oscillator. As the load approaches sensor S1 (typically a back-pressure switch), positive pressure develops at C1, output switches from O2 to O1, and the cylinder retracts. At the other end of its travel, the load activates sensor S2, thus switching the air supply to output O2 and causing the cylinder to extend.

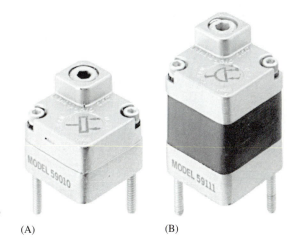

Figure 11–37 Plug-in type fluidic logic devices: (A) OR element; (B) AND element. *(The Aro Corporation, Bryan, Ohio)*

(A)   (B)

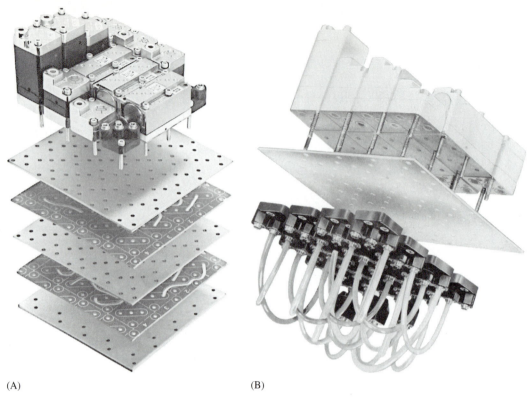

(A)   (B)

Figure 11–38 Plug-in fluidic elements assembled on circuit boards and interconnected by (A) module plates or (B) standard tubing. *(The Aro Corporation, Bryan, Ohio)*

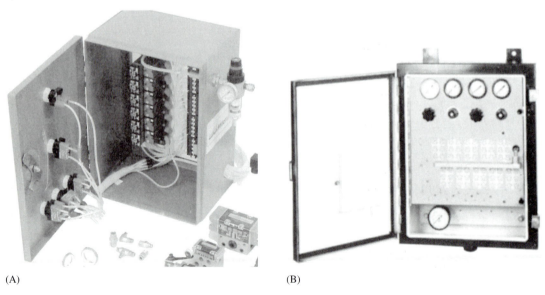

(A)   (B)

Figure 11–39 Typical cabinets for housing programmable air controllers. *((A) Miller Fluid Power, Bensenville, Ill.; (B) Air Logic, Racine, Wis.)*

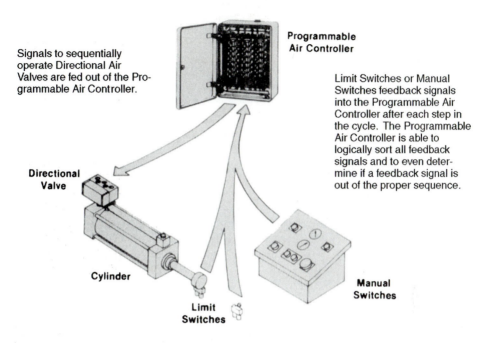

Signals to sequentially operate Directional Air Valves are fed out of the Programmable Air Controller.

Programmable Air Controller

Limit Switches or Manual Switches feedback signals into the Programmable Air Controller after each step in the cycle. The Programmable Air Controller is able to logically sort all feedback signals and to even determine if a feedback signal is out of the proper sequence.

Directional Valve

Cylinder

Manual Switches

Limit Switches

Figure 11–40 Elements of typical system utilizing PAC to control sequential operations. *(Miller Fluid Power, Bensenville, Ill.)*

Although programmable air controllers are not as sophisticated as their electronic counterparts, they are relatively easy to assemble and maintain. Like their individual components, PACs are also durable and reliable, particularly in harsh, demanding environments. For these reasons, they are well suited to a wide variety of industrial process and motion control applications.

Increasing numbers of fluid power systems are controlled by microprocessor-based *programmable logic controllers* (PLCs) such as the electropneumatic unit shown in Figure 11–41. The PLC is first connected to a host computer that is used to program the PLC's microprocessor for the desired system behavior. The computer can then be disconnected and sensors attached to the unit. Input from these sensors is compared with the desired behavior by electronic logic circuits and adjustments are made through electropneumatic valves and actuators. Adjustments to individual operations can generally be made directly from a keypad mounted on the front panel (Figure 11–42); in some applications, the host computer can be used in place of the PLC to provide direct keyboard control of the system (Figure 11–43). Generally, microprocessor-based PLCs are able to continuously monitor system behavior, anticipate future changes before they occur, and thus quickly provide the necessary responses or adjustments.

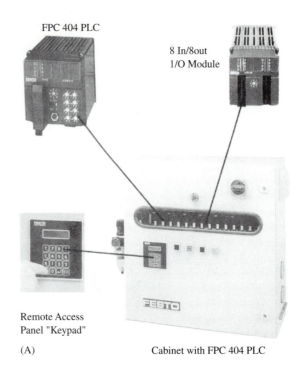

FPC 404 PLC

8 In/8out
1/O Module

Remote Access
Panel "Keypad"

(A)

Cabinet with FPC 404 PLC

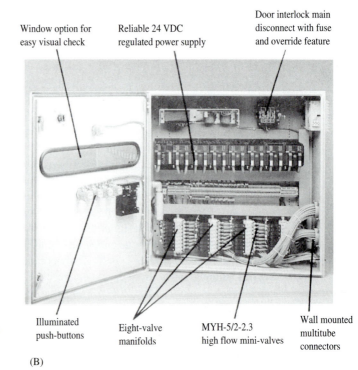

Window option for
easy visual check

Reliable 24 VDC
regulated power supply

Door interlock main
disconnect with fuse
and override feature

Figure 11–41 Exterior
(A) and interior (B) views of an
electropneumatic PLC. *(Festo
Corp., Hauppauge, N.Y.)*

Illuminated
push-buttons

Eight-valve
manifolds

MYH-5/2-2.3
high flow mini-valves

Wall mounted
multitube
connectors

(B)

394

(A)

(B)

Figure 11–42 (A) Front panel of a microprocessor-based PLC used to operate a hydraulic press. (B) A touch screen version is also shown. *(Greenerd Press & Machine Company, Inc., Nashua, N.H.)*

Figure 11–43 Rotation of this programmable robot arm (upper right) about three vertical axes is achieved electrically by one stepper motor and two servo motors; both actuation and vertical travel of the end-of-arm gripper are accomplished pneumatically. Control of the arm is possible either through push button panel or computer keyboard, both visible in center of photo. Here an engineer programs the arm to perform a specific sequence of operations used in an industrial assembly process.

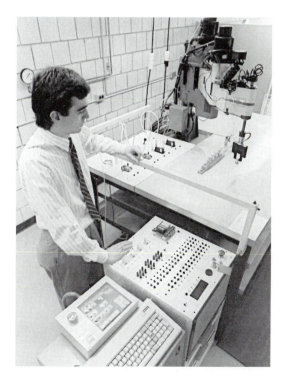

11.4 Actuators

In Chapter 7 we examined simple linear and rotary actuators that are widely used in general-purpose fluid power systems. Also available are a variety of hydraulic and pneumatic actuators designed primarily for use in automated control systems. These include escapements, grippers, multimotion actuators, and powered slides, as well as proportional valves and servovalves.

Since hydraulic systems generally operate at higher pressures than pneumatic systems, they are able to produce greater forces from an actuator of given size. Also, the incompressible liquid allows for more precise positioning of an actuator, though some digitally controlled pneumatic systems claim an accuracy of 0.002 in. On the other hand, hydraulic actuators are always slower and more expensive than their pneumatic counterparts. For all of these reasons, it is not uncommon to find both types of fluid power used within the same system, if not within the same device!

The *escapement* is a pneumatic device that automatically releases individual parts from a hopper, magazine, or vibratory feeder to any manufacturing operation such as assembly, inspection, or packaging (Figure 11–44). This action is accomplished by the sequential extension and retraction of two shafts or rods, as shown in Figure 11–45, which also pictures a typical device.

Grippers may be air- or hydraulic-operated, and, as the name implies, may be used to grip objects for point-to-point transfer and placement as well as for subsequent machining and assembly operations. These devices are classified as angular, parallel, or three-jaw grippers and are actuated by a combination of double- or single-acting pistons and springs. An angular gripper and typical construction details are pictured in Figure 11–46.

Powered slides are nothing more than pneumatic or hydraulic linear actuators that contain precision guide rods and heavy-duty bearings to accommodate side loads while minimizing deflections of the moving shaft. A sturdy tooling plate mounted on the end of this shaft allows for the attachment of various jigs and fixtures. These units can be used for

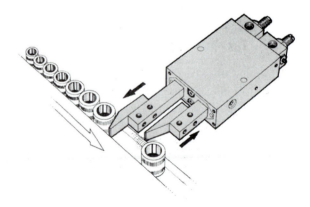

Figure 11–44 Escapement feeds individual parts to any manufacturing operation. (See Figure 11–45 for sequence of operation.) *(PHD, Inc., Fort Wayne, Ind.)*

accurate positioning of tools or workpiece and are available in a variety of configurations as shown in Figure 11–47.

Multimotion actuators combine linear and rotary motions into a single device such as that shown in Figure 11–48. A piston-driven rack-and-pinion in the base section rotates the vertical output shaft, while a piston connected to this shaft provides a linear up-and-down motion. These actuators can be driven either by hydraulic or pneumatic power; many use air-over-oil systems. Applications include repetitive testing (see Figure 11–1) and automated assembly, such as that shown in Figure 11–49.

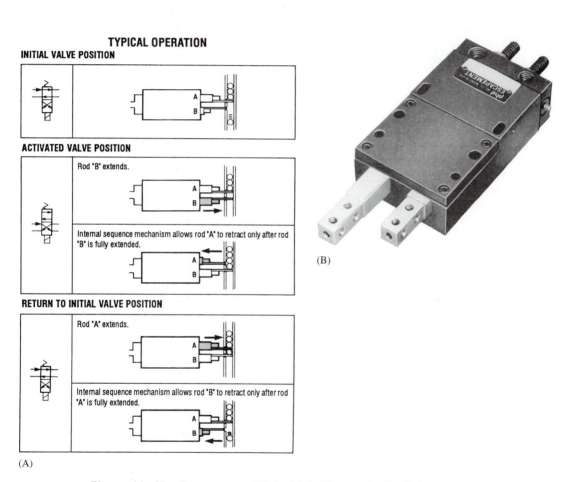

TYPICAL OPERATION

INITIAL VALVE POSITION

ACTIVATED VALVE POSITION

Rod "B" extends.

Internal sequence mechanism allows rod "A" to retract only after rod "B" is fully extended.

RETURN TO INITIAL VALVE POSITION

Rod "A" extends.

Internal sequence mechanism allows rod "B" to retract only after rod "A" is fully extended.

(A)

(B)

Figure 11–45 Escapements: (A) Rod A holds a stack of cylinders or spheres in position (top). Once the device is activated, rod B extends fully, and rod A then retracts (middle). Due to gravity, the supply stack now drops down and rests on rod B. Finally, rod A extends and rod B retracts (bottom), allowing a single cylinder/sphere to "escape" downward. (B) A typical escapement. *(PHD, Inc., Fort Wayne, Ind.)*

(A)

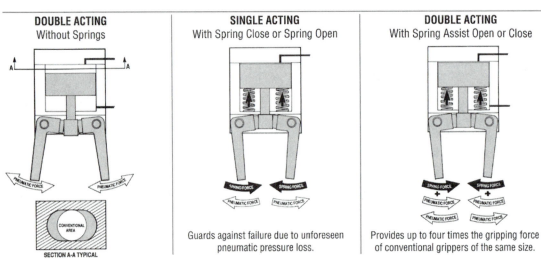

(B)

Figure 11–46 Typical angular grippers (A) and construction details showing piston, springs, and gripping actions (B). *(PHD, Inc., Fort Wayne, Ind.)*

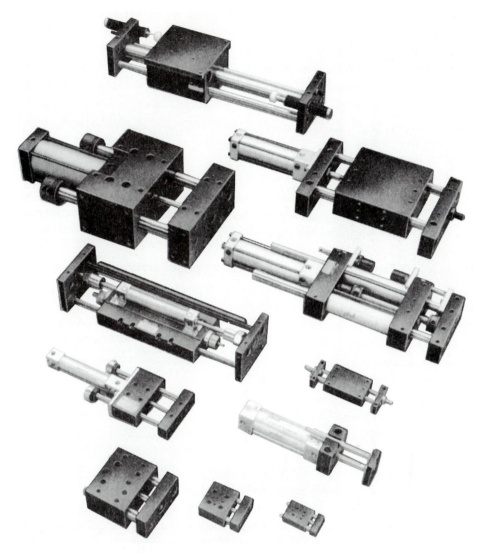

Figure 11–47 Powered slides in various configurations. Each assembly contains a pneumatic or hydraulic cylinder whose motion is held to accurate tolerances by precision guide rods and heavy-duty bearings. A sturdy plate mounted on the end of the cylinder rod allows for attachment of various tools, jigs, or fixtures. *(PHD, Inc., Fort Wayne, Ind.)*

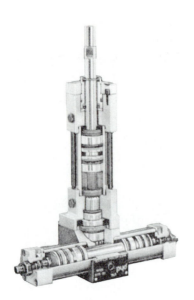

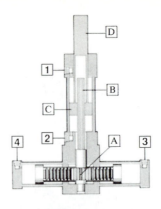

WORKING PRINCIPLE

The main components of the Actuator consist of a cylinder and a rack-and-pinion type rotary actuator. Linear motion of the rod (D) is produced when port 1 or 2 is pressurized. Rotary motion of the rod (D) is produced when port 3 or 4 is pressurized causing pinion gear (A) and spline bar (B), which are coupled together, to rotate coupled piston (C).

Figure 11–48 Operation of a multimotion actuator. *(PHD, Inc., Fort Wayne, Ind.)*

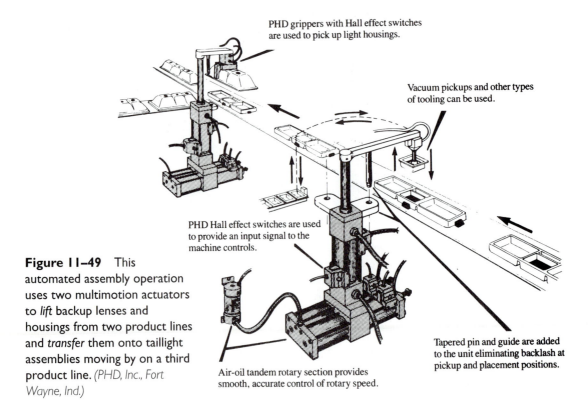

PHD grippers with Hall effect switches are used to pick up light housings.

Vacuum pickups and other types of tooling can be used.

PHD Hall effect switches are used to provide an input signal to the machine controls.

Figure 11–49 This automated assembly operation uses two multimotion actuators to *lift* backup lenses and housings from two product lines and *transfer* them onto taillight assemblies moving by on a third product line. *(PHD, Inc., Fort Wayne, Ind.)*

Air-oil tandem rotary section provides smooth, accurate control of rotary speed.

Tapered pin and guide are added to the unit eliminating backlash at pickup and placement positions.

Figure 11–50   Electrohydraulic motion controls and actuators are used on this FlightSafety International flight simulator in which pilots receive training for certification. *(Moog Controls, Industrial Division, Aurora, N.Y.)*

Control of most fluid power actuators is achieved through the use of multimedia valves that meter the flow rates received by these actuators. Pneumatic control signals, for example, are often used to regulate the flow of liquids or gases within a system. However, electrically controlled valves, both electrohydraulic and electropneumatic, are becoming increasingly popular in response to the trend toward microprocessor-based controllers (Figure 11–50). *Servovalves* deliver an oil or gas flow that is exactly proportional to a small electrical current. They are often used to control position, force, or velocity in *closed-loop* systems such as that pictured in Figure 11–3. Feedback control systems are referred to

Servovalves

Figure 11–51   Typical servovalves. Because they deliver a liquid or gas flow that is proportional to an electrical current, servovalves are frequently used in feedback-control systems. The example of Figure 11–52 shows how a servovalve automatically adjusts flow to an actuator in order that position of the actuator load remains within specified limits. *(Moog Controls, Industrial Division, Aurora, N.Y.)*

as closed loops because their main elements (sensor-controller-actuator) form self-regulating closed systems that function with minimal operator intervention. Figure 11–51 shows several typical servovalves, while Figure 11–52 diagrams a position servo system.

Proportional control valves are similar to servovalves but are used primarily in manually operated or *open-loop* systems. A *joystick* (Figure 11–53) is often used to control this type of system, which includes aerial buckets, cranes, mobile equipment, and industrial manipulators or robot arms.

In Chapter 7 we saw that flow through an axial piston pump could be controlled by varying the swash plate angle (see Figure 7–29). When such a pump is used to supply liquid to a hydraulic motor, the configuration is known as a *hydrostatic drive.* Many types of agricultural machinery, construction equipment, (Figure 11–54), garden tractors, recreational vehicles, and machine tools use this method to manually control the transmission of fluid power. The unit shown in Figure 11–55 provides electrical control of the swashplate angle for a pump or motor and thus represents an additional method of automatically regulating hydraulic systems.

A discussion of fluid power actuators would not be complete without mention of *remote-* or *radio-controlled* devices. The use of such devices to actuate garage door openers, model planes, and toy cars is well known. They are also well suited to some fluid power applications, particularly where convenience or operator safety are involved. One example of

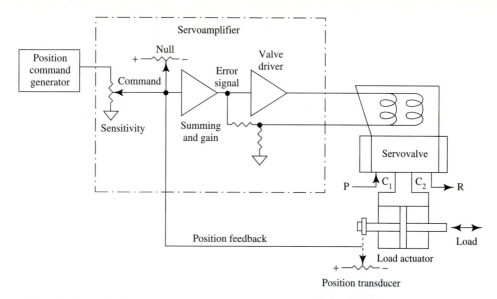

Figure 11–52 A typical closed-loop position control system utilizing a servovalve. A load-positioning servo system is comprised of a servovalve, actuator, position feedback transducer, position command generator, and a servoamplifier. A typical linear position servo system using a double-ended piston is shown (rotary position servo systems can be created by substituting the appropriate rotary components). The valve's two output control ports are connected across the load cylinder. In the servoamplifier, the command input is compared to the present position output of the position transducer. If a difference between the two exists, it is amplified and fed to the servovalve as an error signal. This signal shifts the valve spool position, adjusting flow to the actuator until the position output agrees with the command input. *(Moog Controls, Industrial Division, Aurora, N.Y.)*

Figure 11–53 This dual-axis joystick contains four electrical microswitches per axis. Such units are often used with proportional control valves in manually operated (or open-loop) fluid power systems. *(OEM Controls, Inc., Shelton, Conn.)*

Figure 11–54 Hydrostatic transmissions are commonly used on mobile equipment, ranging from small riding lawnmowers to large and powerful construction machines such as this 192-hp dozer. These hydraulic units offer smooth, continuous speed control over a wide range of operating conditions, without the need for gear-shifting or repeated throttle adjustments. *(John Deere Construction Equipment Company, Moline, Ill.)*

Figure 11–55 This unit electrically controls swashplate angle for certain hydraulic pumps and motors used in hydrostatic drives. *(Moog Controls, Industrial Division, Aurora, N.Y.)*

this occurs in salvage yards, where large crushing machines are used to compact scrap metal for shipment. Typically, a worker using either a front-end loader or large forklift truck places scrap in the crushing machine. Rather than dismounting from the loader and manually activating the crusher (or having a second person on the ground to perform this function), the worker simply keys a radio transmitter in his loader cab to initiate the crushing process. Not only does this allow for one-person operation of both machines, it dramatically decreases the chance of injury to the workers.

11.5 Automated Systems

Automated systems for the control of fluid power can be designed and built using sensors, controllers, and actuators such as those discussed in this chapter. Sequential circuits can range from the simple clamp-and-work configuration presented in Chapter 8 to extensive and very complex systems that automate an entire series of manufacturing activities. Always remember, however, that no matter how complicated an automated system is, it consists of individual operations such as transfer or "pick-and-place" (Figure 11–56), gauging and inspection (Figure 11–57), and machining or assembly (Figure 11–58), which have been combined to form a complete, integrated system. When analyzing such systems, begin by specifying the desired sequence of actions and then proceed to the appropriate hardware necessary to achieve this series of events.

Consider, for example, the automated drilling machine shown in Figure 11–59. Once a workpiece has been placed on the stationary table it must undergo three separate yet interconnected operations: clamping, drilling, and ejection. The part is first clamped in place by activating solenoid C, extending the cylinder that trips limit switch $LS4$. This switch energizes solenoid A, thereby allowing flow to the hydraulic cylinder, which in turn moves the rotating drill bit into the workpiece. At the end of its travel, the movable platform trips limit switch $LS2$; this switch deenergizes solenoid A while energizing solenoid B, causing the drill to retract to its original position. In doing this, the movable platform activates limit switch $LS1$, thereby deactivating solenoid C. Once the clamping cylinder has retracted, $LS3$ energizes solenoid D and a second pneumatic cylinder ejects the drilled part. $LS6$ deactivates solenoid D and returns this cylinder to its original position. The system is now ready for another event-driven work cycle.

For complicated fluid power systems, sequential operations are often described using *ladder diagrams.* Essentially, these represent truth tables for an entire system and indicate what output will be achieved when each sensor, switch, and relay is in a specific state or condition. Ladder diagrams are particularly useful for electropneumatic and electrohydraulic circuits that are controlled by PLCs.

Automated sequential motion can also be achieved through the use of *robots* such as the one shown in Figure 11–60. Although these devices provide functions similar to those of automated assemblies, they are generally self-contained units characterized by the fact that changes in motion can be obtained by reprogramming the control system rather than by rearranging the actuating hardware. Virtually any operation can be performed with

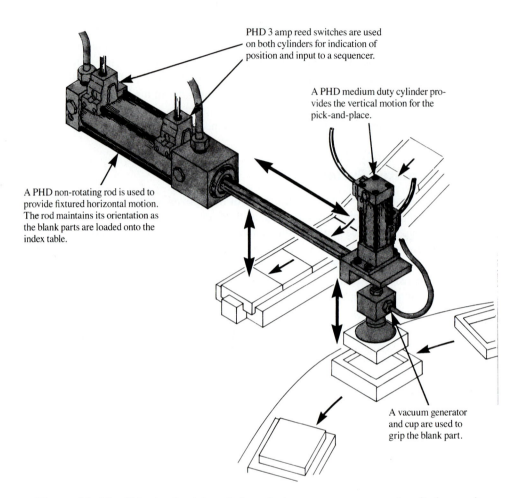

PHD 3 amp reed switches are used on both cylinders for indication of position and input to a sequencer.

A PHD medium duty cylinder provides the vertical motion for the pick-and-place.

A PHD non-rotating rod is used to provide fixtured horizontal motion. The rod maintains its orientation as the blank parts are loaded onto the index table.

A vacuum generator and cup are used to grip the blank part.

Figure 11–56   This simple pick-and-place device uses two pneumatic cylinders and a vacuum cup to lift parts from a supply chute and place them on an adjacent rotary indexing table. Heavy parts might be lifted using pneumatic grippers rather than a vacuum cup. *(PHD, Inc., Fort Wayne, Ind.)*

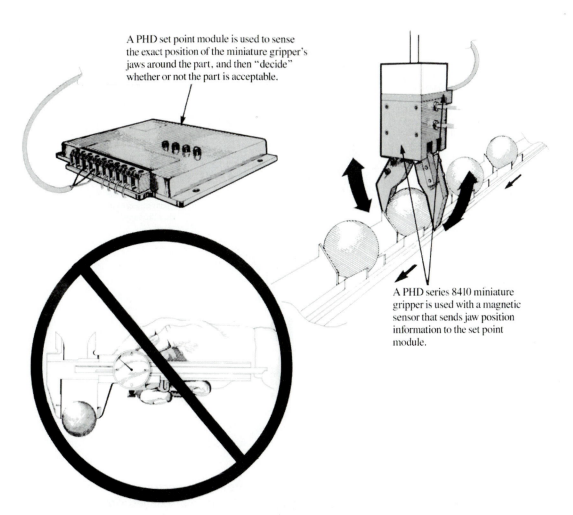

A PHD set point module is used to sense the exact position of the miniature gripper's jaws around the part, and then "decide" whether or not the part is acceptable.

A PHD series 8410 miniature gripper is used with a magnetic sensor that sends jaw position information to the set point module.

Figure 11–57 In this gauging operation, composite spheres on an assembly line are "felt" by a sensor/gripper and their diameters measured to within 0.002 in. Parts whose dimensions fall outside established tolerances are automatically removed from the line by the gripper. This procedure was originally performed manually by an inspector using a micrometer. *(PHD, Inc., Fort Wayne, Ind.)*

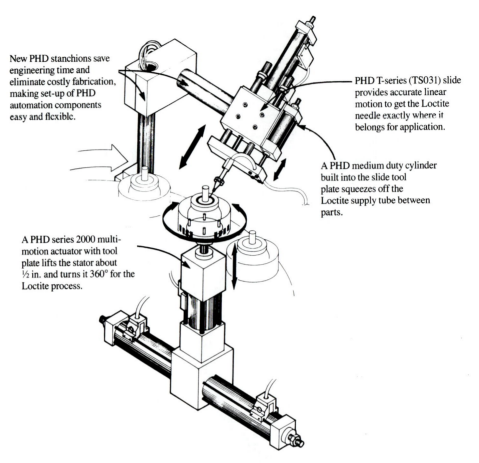

New PHD stanchions save engineering time and eliminate costly fabrication, making set-up of PHD automation components easy and flexible.

PHD T-series (TS031) slide provides accurate linear motion to get the Loctite needle exactly where it belongs for application.

A PHD medium duty cylinder built into the slide tool plate squeezes off the Loctite supply tube between parts.

A PHD series 2000 multi-motion actuator with tool plate lifts the stator about ½ in. and turns it 360° for the Loctite process.

Figure 11–58 The multimotion actuator (bottom) lifts electric motor stator housings and rotates them through 360° while a powered slide (top) extends, allowing a nozzle to apply adhesive to the housing in a circular pattern. At a separate work station bearing races will be pressed onto the adhesive ring. Notice that a small linear cylinder attached to the slide pinches off the adhesive supply tube between applications. *(PHD, Inc., Fort Wayne, Ind.)*

appropriate end-of-arm tooling; typical applications include cutting, spray painting, machining, welding, and lifting (Figure 11–61).

In recent years the trend has been toward electric robots that operate using servomotors and/or stepping motors, with some functions provided pneumatically (see Figure 11–43). Hydraulic systems, however, remain the units of choice for most heavy-duty applications such as those metal working industries that use processes like casting, forging, welding, or any machine tool operations (Figure 11–62). Even electric robots often employ the advantages of fluid power in their specially designed end-of-arm tooling (Figure 11–63).

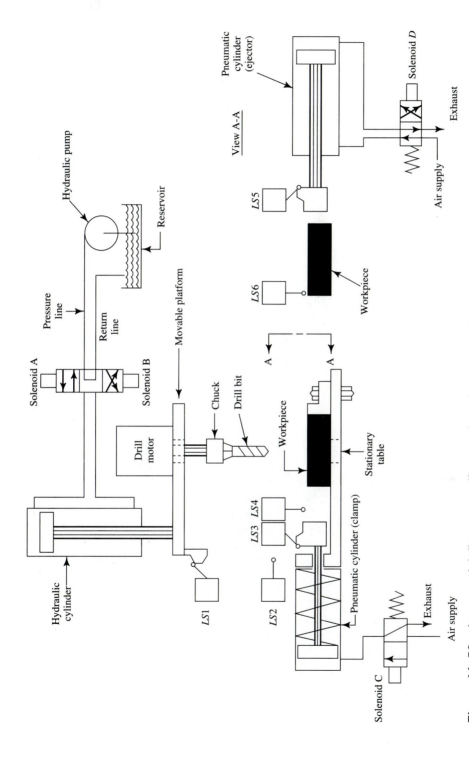

Figure 11–59 Automated drilling system. (See text for a description of system operation.) *(Adapted from Introduction to Control System Technology, 3rd ed., by Robert N. Bateson, Prentice Hall, Upper Saddle River, NJ.)*

409

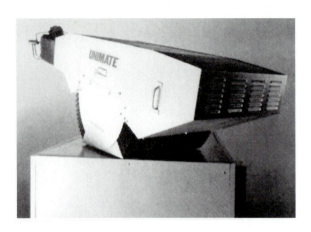

Figure 11–60 A typical hydraulic robot. *(UNIMATE® is a registered trademark of Westinghouse Electric Corporation; these units are manufactured under license by Prab Robots, Inc., Kalamazoo, Mich., who provided this illustration.)*

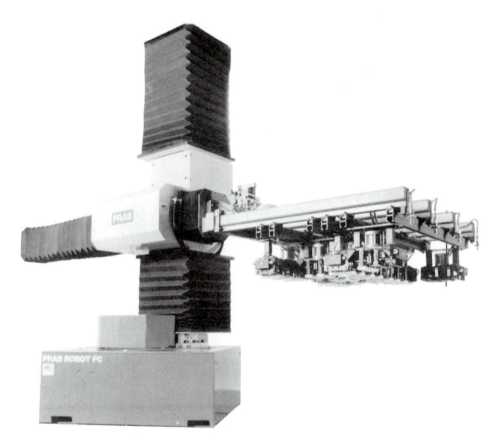

Figure 11–61 This servocontrolled electrohydraulic robot stands more than 12 ft tall and can lift payloads of 6000 lb. *(Prab Robots, Inc., Kalamazoo, Mich.)*

(A)

(B)

Figure 11–62 Industrial robots shown (A) loading hot billet into forging press and (B) spotwelding on automotive assembly line. *(Prab Robots, Inc., Kalamazoo, Mich.)*

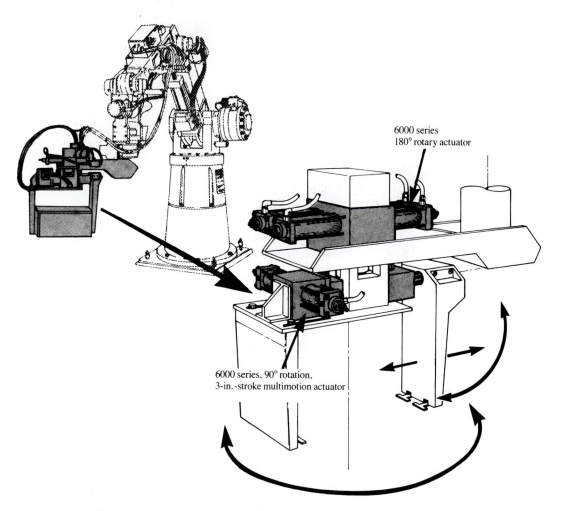

6000 series
180° rotary actuator

6000 series, 90° rotation,
3-in.-stroke multimotion actuator

Figure 11–63 Rotary and multimotion actuators combine to form this end-of-arm tooling for a robot used to remove 70-lb cartons from a conveyor and stack them on a pallet. *(PHD, Inc., Fort Wayne, Ind.)*

Questions

1. List the important components in a feedback, or closed-loop, control system.

2. Activating an automatic garage door opener either fully opens or fully closes the door. How are these extreme positions of the door sensed and how is the opener itself controlled at these limits?

3. Name several advantages and disadvantages of thermocouples.

4. Describe at least six different devices commonly used to measure the flow rates of fluids. Indicate which of these are used only for gas flows, which are used only for liquid flows, and which can be used with either gas or liquid flows.

5. What is a fluidic device?

6. Explain the Coanda effect.

7. Draw the schematic symbols for:
   (a) Interruptable jet sensor.
   (b) Back-pressure switch.
   (c) Flip/flop.
   (d) OR/NOR
   (e) AND/NAND
   (f) Fixed one-shot.
   (g) Schmitt trigger.

8. How may three OR/NOR devices be assembled in order to provide the behavior of an AND gate (one output requires two inputs)?

9. In Figure 11–36, if *S*1 and *S*2 are back-pressure switches, proximity of the load causes positive pressures at *C*1 and *C*2, respectively. If contact sensors such as the one shown in Figure 11–26(B) are used, proximity of the load causes an *absence* of pressures at the control ports. How could such sensors be used to control the circuit of Figure 11–36?

10. What is a Hall-effect sensor?

11. An on-off light switch is the electrical equivalent of a bistable flip/flop. In residential structures, it is common practice to install wall-mounted light switches at both ends of a hall or stairway and near both doorways leading into a room. These dual switches are wired in a manner such that either switch will turn lights on or off, regardless of the (on-off) position of the other switch. Draw the circuit connecting the power supply, switches, and light that allows this type of system behavior.

12. What is a PAC? A PLC? How do these devices differ?

13. Why are hydraulic, rather than penumatic, control systems used when accurate positioning is required?

14. What is an escapement? A powered slide?

15. What is a servovalve? What type of control system might contain such a valve?

16. How does a proportional control valve differ from a servovalve?

17. Describe the operation of a hydrostatic drive. What are its most important operating advantages?

18. Many of today's automated robots are all-electric or electropneumatic. For what type of applications are hydraulic robots ideally suited?

Suggested Activities

1. Tape a scale to the inside of a water supply tank in a bathroom fixture. Record the water level reading and flush the commode. When the tank has refilled, record the water level reading again. Repeat this operation several times. What is the average scale reading when the tank is full? What was the largest observed variation from the average value? Is this simple liquid level control system accurate? Expensive?

2. Check the on-off operation of a thermostat by measuring the actual switching temperatures with a mercury or digital thermometer. What range of room temperatures occurs for a particular preset value of the thermostat?

3. Repeat (2) for any type of regulated compressor by attaching an accurate pressure gauge somewhere in the output line and *slowly* allowing air to escape. What range of pressures is measured at the gauge?

Answers to Odd-Numbered Problems

CHAPTER I

Problem Set I
1. 6560 cm^3
3. 8720 mm^2
5. (a) 0.08048 hp   (b) 156,600 W
7. (a) 42.55 ft to 42.64 ft
    (b) 3.145 ft to 3.154 ft
    (c) 156.5 ft to 157.4 ft
    (d) 203 ft
    (e) 202 ft
    (f) 203 ft
9. (a) 11,200 cm^2 (b) 11,200 cm^2
11. 90.7 ft

CHAPTER 2

Problem Set I
1. 8.34 lb/gal
3. 1.12 g/cm^3
5. 2.26

CHAPTER 3

Problem Set I
1. 10,000 mbg or 11,000 mba
3. 497 kPa
5. (a) slightly greater than 1500 psi
    (b) 1660 psi

Problem Set 2
1. (a) 120 lb   (b) 1.5
3. (a) 2.0 MPa
    (b) 35.3 kN
    (c) 1.36 cm
    (d) 319 kPA; 0.902 kN; 53.1 cm
5. (a) 0.237 in.

(b) 21,000 ft
(c) 22.7 psi
(d) No

Problem Set 3
1. (a) 141,600 ft-lb
    (b) 0.358 hp
    (c) 0.715 hp
3. 3.29 hp
5. 0.369 hp

Review Problems
1. 12,700 lb
3. (a) 90 lb
    (b) 28.7 psi
    (c) 276 lb
5. (a) 70 psi
    (b) 1980 lb
    (c) 366 psi
    (d) 10,300 lb

CHAPTER 4

Problem Set I
1. (a) 15,600 psi
    (b) 5.93 gpm
    (c) 53.9 hp
3. For 14-in. piston: 1.33 gpm; 0.606 hp
    For 19.5-in. piston: 2.59 gpm; 3.03 hp

Problem Set 2
1. (a) 8840 lb
    (b) 13.4 s
    (c) 5.25 hp
    (d) Splitter will perform poorly; most
        commercial machines deliver forces in
        the range of 24,000 lb to 30,000 lb.

Problem Set 3
1. (a) 14.6 hp   (b) 109 rpm

Review Problems
1. $P = 1000 \times q \times p$

CHAPTER 5

Problem Set 1
1. (a) 18.6 psig
  (b) 63.1 ft
  (c) 27.0 ft
3. 5.17
5. 10,000 lb

Problem Set 2
1. Water: 22.0 ft; oil: 31.3 ft
3. -3.92 kPag

Problem Set 3
1. 7.15 psig
3. 0.600

Problem Set 4
1. 1.50
3. (a) 0.0271   (b) 1,946 g

Review Problems
1. (a) 129 psig   (b) 142 psig
3. 2.71 ft

CHAPTER 6

Problem Set 1
1. $v_2 = 15.3$ ft/s; $v_4 = 7.66$ ft/s
3. (a) 9.08 ft/s
  (b) 1.50 lb/s
  (c) 3120

Problem Set 2
1. (a) 270.6 ft-lb/lb
  (b) 270.6 ft
  (c) Yes for (a). 5.00-oz ball: KE $= 84.6$ ft-lb
                 5.25-oz ball: KE $= 88.8$ ft-lb
    No for (b).
3. (a) 2.24 ft   (b) 113 ft
5. 1480 ft

Problem Set 3
1.

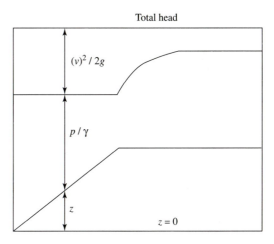

3. $p_3 = 56.2$ psi; $p_6 = 82.0$ psi
5. 9.98 psi

Problem Set 4
1. 73.1 ft/s
3. 610 ft/s (416 mph)

Problem Set 5
1. 5.88 m/s
3. 16.2 ft

Problem Set 6
1. (a) 1 lb requires 85 ft-lb; 1 gal requires 709 ft-lb
  (b) 8500 ft-lb; 1.29 hp
  (c) 0.215 hp
3. 11.3 hp
5. For 3-in. pipe: 87.8 hp
   For 4-in. pipe: 74.3 hp
7. (a) 0.997 hp   (b) 35.9 in.-lb

Problem Set 7
1. (a) 9.33 hp
  (b) Pressure head: 6150 ft;
     velocity head: 0.0328 ft;
     elevation head: 40 ft;
     ideal power: 9.39 hp

Problem Set 8

1. 685 psi; no, almost quadrupled.
3. 160.3 ft

Review Problems

1. (a) 1.77 in.
   (b) 0.148 ft
   (c) 7570

3. For $\frac{1}{2}$-in. pipe: 39.6 psi; for $\frac{3}{8}$-in. pipe:

   38.6 psi

7. 260 kW

CHAPTER 9

Problem Set 1

1. 75.1 psig
3. 686 kPa; 99.7° C

Problem Set 2

1. (a) 1415 N/m^3
   (b) 120
   (c) 1533 N/m^3; 130
3. 38.6° F

CHAPTER 10

Problem Set 1

1. (a) 61.5 acfm   (b) 267 cfm

Index

Table of Conversion Factors

Length

1 ft = 12 in.
1 yd = 3 ft
1 mi = 5280 ft
1 m = 100 cm = 1000 mm

1 in. = 2.540 cm = 25.40 mm
1 ft = 0.3048 m
1 m = 3.281 ft = 39.37 in.

Area

$1\ ft^2 = 144\ in.^2 = 0.09290\ m^2$
$1\ in.^2 = 6.452\ cm^2$
$1\ m^2 = 10.76\ ft^2$

Volume

$1\ ft^3 = 1728\ in.^3 = 0.02832\ m^3$
$1\ in.^3 = 16.39\ cm^3$
$1\ l\ (liter) = 1000\ cm^3 = 1000\ ml\ (milliliter) = 61.02\ in.^3$
$1\ m^3 = 35.32\ ft^3$
$1\ U.S.\ gal = 4\ qt = 231\ in.^3 = 3.785\ l = 0.1337\ ft^3$
$1\ qt = 2\ pints = 32\ fluid\ oz = 57.75\ in.^3 = 0.9464\ l$

Time

1 hour = 1 h = 60 min = 3600 seconds = 3600 s
1 min = 60 s

Force & Weight

1 lb = 16 oz = 4.448 N (newtons)
$1\ N = 0.2248\ lb = 10^5\ dynes$
1 ton = 2000 lb

Torque

1 ft-lb = 1.356 N-m

Velocity

1 mi/h = 1.467 ft/s = 0.4470 m/s = 1.609 km/h
1 ft/s = 0.3048 m/s

Mass

1 slug = 14.59 kg = 32.17 lb (mass)
1 lb (mass) = 0.03108 slug = 0.4536 kg
1 kg = 0.06852 slug = 2.205 lb (mass)

Specific Weight

$1\ lb/ft^3 = 157.1\ N/m^3$